Routledge Revivals

Postwar Trends in U.S. Forest Products Trade

By the end of World War II, the United States had become well integrated into the world markets for forest products. No longer can domestic prices of forest products be viewed as being wholly determined by domestic demand and supply, nor even by North American supply and demand, but must be viewed in a worldwide context. Originally published in 1980, this work provides a comprehensive overview of the nature of global forestry, particularly as it pertains to international trade flows of forest products, and analyses the role of the United States in a global context. This is a valuable resource for any student or researcher interested in environmental studies, global trade relations, and foreign market development.

Postwar Trends in U.S. Forest Products Trade

A Global, National, and Regional View

Roger A. Sedjo and Samuel J. Radcliffe

First published in 1980
by Resources for the Future, Inc.

This edition first published in 2016 by Routledge
2 Park Square, Milton Park, Abingdon, Oxon, OX14 4RN
and by Routledge
711 Third Avenue, New York, NY 10017

Routledge is an imprint of the Taylor & Francis Group, an informa business

Publisher's Note
The publisher has gone to great lengths to ensure the quality of this reprint but points out that some imperfections in the original copies may be apparent.

Disclaimer
The publisher has made every effort to trace copyright holders and welcomes correspondence from those they have been unable to contact.

A Library of Congress record exists under LC control number: 80008886

ISBN 13: 978-1-138-95426-7 (hbk)
ISBN 13: 978-1-315-66700-3 (ebk)
ISBN 13: 978-1-138-95429-8 (pbk)

Research Paper R-22

POSTWAR TRENDS IN U.S. FOREST PRODUCTS TRADE

A Global, National, and Regional View

Roger A. Sedjo
Samuel J. Radcliffe

RESOURCES FOR THE FUTURE / WASHINGTON, D.C.

RESOURCES FOR THE FUTURE, INC.
1755 Massachusetts Avenue, N.W., Washington, D.C. 20036

Resources for the Future is a nonprofit organization for research and education in the development, conservation, and use of natural resources and the improvement of the quality of the environment. It was established in 1952 with the cooperation of the Ford Foundation. Grants for research are accepted from government and private sources only if they meet the conditions of a policy established by the Board of Directors of Resources for the Future. The policy states that RFF shall be solely responsible for the conduct of the research and free to make the research results available to the public. Part of the work of Resources for the Future is carried out by its resident staff; part is supported by grants to universities and other nonprofit organizations. Unless otherwise stated, interpretations and conclusions in RFF publications are those of the authors; the organization takes responsibility for the selection of significant subjects for study, the competence of the researchers, and their freedom of inquiry.

Research Papers are studies and conference reports published by Resources for the Future from the authors' typescripts. The accuracy of the material is the responsibility of the authors and the material is not given the usual editorial review by RFF. The Research Paper series is intended to provide inexpensive and prompt distribution of research that is likely to have a shorter shelf life or to reach a smaller audience than RFF books.

Library of Congress Catalog Card Number 80-8886

ISBN 0-8018-2635-7

Distributed by The Johns Hopkins University Press,
Baltimore, Maryland 21218

Manufactured in the United States of America

Published February 1981. $15.00.

CONTENTS

Page

APPENDIXES

Page

List of Text Tables

Page

Page

List of Text Figures

List of Appendix Tables

Page

Page

Page

ACKNOWLEDGMENTS

In conducting this study, we received cooperation and assistance from numerous individuals and organizations. The Maritime Administration provided a considerable portion of the data found in the appendixes of this volume and used for the analysis of the international trade of the various U.S. regions. We wish to thank Russell Stryker, Howard Norseth, Janis Lawrence and Thomas Hartman for their assistance. Also, we acknowledge the assistance of Gordon Blaney of Statistic Canada.

In addition, we are particularly grateful to those individuals who commented on an earlier draft of the study. They were: David Darr, Hans Gregerson, Perry Hagenstein, Richard Haynes, Jan Laarman, Bruce Lippke, Harold Wisdom, Ken Frederick and particularly Irving Hoch for their reviews of the draft manuscript.

Furthermore, we want to thank Lorraine Van Dine, Avery Gordon, and Marian Lesko for their considerable efforts typing and retyping the manuscript.

Finally, we want to thank the Weyerhaeuser Foundation Company and the Forest Service for their sponsorship of the Forest Economics and Policy Program at Resources for the Future, through which this research was undertaken.

We should note here, however, that the authors accept full responsibility for the manuscript and any errors that it may contain.

GLOSSARY

CONIFER

All trees classified botanically as Gymnospermae. They are generally referred to as softwoods.

FIBER PRODUCTS

Products utilizing wood fiber in their production. Examples include paper, wood pulp, and paperboard.

FIBERBOARD

Board made from compressed and noncompressed wood fiber.

HARDWOOD

Wood derived from trees classified botanically as Angiospermae.

INDUSTRIAL ROUNDWOOD

Roundwood other than fuelwood and wood used for charcoal. The commodities include sawlogs, veneer logs, pulpwood, and also--in the case of trade commodities--wood chips, particles, and wood residuals.

LUMBER

Sawnwood, timber sawn into beams, planks, boards, and so forth, of convenient size.

NEWSPRINT

Uncoated paper of the type used mainly for the printing of newspapers.

PAPER AND PAPERBOARD

Includes newsprint, printing and writing paper, and other paper and paperboard.

PAPERBOARD

Rigid paper made from wood fiber commonly used for boxes and packaging material.

PARTICLE BOARD

A sheet material manufactured from small pieces of wood or other cellulosic materials agglomerated by use of an organic binder.

PLYWOOD

A sheet constructed by bonding together veneer sheets or bonding veneer sheets with other sheets, for example, with a solid core.

PULPWOOD

Roundwood used to make wood pulp. The aggregate includes chips, particles, and residuals.

ROUNDWOOD

Wood in its natural state as felled or otherwise harvested.

RECONSTITUTED WOOD

Includes both particle board and other improved wood densified and/or impregnated with resin or resinlike materials.

SAWNWOOD

Timber unplaned, planed, grooved, sawn lengthwise or produced by a profile chipping process, for example, planks, beams, boards, boxboards, and lumber.

SOFTWOOD

Wood from conifers.

WOOD CHIPS

Wood that has been deliberately reduced to small pieces from wood in the rough or from industrial residuals.

WOOD-BASED PANELS

Panels made of wood including veneer sheets plywood, particle board, and fiberboard.

WOOD PULP

A mixture of cellulosic material, which has been grooved or chemically pressed, from which paper products are made.

Chapter 1

INTRODUCTION

International trade in forest products has a long history. One of the earliest recorded accounts of a seagoing voyage documents the delivery of forty shiploads of cedar from the coast of Lebanon on the eastern Mediterranean to Egypt in about 2600 B.C.(1). The Old Testament records King Solomon's importation of large quantities of cedar from Lebanon, which was used to build the great temple of Israel nearly one thousand years before the birth of Christ.

However, despite ancient references to such international trade, this topic has not typically elicited a great deal of interest. The reasons for this are numerous. Some form of wood has generally been available in most areas of the globe. Also, wood substitutes, for particular uses, are plentiful. Thus, building in much of the ancient world utilized materials other than wood for construction. In addition, in areas where fuelwood was not readily available, other fuels were commonly used. Furthermore, the value per unit of wood is relatively low, implying high transportation costs relative to its value. These factors suggest that traditional wood markets tended to be localized and not well integrated internationally.

In North America the early settlers almost certainly viewed wood as an inexhaustible resource and one that was more of an obstacle than an asset (2). It is true that in some areas, such as the Great Plains, the sod house replaced the log cabin that was so common on the heavily wooded East Coast. But generally, as the population migrated, new areas of forest provided the necessary wood supplies, and logging activity generally shifted to accommodate the movement of population and markets.

Given this environment, it is not surprising that international and interregional trade volumes in forest products were modest, and trade in forest products was not very important. Therefore, historically, neither foresters, forest economists, the forest product industry, nor public agencies charged with the stewardship of the public forest resources have shown much interest in the international, or even interregional, aspects of the trade in forest resources. Consistent with the historic lack of interest in the world forest product markets was an apathy regarding the contributions of the various U.S. regions to the totality of U.S. forest products trade.

Gradually, but continuously, however, U.S. interactions with foreign markets for forest products have been increasing, and the view that U.S. market conditions are determined by strictly local or domestic forces is becoming increasingly unrealistic. By the end of World War I the United States had become a net importer of pulp and paper products, a situation which has continued to prevail without exception for over sixty years. During World War II the United States became a net importer of softwood lumber, and this situation has continued unaltered, except for the year 1947 (3), to the present. In the early 1960s the great Columbus Day storm swept the forests of the Pacific Northwest. The resulting salvage operation stimulated the then-fledgling log export market to Japan so that within a decade earnings on conifer log exports from the Pacific

Northwest alone amounted to more than $750 million. In recent years, the United States, as a major importer of both wood pulp and newsprint, has also experienced earnings of over $1 billion from the exports of paper and paperboard products. Today the United States is generally well integrated into world markets for forest products. Domestic prices of forest products can no longer be viewed as being wholly determined by domestic demand and supply, nor even by North American supply and demand, but must be viewed in a worldwide context.

In response to the evolving environment, there has been additional interest in the international dimension of forest product trade and resources, and research in this topic has been on the increase in recent years. The U.S. Forest Service has gradually expanded its research and informational activities related to international aspects of forestry and the forest products trade. For example, the Pacific Northwest Forest and Range Experiment Station is now actively documenting and publishing trade data and examining trade issues that are of special interest to that region. On the national level, the 1980 Forest Service Timber Assessment Model has incorporated international dimensions into its structure, again reflecting the increased role perceived for international trade. Also, the forest products industry--particularly that segment involved in international trade--now looks with increased interest toward international markets, recognizing the importance of having a more sophisticated understanding of international forest product trade relationships. Finally, our study, supported by both the Forest Service and the Weyerhaeuser Company Foundation, is also a testimonial to the increased interest in the international dimensions of forestry.

Background to the Study

In the summer of 1977 Resources for the Future (RFF) began its formalized and expanded Forest Economics and Policy Program. Forest policy issues were not new to RFF, research in this topic area having been undertaken at various times by numerous researchers beginning in the late 1950s. However, the creation of a formalized program reflected a commitment by RFF to a continuing interest in research in forest economics and policy issues.

Several program objectives were established at the inception of the program. These included the desire to "undertake research in important areas which had been neglected" and to "contribute to the establishment of a comprehensive professional literature." At that time several broad, research topic areas were selected for extensive study including international forestry. The inclusion of this topic area reflected the judgment that this was an important area that had not been well researched. Within international forestry, we subsequently chose as the principal thrust the broad question of the future U.S. role in the world forest economy. However, in order to adequately examine this question, we needed a clearer view of the current U.S. role as well as the U.S. role in the recent past.

A literature search revealed that little had been written on this question. Within the United States, John Zivnuska had done a report on U.S. Timber Resources in a World Economy for RFF in 1967. Also, Irving Holland had an appendix on international trade in The Report of the President's Advisory Panel on Timber and the Environment (1973). In addition, the Forest Service's The Outlook for Timber in the United States

(1973) had a section devoted to international trade. Beyond these, little had been done in the United States.

With the exception of Zivnuska's work, these studies were not very satisfactory for us since they tended to take a two-country view of forest product trade between the United States and the rest of the world, while we were primarily interested in understanding the overall structure of world forest resource production and trade. In addition, to properly understand the role of the United States, it was necessary to investigate the forest product trade relationships between particular countries, especially the United States and Canada. Finally, it had been recognized that there were vast differences within the United States in the contributions to total U.S. international forest products trade of the various U.S. regions. However, very little research had been undertaken on the regional trade contributions since the data, although they did exist, were not readily accessible.

In this context, the first priority of the newly founded research activities in international forestry was to examine the global structure of forest products production and trade and recent U.S. experience within that structure. The second priority was to investigate the contributions of the individual U.S. regions to that experience. This manuscript is the result of that effort.

Objectives of the Study

The study fulfills several important objectives. First, in the tradition of earlier RFF studies [for example, Potter and Christy (4), and Manthy (5)], it is designed to organize available data that has not previously been assembled in an easily usable form. For example, the data for the various U.S. regions appearing in this report fall into this category. The U.S. Census has maintained data on forest products exports and imports by commodity and by customs districts for an extended period of time. However, the forest products portions of this data were never assembled separately, and only data for the most recent five years are available on computer tapes. Fortunately, the U.S. Maritime Administration (MarAd) has classified oceangoing cargoes by their U.S. region of origin. These data are available on computer tapes for the most recent ten-year period. Also, fortuitously, the MarAd regional groupings are quite compatible with the major U.S. forest-producing regions. Therefore, the MarAd data provide the basic source from which our regional series for international trade in forest products has been developed. Of course, oceangoing trade, while dominant for forest products, does not exhaust the mode of international transport. Particularly for the U.S.-Canadian and U.S.-Mexican trade, large volumes of forest products are transported overland. These flows were captured by using U.S. Census data and integrating these into the MarAd data.

Our second objective is to provide a comprehensive overview of the nature of global forestry, particularly as it pertains to international trade flows of forest products. We describe the forest inventory and how the existing world forest resources are distributed, identify the major forest product-producing countries, summarize important trading patterns by country and commodity, and briefly discuss changes in production and trade patterns that have occurred following World War II. In addition to the U.S. data, we also make use of various international data sources, particularly the Food and Agricultural Organization (FAO) of the United Nations. For the analysis of changing trends, we focus principally on three time periods--1950, 1961-65, and 1976. This analysis is largely descriptive and designed to provide the reader with a comprehensive overview of the contributions of the traditional wood products producers--North America, the USSR, and Europe--to world production and trade during the postwar period, and it also documents the increasing role being assumed by the nontraditional producers.

As a third objective, we describe comprehensively and analyze the role of the United States in a global context. Particular attention will be given to the complementary nature of the United States and Canada within North America. In addition, the trade patterns of the United States are analyzed separately vis-à-vis both the rest of the world and U.S. individual trading partners.

Our fourth objective is to analyze the role of the regions within the United States in world trade. We utilize the newly compiled U.S. regional data to describe and compare the roles of the various U.S. regions in the forest products trade patterns and trends of the United States for the period 1967-76.

Our final objective is to provide other researchers with a reference source as a point of departure for further work. The report's appendixes are extensive and a detailed table of contents is provided to assist the user in determining what sections of the report or the appendixes may be useful. Although the report's appendixes include data only for the years 1967 to 1976, the Receiving and Evaluation Division, Foreign Market Development of the Foreign Agricultural Service, in the U.S. Department of Agriculture has indicated its intention to maintain and update the data series developed in this study. Thus, a growing data base will become available to researchers.

Appendix A provides a detailed documentation of the regional data. It is intended to provide researchers with sufficient information on the data sources, definitions, methods used, and so forth, so that the data can be updated with consistency or modified to fit the requirements of individual researchers. Appendixes B and C provide detailed data of national U.S. forest products trade by commodity and relate to the discussion of chapters 5 and 6. Appendixes D through M examine ten U.S. regions over the period 1967-76. The nature of each region's forest products trade is briefly described followed by considerable amount of regional specific data. Most of the data have not previously been published in a comprehensive easily accessible form.

References

1. Gerhard Herm, The Phoenicians (New York, William Morrow, 1975).
2. Marion Clawson, "Forests in the Long Sweep of American History," Science vol. 204, no. 4398 (1979) pp. 1168-1174.
3. U.S. Department of Agriculture, Forest Service, "Historical Forestry Statistics of the United States," Statistical Bulletin no. 228 (Washington, D.C., USDA, Division of Forest Economics Research, 1958).
4. Neal Potter and Francis T. Christy, Jr., Trends in Natural Resource Commodities: Statistics of Prices Output, Consumption, Foreign Trade, and Employment in the United States, 1870-1959 (Baltimore, Md., Johns Hopkins University Press for Resources for the Future, 1962).
5. Robert S. Manthy, Natural Resource Commodities--A Century of Statistics: Prices, Output, Consumption, Foreign Trade, and Employment in the United States, 1870-1973 (Baltimore, Md., Johns Hopkins University Press for Resources for the Future, 1978).

Chapter 2

SUMMARY AND FINDINGS

The world's forestry resources are substantial and broadly distributed. As measured in either land area or growing stock, the USSR (with 33 percent of the world's volume) is the dominant region. However, North America (with 16 percent of the growing stock), South America (with 27 percent), and Asia (with 15 percent) also have a large portion of the world's total forest resources, while Europe's portion is relatively modest (5 percent). Africa and Oceania make up the remainder. The world conifer (softwood) and broadleaf (hardwood) forests are less evenly distributed, with the USSR and North America accounting for about 85 percent of the world's total; adding Europe brings the percentage to over 90 percent. Broadleaf forests are more evenly distributed, with South America and Asia accounting for over 70 percent by volume and the USSR and North America accounting for another 18 percent. Within the global context, North America has about 16 percent of the total volume, with the United States and Canada having roughly equal shares of this inventory.

In 1976 the traditional world timber producers--North America, Europe, and the USSR--accounted for 46 percent of the world's roundwood production. Their dominance in 1976 was even greater for industrial roundwood (76 percent) and still greater for conifer industrial roundwood (88 percent). This dominance, as reflected in the share of worldwide production, however, has eroded considerably since 1959; the earliest date for which data are available. In 1959 the traditional producers' share was 66 percent, 83 percent, and 92 percent for roundwood, industrial roundwood, and conifer industrial roundwood, respectively.

While North America's portion of total roundwood production also declined over the 1959-76 period, her conifer industrial production share moved against the trend and rose from 38.4 percent in 1959 to about 40 percent in 1976. Within this context, the U.S. share of both total and conifer roundwood production fell. However, the decline in the U.S. conifer roundwood production share was modest (from 28.3 percent to 27 percent).

To summarize, the traditional producing regions have a greater share of the world's conifer resources than of its nonconifer resources. They also dominate all types of forest production; however, they have a greater impact on industrial conifer production and a lesser one on total roundwood production. The traditional producers' share of world production, both total and conifer, declined between 1959 and 1976. North America had a substantial share of both inventories and production and has modestly expanded its share of the world's industrial conifer production.

In 1976 only two of the world's seven continental regions had net out-flows (surpluses)[1] of forest products. North America accounted for over three-fourths of the total intercontinental surplus and the USSR had the remainder. Almost 90 percent of the intercontinental deficits were accounted for, in about equal proportions, by Europe and Asia, with the

---

[1]A surplus occurs when exports exceed imports while a deficit is defined as imports exceeding exports.

remainder of the deficit distributed among Africa, Latin America, and Oceania. Between 1961-65 and 1976 the share of the intercontinental deficit accounted for by both Latin America and Oceania declined, as these regions moved toward relatively greater self-sufficiency in forest products.

The disaggregation of the North American and European forest product trade reveals that Canada and the Nordic countries (Finland, Norway, and Sweden) are the largest exporters of forest products. During the 1950-76 period, Canadian forest product trade surplus grew almost continuously, whereas U.S. forest product trade experienced a continuous deficit that, while volatile, declined modestly in its real value. The Canadian surplus far exceeded the U.S. deficit so that a large and growing North American surplus resulted. By contrast, in Europe, Nordic forest product flows went largely to the rest of Europe and were supplemented by large inflows from outside Europe. Thus, the Nordic surplus was undercut by the deficit in the rest of Europe, leaving Europe as a whole with a large trade deficit in forest products.

Trends in world forest products trade are discussed. It is estimated that between 1950 and 1976 the physical volume of solid wood trade flows grew at an average annual rate of about 5.8 percent per annum while the wood fiber products growth rate was about 4.9 percent. A brief description of the global pattern of major forest products trade is presented.

While North America was the dominant net exporting region, within North America the United States and Canada had complementary roles. The United States had high output levels and high domestic demand. A large portion of Canadian exports of forest products went to the United States. For many items, the United States consumed a large portion of Canadian domestic production, indicating Canada's great reliance on the United States as a market. Also, while U.S. domestic production relied to a lesser extent on Canadian markets, the United States relies extensively on Canada for certain products (for example, lumber and newsprint).

The North American forest products trade balance is examined in both value and physical volume terms. In the period 1950-76 the North American net value of forest products exports (given in 1976 constant U.S. dollars) increased from a small surplus of $139.3 million in 1950 to $4.2 billion in 1976. In physical units, both solid wood and fiber products net exports from North America increased markedly between 1950 and 1976.

The most dramatic increase was in solid wood products because of unprecedented increases in industrial roundwood exports (largely conifer sawlogs) together with the large growth experienced in lumber net exports. This more than offset the increases experienced in net imports of wood panel products, caused by rising imports of tropical hardwood panels.

For fiber products, North America moved from a small trade deficit in both physical volume and value in 1950 to a substantial trade surplus in 1976. Wood pulp and paper and paperboard (including newsprint) contributed approximately equally to the surplus of 1976.

A detailed analysis of U.S. forest products trade and the U.S. forest products trade balance is undertaken. In 1950 the United States had trade deficits in all five categories--industrial roundwood, sawnwood, wood panels, wood pulp, and paper and paperboard--as measured by both volume and value. By 1976 the United States achieved a trade surplus in only the industrial roundwood category. However, in value terms (1976 constant dollars) the U.S. deficit exhibited an improvement, falling from

$2.3 billion in 1950 to $1.4 billion in 1976. However, it should be noted that, subsequently, the deficit increased markedly through 1979.

The study also describes the U.S. trading experience and patterns for thirteen major commodities. Major trading partners are identified as well as are the principal U.S. regions involved in the commodity trade. The data presented for thirteen commodities indicate the quantity and value of goods traded internationally by each region. In addition, the major foreign countries and world areas involved in U.S. trade are identified.

Although there are 130 possible pairings of ten trading regions with thirteen forest products, the bulk of U.S. international trade is concentrated only in a few regions and commodities. For both imports and exports, the ten most important region-commodity pairs constitute about 70 percent of total U.S. trade.

On the import side, the Great Lakes region is a heavy importer of newsprint, wood pulp, and softwood lumber from Canada. This region alone accounted for 46 percent of total U.S. imports of forest products in 1976.

Relatively large shares of the total were also captured by the North Atlantic (19 percent of the 1976 value) and the Pacific Northwest (14 percent) regions. Again, the important commodities were newsprint, wood pulp, and softwood lumber, imported principally from Canada.

In 1976 nearly half the U.S. forest product exports were of paper and paperboard (other than newsprint) and wood pulp, but the largest single flow was softwood logs exported from the Pacific Northwest to Japan. The region also exported considerable amounts of softwood lumber, wood pulp, and paper and paperboard; in 1976 the Pacific Northwest accounted for almost 40 percent of total U.S. exports, by value.

The North Atlantic, South Atlantic, Gulf, and Great Lakes regions were all important forest product exporters, predominantly of wood pulp, and paper and paperboard (excluding newsprint). Canada and the western European nations were the chief importers of these U.S. products.

Attention is also given to the comparative export performance of the ten U.S. regions, with the focus being upon the relative performance of the Pacific Northwest region and the regions comprising the U.S. South. Over the 1967-76 period the Pacific Northwest region increased its dominance in forest products exports, which are largely softwood, even as the South sharply expanded its relative share of total softwood product production. This suggests that increased southern production is being marketed domestically, perhaps displacing Pacific Northwest production while Pacific Northwest products are increasingly being marketed abroad.

Chapter 3

WORLD FOREST RESOURCES AND FOREST PRODUCTS TRADE--AN OVERVIEW

The world's forest products trade has many dimensions. Total trade and the rate of increase in trade are important, but also are the distribution of forestry resources, the commodity composition of forest trade, the existing interregional structure of forest trade flows, and recent changes in that structure.

In this chapter we will present an overview of world forestry resources, world timber production, and some postwar regional trends and structural changes in international trade of aggregate world forest products, and discuss briefly recent international trade flows in important forest products. Our purpose is to provide the reader with a descriptive overview of the current global situation in forest resources and forest products trade, as well as a perspective on some of the important changes that have occurred in recent years. This perspective will be useful later when the roles of the United States and North America are examined in the context of global trade in forest products.

World Forestry Resources

Economists believe that patterns of international trade flows generally reflect the comparative economic advantage of a specific region.

Therefore, a reasonable starting point for such an analysis is with an assessment of the existing inventories of world forest resources.

Forests are a renewable resource. Not all forest inventories or forestlands are of equal biological or economic productivity. The existing inventory base per se tells us little about the rate at which the harvested forest can be regrown and, therefore, little about the potential inventory of a region over the long term. To a large extent, existing inventories are the result of uncontrolled natural forces and historical accident. Thus, in the long run, we may well expect that both the locational distribution and the level of such inventories are likely to change. Many of these changes will reflect economic considerations and hence will be directed to sites whose economic potential for forest production appears to exceed their value for alternative uses. Other noneconomic factors also influence the location of forests and have implications for future production. However, the task of this report is to focus upon existing and near-term forest product trade flows and changes in those flows. Our concern here, therefore, is not to determine where new forests may be developed. Rather we hope that by evaluating existing forest inventories we will be able to determine recent, current, and even near-term future forest product flows.

In 1968 the Food and Agricultural Organization (FAO) of the United Nations began the fifth in a series of world forest inventory reviews. However, because of financial considerations, the 1968 study was never completed. Fortunately, Reidar Persson, the compiler of the study, arranged for independent publication of the incomplete work (1).

Tables 3-1 and 3-2 present information on the world's inventory of forests, as developed by Persson. As can be seen, the world's total forestry resources are broadly distributed, with the USSR having the greatest land area and the highest volume of growing stock. However, North America, South America, and Asia possess large portions of the world's total forest resource.

The distribution between coniferous (softwood) and broadleaf (hardwood) forests, an important distinction for discussions of forest products and forest product trade flows, reveals a much less evenly distributed pattern. The USSR and North America account for almost 85 percent of the world's coniferous forests by both the land area and volume measures, and with Europe account for over 90 percent. Broadleaf forests are more evenly distributed worldwide than are conifers. South America and Asia account for over 50 percent of the broadleaf forest by land area and over 70 percent by volume. The USSR and North America account for about 18 percent of the world's hardwood volume. Europe, historically a major producer of forest products, especially of conifer products, accounts for less than 8 percent of the world's conifer forests by either land or volume measures and for only about 3 percent of broadleaf forests.

Within the global context the United States and North America are obviously important holders for some 22.3 percent of the land area covered by forests, and about 16.1 percent of the global volume of growing stock. The distribution of inventories is roughly equally divided between the United States and Canada.

North America accounts for over 32 percent of the coniferous forest-land area and for 25.8 percent of the world's total volume of coniferous

Table 3-1. Land Area of World Forest Resources by Region and Type

(million hectares)

	Coniferous		Broadleaf		Combined coniferous and broadleaf forests[a]	
Region	Land area	%	Land area	%	Land area	%
North America	400	35.2	230	14.1	630	22.3
Central America	20	1.8	40	2.4	60	2.1
South America	10	0.8	550	33.6	560	19.8
Africa	2	0.2	188	11.5	190	6.7
Europe	75	6.6	50	3.0	140	5.0
USSR	553	48.7	175	10.7	765	27.1
Asia	65	5.7	335	20.5	400	14.2
Oceania	11	1.0	69	4.2	80	2.8
Total world	1136	100.0	1637	100.0	2825	100.0

Note: Closed forests.

Source: Reidar Persson, World Forest Resources: Review of the World's Forest Resources in the Early 1970s (Stockholm, Royal College of Forestry) no. 17 (1974).

[a] The totals for combined coniferous and broadleaf forests do not always add because no breakdowns have been given for areas in Europe and the USSR excluded by law for exploitation.

Table 3-2. World Growing Stock: Volume in Closed Forest

(100 million cubic meters)

Region	Coniferous		Broadleaf		Combined coniferous and broadleaf forests[a]	
	No.	%	No.	%	No.	%
North America	265	25.8	95	7.9	360	16.1
Central America	7	0.7	15	1.2	22	1.0
South America	5	0.5	595	49.6	600	26.8
Africa	1	0.1	51	4.3	52	2.3
Europe	80	7.8	40	3.3	120	5.4
USSR	612	59.5	120	10.0	733	32.7
Asia	55	5.3	285	23.7	340	15.2
Oceania	3	0.3	0	0	13	0.5
Total world	1028	100.0	1201	100.0	2240	100.0

Source: Reidar Persson, World Forest Resources: Review of the World's Forest Resources in the Early 1970s (Stockholm, Royal College of Forestry) no. 17 (1974).

[a]Sum of coniferous and broadleaf forests do not always add to total.

growing stock. However, North America's share of the world's economically recoverable conifer forests is probably larger than her percentage of total volume might suggest since much of the larger USSR stock is located in very inaccessible regions and it never may be economic to harvest it (2).

World Timber Production

As noted, North America, Europe, and the USSR hold more than half the world's total forest resources and more than 90 percent of the world's coniferous forests. Because these regions control most of the world's resources and have a high level of economic development, it is not surprising that the vast majority of forest products, both primary and processed, originated in these regions. This study will refer to these as the "traditional" producers.

In recent years, however, substantial changes have occurred. Between 1959 and 1976 total world roundwood production, the broadest category of wood production, is estimated to have increased in volume by 51.2 percent (table 3-3) (3). However, the total roundwood production of the traditional forest-producing regions--North America, Europe, and the USSR--increased only 5.8 percent during that seventeen-year period, leading to substantial declines in the share of total roundwood production in these traditional producing regions. This is caused in part by absolute production declines in non-Nordic Europe and the USSR. The traditional regions' share of total world production fell from 65.9 percent in 1979 to 46.1 percent in 1976. All the traditional regions experienced decreased shares with the exception of Canada, whose share remained essentially unchanged. The United States, whose production increased only by 6.9 percent during

Table 3-3. World Roundwood Production for 1959 and 1976

(1,000 cubic meters)

Region	1959		1976	
	No.	Percentage of total	No.	Percentage of total
North America	405,627 (100.0)	24.3	473,790 (116.8)	18.7
United States	319,421 (100.0)	19.1	341,397 (106.9)	13.5
Canada	86,206 (100.0)	5.2	132,393 (153.6)	5.2
Europe (excluding Nordic countries)	210,545 (100.0)	12.6	195,612 (92.9)	7.7
Nordic countries[a]	87,215 (100.0)	5.2	110,206 (126.4)	4.5
USSR	397,000 (100.0)	23.8	384,534 (96.9)	15.2
Other	569,728 (100.0)	34.1	1,360,077 (238.7)	53.9
Total world	1,670,115 (100.0)	100.0	2,524,219 (151.2)	100.0

Note: Parenthetical figures give percentage change in output with 1959 = 100.

Source: Food and Agricultural Organization, Yearbook of Forest Products, selected issues (Rome, FAO).

[a]Finland, Norway, and Sweden.

the seventeen-year period, saw a decline in its share of world production from 19.1 percent to 13.5 percent. North America (the United States and Canada) of course also experienced a decline, from 24.3 percent in 1959 to 18.7 percent in 1976.

However, industrial roundwood, which excludes fuelwood and charcoal that are included in the "roundwood" definition, is a better indication of a country's industrial performance in forest resource production, since it does not include the large fuelwood volumes so common in developing countries. Table 3-4 summarizes comparative worldwide production of industrial roundwood for 1959 and 1976. Here, the traditional producers supplied 76.3 percent of total world industrial roundwood, down somewhat from their 82.9 percent in 1959. Of the traditional producers, Canada increased production 64 percent to raise her total share of industrial roundwood from 8 percent in 1959 to 9.6 percent by 1976. The decline in the U.S. share of world production from 27.7 percent to 24.4 percent because of the slower than average growth of U.S. production, resulted in a slight decline of the total North American share from 35.7 percent to 34 percent. During this period, the shares of total world production of both the USSR and the Nordic countries also declined despite increases in their production.

Given their dominance over conifer inventories, traditional producing regions not unexpectedly increased their volume output of industrial conifer roundwood at almost the same rate as the increases in worldwide production (table 3-5). While world production increased 30.1 percent in the 1959-76 period, the traditional producers increased production by 26 percent. During this period, Canada increased her production by 62.6

Table 3-4. World Production of Industrial Roundwood

(1,000 cubic meters)

Region	1959 No.	1959 Percentage of total	1976 No.	1976 Percentage of total
North America	350,108 (100.0)	35.7	455,643 (130.1)	34.0
United States	271,652 (100.0)	27.7	326,991 (120.4)	24.4
Canada	78,456 (100.0)	8.0	128,652 (164.0)	9.6
Europe (excluding Nordic countries)	125,940 (100.0)	12.8	176,168 (139.9)	13.1
Nordic countries[a]	68,225 (100.0)	7.0	88,125 (129.2)	6.6
USSR	269,000 (100.0)	27.4	302,932 (113.5)	22.6
Other	167,422 (100.0)	17.1	317,261 (189.5)	23.7
Total world	980,695 (100.0)	100.0	1,340,129 (136.7)	100.0

Note: Parenthetical figures give percentage change in output with 1959 = 100.

Source: Food and Agricultural Organization, *Yearbook of Forest Products*, selected issues (Rome, FAO).

[a]Finland, Norway, and Sweden.

Table 3-5. World Production of Industrial Conifer Roundwood

(1,000 cubic meters)

Region	1959 No.	1959 Percentage of total	1976 No.	1976 Percentage of total
North America	279,292 (100.0)	38.4	375,191 (134.3)	39.6
United States	205,830 (100.0)	28.3	255,729 (124.1)	27.0
Canada	73,462 (100.0)	10.1	119,462 (162.6)	12.6
Europe (excluding Nordic countries)	87,190 (100.0)	12.1	114,040 (130.8)	12.1
Nordic countries[a]	65,250 (100.0)	8.9	79,367 (121.6)	8.4
USSR	232,900 (100.0)	32.1	268,954 (115.5)	28.4
Other	62,088 (100.0)	8.5	108,193 (174.3)	11.5
Total world	726,720 (100.0)	100.0	945,745 (130.1)	100.0

Note: Parenthetical figures give percentage change in output with 1959 = 100.

Source: Food and Agricultural Organization, Yearbook of Forest Products, selected issues (Rome, FAO).

[a]Finland, Norway, and Sweden.

percent, while that of the United States increased by 24.2 percent. This resulted in a modest increase in North America's share of total industrial conifer production, from 38.4 percent in 1959 to 39.6 percent in 1976, despite a modest decline in the U.S. share.

Also experiencing modest declines in production were the USSR (from 32.1 percent to 28.4 percent) and the Nordic countries (from 8.9 percent to 8.4 percent). Meanwhile, Europe (excluding the Nordic countries) maintained an unchanged share (12.1 percent), whereas the rest of the world, the nontraditional producers, increased their share from 8.5 percent to 11.5 percent.

To summarize, total world roundwood production increased by 51.1 percent between 1959 and 1976, industrial roundwood increased by 36.7 percent, and industrial conifer roundwood by 30.1 percent. North America, Europe, and the USSR increased their total production by 5.8 percent for roundwood, 25.8 percent for industrial roundwood, and 26 percent for industrial conifer roundwood. They have almost maintained their 1959 share of world conifer production while, at the same time, experiencing declines in their share of industrial roundwood production, the two categories that have experienced rapid worldwide expansion.

North America increased production in all these categories between 1959 and 1976. Roundwood production increased 16.8 percent, industrial roundwood production increased 30.1 percent, and industrial conifer roundwood increased 34.3 percent.

Despite absolute production increases, however, North America's share of total world output dropped for both roundwood and industrial roundwood but increased slightly for industrial conifer roundwood. Against the trend, both Canadian production and shares increased in all three categories, while the U.S. share in each declined despite modest output increases.

Trends in World Forest Products Trade

Worldwide trade in forest products has increased markedly in the period following World War II. However, it has not increased symmetrically across products and regions. In this section, we will examine the growth of forest product trade flows in the aggregate, by major commodity and by major world region. Throughout, we will focus on changes in both the level and the structure of forest product trade flows. Since we are concerned with long-term changes in the level and structure of trade, we will not examine year-to-year changes. Rather, our approach has been to choose representative years, usually well over a decade apart, and to focus upon the changes which have occurred between these time periods.[1]

Trade Flows in Major Forest Products

Table 3-6 presents the volume trade levels by major forest product commodity groups for 1950 and 1976. As the table shows, over the twenty-six-year period, the physical volume of traded forest products increased by an average compounded rate of about 5.8 percent for solid wood products

---

[1]The U.S. forest products trade appears to be highly sensitive to the business cycle. Our purpose here is to examine structural changes that reflect secular rather than cyclical phenomena. The 1961-65 period was chosen because the several years represented various stages of the business cycle. The year 1976 was chosen for the terminal year, not only because it was the latest available when this study began, but also because it is viewed as "representative" in that it avoids the atypical characteristics of 1975, the trough of a business cycle, and also of 1978, a year that is near the peak of a cycle.

Table 3-6. World Forest Product Trade Flows

Product (1,000 cubic meters)	1950	1976	1976 as a percentage of 1950
Solid Wood			
Sawlog, conifer	2,670	28,420	1,064.4
Sawlog, nonconifer	2,750	43,419	1,578.9
Pulpwood	6,850	15,886	231.9
Sawnwood, conifer	21,420	53,639	250.4
Sawnwood, nonconifer	3,050	10,212	334.8
Veneer sheets	155	1,382	891.6
Plywood	406	6,257	1,541.1
Particleboard[a]	128	4,494	3,510.9
Fiberboard	392	1,979	504.8
Total Solid Wood	37,821	165,688	438.1[b]
Fiber			
Paperboard	1,945	15,866	815.7
Newsprint	5,540	10,746	194.0
Pulp	5,730	16,358	285.5
Total fiber	13,215	42,970	325.2[c]

Source: Food and Agricultural Organization, Yearbook of Forest Products, selected issues (Rome, FAO).

[a]Particleboard data reflect 1959 data since earlier data are unavailable.

[b] $\cong$ 5.8 percent per annum.

[c] $\cong$ 4.87 percent per annum.

and 4.8 percent for paper and pulp products. These rates are comparable to the real growth rates in GNP experienced by the developed countries during that period. It should be noted, however, that the volume of industrial roundwood production grew at an annual rate of only 1.9 percent worldwide between 1959 and 1976 (see table 3-4). Unfortunately, we do not have accurate world production figures for the postwar period before 1979. However, if the 1959-76 production experience is representative of the 1950-59 period, this suggests that the growth in the volume of forest products entering international trade was considerably more rapid than the growth rate of total physical output for the entire period 1950-76. Thus, for forest products, the volume of international trade growth may have increased 2.5 to 3 times as rapidly as overall production.

An interesting aspect of the growth in volume of international trade in forest products involves the disparities in growth rates among major commodities. The slowest-growing commodities are largely the traditional ones, for example, pulpwood, sawnwood, newsprint, and pulp. However, both conifer and nonconifer sawlogs, particularly board and plywood, reached international traded volumes in 1976 in excess of 1,000 percent of their 1950 traded levels. The growth of panel products is certainly related to the fact that, as new commodities, they also experienced some decline in real price as the technology became perfected. Particleboard, an even more newly developed product, achieved its dramatic increase in trade (3,510.9 percent) between 1959 and 1976.

The rapid growth in the international trade of sawlogs is an anomaly within the overall context of international trade and is caused largely by the influence of one importing country--Japan--on the log market. The importation of hardwood sawlogs, which are imported mainly from developing countries in Southeast Asia, may be explained by the limited processing facilities that these countries have had. However, such an explanation is inadequate for the conifer sawlogs which are imported from some of the more industrialized countries, with 90 percent coming from the United States and the USSR. The log flow into Japan apparently reflects unique Japanese lumber specifications and national policy.

Regional Contributions to Aggregate Forest Product Trade Flows

In the previous section, we compared the 1950 and 1976 physical volumes of trade worldwide for major forest products. In this section we will examine the values of aggregate forest product trade flows by major world regions for a shorter time period. The level and share of each major region in the aggregate exports and import values of forest products are determined for 1976 and are averaged for 1961-65. Our principal focus will be on the structural changes that occurred during that period.

Following our earlier regionalization format, we will examine "intercontinental" trade flows using seven regions--the five continents (with Central America aggregated with South America) plus the USSR and Oceania. Subsequently, North America is divided into the United States and Canada, and Europe is divided into Nordic and non-Nordic components. We then will examine the nine-region trade flows. It should be noted that the portion of total international trade defined as <u>interregional</u> increases as the regions are disaggregated.

Table 3-7 compares the total value and share of industrial forest product gross international exports, both interregional and intraregional, by major world region for 1976 with the average of the 1961-65 period. While the total nominal value of the world's forest product trade increased by over 400 percent during this period, we will focus on changes in the regional structure of those exports. As with production, the traditional producers' share of gross world exports declined over the period. The 1976 share was 83 percent as compared with 88.5 percent in the 1961-65 period. Of the worldwide total for gross exports of forest products, the European countries' share dropped from 45.7 percent for 1961-65 period to 41.5 percent in 1976. The Nordic countries' share fell sharply from 29.4 percent in 1961-65 to 22.1 percent in 1976, and the share of the non-Nordic European countries rose from 16.3 percent to 21.4 percent. North America's share of gross exports also declined, dropping from 35.9 percent to 33.0 percent over the same period. However, the U.S. share rose from 10 percent in 1961-65 to 12.9 percent in 1976, and Canada's share declined from 25.9 percent to 20.1 percent. The share losses of the traditional exporters were almost exactly captured by the expanding share of exports captured by Asia. By 1976 Asian exports had increased to 11.6 percent of the world gross from a 5.9 percent share in 1961-65. Oceania also experienced an increase from 0.6 percent to 1.1 percent.

Table 3-8 compares the total value and share of gross imports of forest products by major region for the same time periods. Again, the nominal value increased over 400 percent. However, North America's share of gross world imports of forest products dropped from 23 percent in

Table 3-7. World Forest Product Exports by Region for Selected Years (million current $)

Region	1961-65 Average		1976	
	Value	Share of total (%)	Value	Share of total (%)
North America	2,296.2	35.9	10,235.8	33.0
United States	636.5	10.0	3,994.4	12.9
Canada	1,659.7	25.9	6,241.4	20.1
Europe	2,915.3	45.7	13,504.5	43.5
Nordic countries[a]	1,875.4	29.4	6,868.2	22.1
Non-Nordic countries	1,039.9	16.3	6,636.3	21.4
USSR	435.7	6.8	2,008.1	6.5
Africa	232.1	3.6	853.8	2.8
Asia	374.7	5.9	3,604.9	11.6
Latin America	96.9	1.5	476.1	1.5
Oceania	38.5	0.6	339.2	1.1
World total	6,389.4	100.0	31,022.4	100.0

Source: Food and Agricultural Organization, *Yearbook of Forest Products*, selected issues (FAO, Rome).

[a]Finland, Norway, and Sweden.

Table 3-8. World Forest Product Imports by Region for Selected Years (million current $)

Region	1961-65 Average		1976	
	Value	Share of total (%)	Value	Share of total (%)
North America	1,770.7	23.0	6,053.1	17.5
United States	1,630.0	21.2	5,431.0	15.7
Canada	140.7	1.8	622.1	1.8
Europe	4,223.8	54.8	18,411.4	53.1
Nordic countries[a]	136.5	1.8	857.7	50.6
Non-Nordic countries	4,087.3	53.0	17,553.7	2.5
USSR	98.9	1.3	519.1	1.5
Africa	248.3	3.2	1,072.3	3.1
Asia	858.0	11.1	7,003.0	20.2
Latin America	343.4	4.5	1,128.2	3.2
Oceania	167.7	2.1	486.1	1.4
World total	7,710.8	100.0	34,673.2	100.0

Source: Food and Agricultural Organization, Yearbook of Forest Products, selected issues, (FAO, Rome).

[a]Finland, Norway, and Sweden.

1961-65 to 17.5 percent in 1976. This decrease can be accounted for by the decline in the U.S. share of gross world imports--from 21.2 percent in 1961-65 to 15.7 percent in 1976. Canada's share remained constant at 1.8 percent. Europe's share also decreased slightly, from 54.8 percent in 1961-65 to 53.1 percent in 1976. Asia experienced an increase, from 11.1 percent in 1961-65 to 20.2 percent in 1976.

Net Interregional Trade Flows

As an extension of the foregoing analysis, we will examine the net interregional trade flows. In order to obtain net values, it is usual to subtract import values from export values. As can be seen from the totals given in tables 3-7 and 3-8, the aggregate data is not entirely consistent. Since the import total includes cost, insurance, and freight (c.i.f.) and the export total is free on board (f.o.b.), the import total is systematically higher. Therefore, imports have been adjusted downward by a factor that equated the export and import totals while leaving the proportions unchanged. The adjusted imports were then subtracted from the exports (tables 3-9 and 3-10).

Table 3-9 provides information on intercontinental net trade flows for forest products, with the USSR included separately. It will be noted that the volume of trade flows reported here are only those net flows that are involved in intercontinental trade, and therefore, the values are less than total trade. Again, the two periods examined are 1961-65 and 1976.

The structure of intercontinental net flows for 1976 can be compared with those of the 1961-65 average. In 1961-65 net outflows (exports) equaled about 19 percent of worldwide trade in forest products by value

Table 3-9. Net Interregional Trade in Forest Products (million $)

Region	1961-65				1976			
	Surplus	% of total surplus	Deficit	% of total deficit	Surplus	% of total surplus	Deficit	% of total deficit
North America	828.9	68.6			4820.0	75.7		
Europe			(584.7)	48.4			(2968.3)	46.6
USSR	353.8	29.3			1543.7	24.3		
Africa	26.4	2.1					(105.7)	1.7
Asia			(336.3)	27.8			(2660.7)	41.8
Latin America			(187.7)	15.5			(533.3)	8.4
Oceania			(100.4)	8.3			(95.7)	1.5
TOTAL	1209.1	100.0	(1209.1)	100.0	6363.7	100.0	(6363.7)	100.0

Source: See table 3-10.

as compared with 20.5 percent in 1976. Thus, the intercontinental category had increased its trade in forest products only very slightly over the intervening years. However, the contribution of the various continents shifted.

In 1961-65 North America accounted for about 68.6 percent of the world's total exports in forest products, the USSR accounted for 29.3 percent, and Africa for 2.1 percent. By 1976 the North American share rose to 75.7 percent, but the USSR's share declined slightly to 24.3 percent, and Africa had modest net inflows (net imports) of forest products.

The major intercontinental net importers in both periods were Europe and Asia. Together they constituted 76 percent of total intercontinental net imports of forest products in 1961-65. In 1976 this grew to 88.6 percent, with Asia experiencing a large increase in its share and Europe's share declining slightly. In absolute value, of course, both continents experienced large increases in net imports of forest products. Simultaneously, the net import share of both Latin America and Oceania fell, with Oceania's net imports declining in absolute terms as well.

A further degree of regional disaggregation is shown in table 3-10. North America is subdivided into the United States and Canada, and Europe is separated into Nordic and non-Nordic countries. Several observations can be made utilizing this regional configuration.

First, North America's continental net exports clearly originate from Canada, whereas the source of Europe's net imports clearly resides with the non-Nordic countries. Second, Canada and Nordic Europe are the dominant net exporters, together accounting for over 88 percent of

Table 3-10. Net Trade in Forest Products for Nine World Regions (millions $)

Region	1961-65				1976			
	Surplus	% of total surplus	Deficit	% of total deficit	Surplus	% of total surplus	Deficit	% of Total deficit
United States			(714.2)	19.4			(864.8)	6.5
Canada	1,543.1	41.9			5,684.8	42.6		
Nordic countries[a]	1,762.3	47.8			6,100.8	45.8		
Europe			(2,347.0)	63.7			(9,069.1)	68.0
USSR	353.8	9.6			1,543.7	11.6		
Africa	26.4	0.7					(105.7)	0.8
Asia			(336.3)	9.1			(2,660.7)	20.0
Latin America			(187.7)	5.1			(533.3)	4.0
Oceania			(100.4)	2.7			(95.7)	0.7
TOTAL	3,685.6	100.0	(3,685.6)	100.0	13,329.3	100.0	(13,329.3)	100.0

Source: Tables 3-7 and 3-8. To construct this table, the figures in table 3-8 were adjusted proportionately downward to allow total imports to equal the total exports of table 3-7.

[a]Finland, Norway, and Sweden.

nine-region net outflows for both 1961-65 and 1976. The USSR accounted for most of the remaining net outflows in 1961-65 and for all the remainder of 1976. Finally, the most dramatic changes observed over the period are the decline in the U.S. share of net imports and the simultaneous increase of Asia's share.

Several observations can be made based on the above analysis. First, the dominant continental forest product supplier to world markets for both the 1961-65 and 1976 periods was North America--and within North America, it was Canada. North America's contribution to the intercontinental net exports decreased between 1961-65 and 1976. Second, the dominant continental net importers were Europe and Asia, with Asia's share of net imports increasing substantially. Within Europe, however, the exports from the Nordic countries provided a large portion of forest products imports to the rest of Europe. Third, the United States, while a large net importer, experienced a marked reduction between 1961-65 and 1976 in its share of the deficit. Finally, the import dependency of both Latin America and Oceania experienced a substantial relative decline over the period.

World Trade Flows in Major Commodities for 1976

To provide the reader with a quick overview of major world forest product trade flows, this section sketches briefly major flows for 1976. The data used are those available in selected issues of the FAO's *Yearbook of Forest Products*.

Conifer Sawlogs and Veneer Logs

The dominant 1976 world flow of conifer sawlogs was to East Asia from the United States and the USSR.

Japan was easily the major importer of conifer logs, accounting for about 66 percent of the total volume of world imports in 1976. Other large importers were Canada (6.1 percent), Italy (4 percent), and South Korea (2.6 percent).

The principal exporters of conifer sawlogs were the United States, which exported 50.5 percent of the world's volume in 1976, followed by the USSR with 32.1 percent.

Nonconifer Sawlogs and Veneer Logs

East Asia was the major destination of world nonconifer sawlog flows, with the most sawlogs originating in Southeast Asia. A much smaller secondary flow went from West Africa to Europe.

Japan was the dominant importer of nonconifer sawlogs, accounting for 50.8 percent of the 1976 volume imported. Other major importers were South Korea, with 12.8 percent of world imports, Taiwan with 10.8 percent, and Italy with 6 percent.

Major exporter countries in 1976 were primarily in Southeast Asia, with Indonesia accounting for 39 percent of total world export volume, Malaysia 34.6 percent, and the Philippines 5.2 percent. The Ivory Coast was also important, accounting for 7.3 percent of world's nonconifer sawlog exports in that year.

Pulpwood

The major portion of pulpwood (excluding chips and particles) trade in 1976 involved European countries, with a large segment of this being intra-European trade. The USSR was the major world exporter, supplying both Europe and Japan. Small intra-North American flows also occurred.

The Nordic countries were the world's major pulpwood importers, accounting for 41.7 percent of the total 1976 import flows, with Finland accounting for 22 percent of total pulpwood imports in 1976, and Sweden and Norway accounting for 13.3 percent and 6.4 percent, respectively. Other European countries with substantial portions of world pulp imports included Italy (7.7 percent), Belgium (7.5 percent), the Federal Republic of Germany (5.8 percent) and the Democratic Peoples Republic of Germany (5.2 percent). Japan imported 4.9 percent of the total.

The USSR accounted for 38.4 percent of 1976 world pulpwood exports. The Federal Republic of Germany and Czechoslovakia each accounted for 9 percent, with France and Sweden having 7.7 percent and 7.2 percent, respectively.

Chips and Particles

Although chips and particles are readily substitutable for pulpwood, the international trade patterns are somewhat different. Japan imported 86.4 percent of the 1976 total volume of wood chips. The United States was the major world exporter, accounting for 55.5 percent of the 1976 total volume. Australia was next, with 22.7 percent of the world's exports.

Conifer Sawnwood

In 1976 two major conifer sawnwood trade flows existed--one within North America (largely from Canada to the United States) and the other within Europe (from the USSR and the Nordic countries to other European countries). Also, there were secondary flows from North America to both Europe and Japan.

The United States accounted for 33 percent of world imports of conifer sawnwood. The European countries, excluding the USSR, accounted for 49.9 percent of the import total, with the United Kingdom having 13.4 percent, Italy 6.6 percent, France 5.7 percent, and the Federal Republic of Germany 5.4 percent. Japan accounted for 5.6 percent of the world's imports.

Canada was the major exporter of conifer sawnwood and accounted for 27.1 percent of the 1976 total, with most of this going to the United States. Canada was followed by the USSR (13.8 percent), Sweden (9.4 percent), and the United States (5.7 percent).

Nonconifer Sawnwood

The principal world flows of nonconifer sawtimber originated in Southeast Asia and flowed principally to Europe. Much smaller export flows originated in Africa, the United States, and Europe.

Europe imported 50 percent of total world imports of nonconifer sawnwood (by volume). This volume, however, was widely dispersed among the European countries. Italy (10.7 percent), the Federal Republic of Germany (7.3 percent), and the Netherlands (7.1 percent) were the largest European importers. Other major world importers included Singapore (10.7 percent) and the United States (6.6 percent).

The primary world exporter of nonconifer sawnwood was Southeast Asia, which accounted for 50.4 percent of total exports. In 1976 the major exporting countries were Malaysia (27.7 percent), Singapore (10.3 percent), Indonesia (6.0 percent), and the Philippines (4.5 percent).

Plywood

Plywood in world trade consists of two different commodities--softwood plywood, common largely to North America, and hardwood plywood, the dominant type outside of North America. The lack of worldwide data distinguishing plywood by type makes it difficult to separate the two essentially different flows. However, to a large degree the type of plywood can be determined by its country of origin. The dominant flow was hardwood plywood from East and Southeast Asia to North America and, to a lesser extent, Europe. A secondary flow, largely hardwood, went from the Nordic countries and the USSR to Europe. Also, there was a secondary softwood flow, originating in the United States and Canada, and flowing largely to Europe.

Principal Asian exporting countries in 1976 were South Korea (27.1 percent), Taiwan (14.4 percent), Singapore (7.7 percent), Malaysia (6.8 percent), and the Philippines (4.3 percent). The Asian exports were undoubtedly almost exclusively hardwood. In addition, the United States and Canada exported 7.7 percent and 4.2 percent, respectively, of total wood exports, and these were predominantly softwood plywood. (See chapter 5 for a fuller discussion of plywood.) Also, the USSR (5.3 percent), Finland (5.5 percent), and other European countries exported what was probably predominantly hardwood plywood.

Europe (with 44.8 percent) and North America (with 39.3 percent) were the principal importers of plywood, with the United States importing 30.9 percent and the United Kingdom 16.9 percent. Most North American imports were hardwood, with the principal exporters being Korea and Taiwan. Europe exported from both Asia and North America, and thus undoubtedly included substantial quantities of both hardwood and softwood plywood imports.

Wood Pulp

The major exporters of wood pulp originated in Europe and North America, with European flows going largely from the Nordic countries to the rest of Europe and North American flows going from Canada to the United States and also out of North America.

European imports accounted for 61.1 percent of 1976 world imports of wood pulp. The major importers were the United Kingdom (12.4 percent), the Federal Republic of Germany (9.5 percent), France (7 percent), and Italy (6.1 percent). The United States was the world's dominant importer, accounting for 16.9 percent of world wood pulp imports in 1976. Japan was also a major importer, accounting for 6.3 percent of the total.

North America was the major exporting region, accounting for 49.6 percent of the total volume flow. Canada, the world's largest exporter of wood pulp, dominated this flow and accounted for 36.2 percent of the world's total. Europe accounted for 36.9 percent of exports, with the principal exporting countries being Sweden (19.1 percent), Finland (5.6 percent), and Norway (3.5 percent).

Newsprint

World newsprint flows in 1976 were similar to those of wood pulp. North America and Europe were both the major importing and exporting regions, with the North American flow going from Canada to the United States and the European flow going largely from the Nordic countries to the others. Again, North America produced a small residual flow for European and other non-North American destinations.

North America (58 percent) and Europe (28.8 percent) dominated newsprint imports in 1976. Of this, the United States accounted for 55.4 percent of the total. Major European importers were the United Kingdom (10.6 percent), Federal Republic of Germany (6.4 percent), France (3.1 percent), and the Netherlands (2.4 percent).

Exports originated mainly in Canada (68.6 percent), the Nordic countries (20.7 percent), and the USSR (2.8 percent).

Paper and Paperboard

Europe was the major importer of paper and paperboard (excluding newsprint) in 1976. As with the other pulp and paper products, these originated largely in the Nordic countries and flowed to the rest of Europe. North America, principally the United States, also experienced large outflows with a substantial portion of these going to Europe.

Europe imported 66.4 percent of the 1976 world volume of paper and paperboard production. The major importers were the Federal Republic of Germany (20.4 percent), the United Kingdom (17.4 percent), France (11.9 percent), the Netherlands (7.4 percent), and Belgium (6.5 percent). Others were the USSR (4.8 percent), Canada (4.7 percent), and the United States (4.1 percent). The Nordic countries were the major world exporters of paper and paperboard, accounting for 51.6 percent in 1976, followed by the United States with 23 percent. Other major exporters included the Federal Republic of Germany (9.4 percent), Canada (8.8 percent), Austria (6.4 percent), the Netherlands (6.1 percent), and the USSR (5.5 percent).

Summary

This chapter has shown that, while worldwide production of major forest products has been increasing, the traditional forest-producing regions have experienced some declines in their share of production and trade in forest products, and the importance of the United States and North America as a whole has grown in the post-World War II period. The United States and North America now hold large volumes of the world's forestry resources, particularly conifers, and have been increasing their aggregate value of world conifer industrial roundwood production. North America has been the major international supplier of forest products, and its share increased substantially over the period 1961-65 to 1976 while the U.S. forest product trade deficit declined over the 1961-65 to 1976 period. The USSR is the other major continental supplier.

References

1. R. Persson, World Forest Resources: Review of the World's Forest Resources in the Early 1970s (Stockholm, Royal College of Forestry) no. 17 (1974).
2. W. R. J. Sutton, "The Forest Resources of the USSR: Their Exploitation and Their Potential," Commonwealth Forestry Review vol. 54 (1975) pp. 109-138.
3. Food and Agricultural Organization, Yearbook of Forest Products, selected issues (Rome, FAO).

Chapter 4

NORTH AMERICAN FOREST PRODUCTS TRADE

In the previous chapter, we examined the volume of world forest resources, recent changes in the contribution of major regions to output and trade flows, post-World War II trends in world trade given by major forest product, and described recent interregional forest product trade flows by major commodity groupings. This chapter will focus on recent trends in North America's international trade in forest products. A hypothetical trade flow pattern for North America will be articulated, based upon U.S. and Canadian resource endowments and market locations. Against this backdrop, the changing structure of the North American international trade flows and trade balance will be investigated. We will examine both aggregate and disaggregated forest product flows. Changes in North America's trade balance for forest products over the past twenty-six years will be analyzed, and the respective contributions of the United States and Canada will also be examined. Also, intra-North American trade flows are examined, and the nature of the U.S.-Canadian forest product trade relationship is discussed. As in other chapters, forest product trade patterns are examined both in terms of physical volume and value.

North America as Supplier of Forest Products to the World

The Free Trade Paradigm

In a world free of trade barriers, where international trade patterns were determined by underlying comparative advantage, a *prima facie* judgment would probably state that North America was a major world supplier of forest products because it is rich in forest resources, particularly conifer resources. However, the mere existence of resources does not necessarily ensure the region's comparative advantage in production and export of these resources. Not only must the resource base be considered, but the entire spectrum of costs associated with the transformation of the raw unexploited resource (in this case, forestland and stumpage) into economic commodities--for example, sawlogs, pulpwood, and chips--must also be noted. In addition, markets (demand) must exist with prices sufficiently attractive to justify incurring the costs of access, exploitation, transportation, and so forth.

Furthermore, comparative economic advantage implies not only economic production, but also a production level, given market prices, in excess of domestic demand, thereby generating a surplus or "excess supply" for foreign markets. This condition must obtain since it is conceivable that domestic production, even though considerable, could be wholly absorbed into domestic markets, thus allowing for no net forest products surplus for export such as is currently the case in Europe (excluding the USSR).

However, the relationship in North America between forest resources (making up 30 percent of the world's total resources) and market size (as indicated roughly by North America's 6 percent of the world's population) suggests a potentially important role for North America as a major world forest product exporter. As we will subsequently show, the actual post-World War II performance of forest product exports is consistent with a North American comparative advantage in forest product production.

Although North America might be expected to be a major net exporter of forest products, not all regions within North America would be expected to participate equally in either domestic or foreign markets. Even though the forest resources of the United States and Canada are roughly comparable, domestic demand levels are certainly not uniform throughout North America. For example, about 75 percent of North America's population is located in a triangle roughly encompassed by Boston, Chicago, and Washington, D.C. Other large North American markets are in the South Pacific Coast region and the Southeast. Also, the forest resource endowments of the various regions within North America differ considerably. In the absence of trade restrictions, wood-surplus regions within North America would supply wood-deficit regions both within and outside of North America based upon economic considerations, regardless of political boundaries. This suggests that the United States and Canada would be expected to play complementary roles with regard to production and consumption of forest products, with U.S. consumption being far greater than Canadian, resulting in net forest products trade flows from Canada to the United States.

Forest-producing regions in the United States and Canada, with good access to major North American markets, would focus their attention on those domestic markets, be they American or Canadian. On the other hand, less advantageously located regions--particularly those with good access to foreign markets--would tend to look offshore. Since U.S. regions are generally somewhat better located with respect to the major North American markets than are Canadian regions, this tendency would be reinforced, and Canadian production would likely account for a disproportionate share of North America's foreign exports.

The foregoing reasoning suggests that, to the extent that restrictions do not offset underlying comparative advantage, North America is likely to be a major net exporter of forest products with each region's activity in foreign markets varying considerably, depending upon location and resource base. Canada is expected to be the dominant North American exporter, with large quantities reaching the United States and other foreign markets, whereas U.S. production is more domestically oriented.

An Important Trade Distortion

Although restrictions in forest products trade between the United States and Canada exist, they are generally small. Thus, the geographic proximity of the United States and Canada, their long common border, and the locations of the major North American forest resources and markets have all contributed to a trade pattern that generally conforms to the hypothetical pattern described above. There is, however, one notable restriction that has distorted the North American trade pattern away from that which would be predicted in an unrestricted environment.[1] This is the Canadian restriction against the export of unprocessed forest resources

---

[1]A second restriction, the U.S. Jones Act, requires that interstate coastal shipping be in U.S. vessels. Although ostensibly to apply only to interstate trade, the act will affect international forest products flows to the extent that (1) waterborne transportation is potentially least costly, and (2) costs on U.S. vessels are higher than on foreign carriers. This is particularly germane when assessing the competitive position of U.S. and Canadian regions in competing in major North American markets. Since the Jones Act applies only to U.S. shipping, a higher transport cost associated with U.S. waterborne carriers may well provide Canadian forest products with a competitive advantage in major U.S. markets. To the extent that such an advantage is realized, trade will be distorted from the free trade paradigm, with Canadian forest products displacing some U.S. products on the U.S. East Coast, while the displaced U.S. products may attempt to move to offshore markets.

from Crown Lands (lands which make up most Canadian forestlands) which thereby promotes increased domestic wood processing.

We hypothesize that such a policy has implications for both Canadian and U.S. forest products trade. For Canada, production and exports shift toward more processed commodities, thereby skewing the export pattern toward processed rather than unprocessed products. For the United States, the effect is to provide some export markets that are relatively free from Canadian competition, for example, conifer sawlogs, while at the same time substantial Canadian competition exists in processed forest product markets. The effect, therefore, is expected to induce a large volume of U.S. forest product exports of unprocessed wood into foreign markets with no important Canadian competition. The result is to promote specialization in export commodities by the United States and Canada to a degree that is not based upon their underlying comparative advantage. As we will show later in this chapter, the current North American trade structure exhibits the hypothesized U.S.-Canadian specialization in export trade.

Recent Patterns of U.S.-Canadian Forest Product Trade

The free trade paradigm introduced earlier suggests that, given the geographic proximity of the United States and Canada, the roughly similar magnitudes in the size of the U.S. and Canadian forest inventories, and the absence of significant trade restrictions for most forest products, it would be expected that U.S. and Canadian consumption of forest product resources would be drawn from many North American supply regions. However, given also the vastly larger U.S. market and the relative equality in the forest resource endowment, it would also be anticipated that the much larger U.S. domestic market would result in the United States becoming a net importer of Canadian forest resources, while Canada became a net exporter to the United States.

To the extent that some North American producing regions had superior access to the large U.S. market, a greater proportion of these regions' forest resource output would be consumed domestically, whereas other regions would look toward offshore markets to consume large portions of their production.

In this section we will examine very briefly the recent structure of U.S.-Canadian mutual trade in forest products. The hypothesized paradigm presented above will be examined to determine whether it is a reasonable approximation of recent U.S.-Canadian mutual trade patterns in forest products.

Although the data included in this section reflect only the 1976 situation, preliminary investigation of trade flows in the early 1960s, not reported here, has not revealed any major differences in the nature or structure of the U.S.-Canadian forest products trade from what is found in the 1976 data.

Production and Trade

In table 4-1 , U.S. and Canadian production and trade in six major forest products--pulpwood, conifer sawlogs, conifer sawnwood, wood pulp, newsprint, and paper and paperboard excluding newsprint--are analyzed. These six constitute the major portion of aggregate forest product production and trade. U.S. trade restrictions for five of these commodities--pulpwood, conifer sawlogs, sawnwood, wood pulp, and newsprint--are essentially nonexistent. Canada is only slightly more restrictive for sawnwood,

Table 4-1. Major U.S. and Canadian Forest Product Exports, 1976

	North American	United States		Canadian					
Commodity	Production	Production	% of North American total	Production	% of North American total	U.S. exports	% of U.S. production	Canadian exports	% of Canadian production
Pulpwood (1,000 m^3)	149,486	104,217	69.7	45,269	30.3	372	0.4	899	2.0
Sawlogs, conifer (100 m^3)	253,829	176,515	69.5	77,314	30.5	14,295	8.1	513	0.7
Sawnwood, conifer (1,000 m^3)	107,892	72,782	67.5	35,110	32.5	3,715	5.1	22,613	64.4
Wood Pulp (100 metric tons)	58,702	41,130	70.1	17,572	29.9	2,283	5.6	6,132	34.9
Newsprint (1,000 metric tons)	10,999	3,104	28.2	7,895	71.8	115	3.7	6,997	88.6
Paper and paperboard (excluding newsprint) (1,000 metric tons)	53,541	50,086	93.5	3,455	6.5	2,784	5.6	1,066	30.9

Source: Food and Agricultural Organization, Yearbook of Forest Products, 1964-76 (Rome, FAO, 1977).

charging a small import duty. As noted previously, however, Canada has restrictions on the export of most unprocessed wood, including sawlogs. Both countries have higher import duties on paper products excluding newsprint, although for many major products in this group the United States imposes no duty. In addition, both countries have significant import duties on major wood panel products, and we have not included these in our discussion. Although the United States is the dominant producer for each commodity except newsprint, Canadian total exports exceed those of the United States for four of the six commodities (sawlogs and paper and paperboard being the exceptions). For sawlogs, of course, the Canadian export prohibition obtains. Although the United States did not export more than 7 percent of the production of any commodity Canada exported over 30 percent of her domestic production for four of the six commodities. The remaining two commodities were sawlogs, for which the export prohibition applies, and pulpwood, which historically is not traded extensively because of its low value/weight ratio.

To summarize, whereas Canada relies upon the United States as an outlet for her forest product exports, the United States relies on Canada to provide substantial quantities of forest products, particularly newsprint and conifer sawnwood, for U.S. domestic consumption.

Mutual Trade Between the United States and Canada

The free trade paradigm suggests that geographic considerations will be important determinants of trade patterns. Specifically, in the absence of trade restrictions, regions with large markets will provide marketing outlets for surplus regions regardless of national boundaries. The detailed regional analysis of chapters 8 and 10 reveals that the Great Lakes and Northeast regions accounted for about two-thirds of total imports of U.S. forest products in 1976. Doubtless, this is partially because of their large regional markets and their location adjacent to major sea routes. Also, both regions share a common boundary with the wood-producing regions of eastern Canada.

As shown in tables 4-2 and 4-3 for each of the six commodities, Canadian exports to the United States made up at least 39 percent of her total exports. For three--sawnwood, newsprint, and pulpwood--the portion exported to the United States was more than 80 percent. For four of the six commodities, the United States absorbed over 10 percent of Canada's _total_ production. For newsprint, the United States absorbed over 70 percent of total Canadian production; and for conifer sawnwood, it was over 50 percent. In contrast, a much smaller share of U.S. forest products is exported, and generally only a relatively modest share goes to Canada.

Conclusions

The foregoing brief investigation reveals the hypothesized asymmetry in the structure of U.S.-Canadian forest products trade for a recent year. The net flow of forest products within North America is from Canada to the United States. Canada exports both a much higher fraction of her forest products production and a larger absolute volume than does the United States for most commodities. The only important commodity deviation is found in sawlogs where Canada has log-export restrictions. Furthermore, Canadian exports rely on the United States as her major market. The United States, by contrast, exports a far smaller fraction of her total forest products production and also a smaller absolute volume. Also,

Table 4-2. Major Canadian Forest Products Trade with the United States, 1976

Commodity	Canadian production	Canadian exports		Canadian exports to U.S. as a % of total exports	Canadian exports to U.S. as % of Canadian production	Canadian exports to U.S. as % of U.S. production
		Total	To U.S.			
Pulpwood (1,000 m^3)	45,269	899	740	82.3	1.6	0.7
Sawlogs, conifer (1,000 m^3)	77,314	513	200	39.0	Negligible	Negligible
Sawnwood, conifer (1,000 m^3)	35,110	22,613	18,474	81.7	52.6	25.4
Wood pulp (1,000 metric tons)	17,572	6,132	3,218	52.5	18.3	7.8
Newsprint (1,000 metric tons)	7,895	6,997	5,676	81.1	71.9	182.9
Paper and paperboard (exluding newsprint) (1,000 metric tons)	3,445	1,066	454	42.6	13.1	0.8

Source: Food and Agricultural Organization, Yearbook of Forest Products, 1964-76 (Rome, FAO, 1977).

Table 4-3. Major U.S. Forest Products Trade with Canada, 1976

Commodity	U.S. Production	U.S. exports		U.S. exports to Canada as % of total exports	U.S. exports to Canada as % of U.S. production	U.S. exports to Canada as % of Canadian production
		Total	to Canada			
Pulpwood (1,000 m^3)	104,217	372	343	92.2	Negligible	0.8
Sawlogs, conifer (1,000 m^3)	176,575	14,295	1,450	10.1	0.8	1.8
Sawnwood, conifer (1,000 m^3)	72,782	3,715	969	26.1	1.3	2.7
Wood pulp (1,000 metric tons)	41,130	2,283	73	3.1	Negligible	Negligible
Newsprint (1,000 metric tons)	3,104	115	0	0	Negligible	Negligible
Paper and paper-board (exluding newsprint) (1,000 metric tons)	50,086	2,784	642	23.1	1.3	18.6

Source: Food and Agricultural Organization, Yearbook of Forest Products, 1964-76 (Rome, FAO, 1977).

only a relatively small fraction of total U.S. forest product exports are for Canadian markets. Major U.S. forest product markets are elsewhere than Canada.

Assessment of North America's Trade Balance

In this section we will examine North America's international trade balance in forest products and the changes it experienced between 1950 and 1976. We will analyze the trade balance in terms of value and physical wood volume at the aggregate level and later will disaggregate it into five commodity groups. Changes in the total North American balance will be calculated, and the respective U.S. and Canadian contributions (net of intra-North American flows) to the total North American trade balances will be determined.

Approach and Aggregate Results

The forest products trade balance is defined as forest product exports minus forest product imports. When exports exceed imports, the trade balance is said to be in surplus. For the reverse situation, the trade balance is said to be in deficit. The principal focus of our analysis will be upon changes in the trade balance, where the magnitude and direction of the changes are used to indicate an improvement in or deterioration of the trade balance. A larger surplus (or smaller deficit) implies a trade balance improvement, whereas a larger deficit (or smaller surplus) indicates a deterioration.

Two basic types of trade balances will be investigated. The first is the physical wood balance, where the wood trade balance is compared, using cubic meters (for solid wood) and metric tons (for fiber products).[2] Next, the dollar-value trade balance is investigated. Both current and 1976 constant prices are used, with much of the analysis utilizing the constant price values. It is necessary to use constant prices in order to avoid the distortions that would occur if current prices were utilized for the twenty-six-year inflation-plagued period.

Tables 4-4 and 4-5 present the North American aggregate net international trade balance for the period 1950-76. The balance is presented in both current and constant dollars. North America's net trade balance is the difference between the U.S. and the Canadian net trade balances. For this period, by either measure, the North American trade balance had a trade surplus, which exhibited a strongly increasing trend throughout the period. This was the result of a strongly increasing Canadian surplus together with a fluctuating, but rather trendless, U.S. trade balance. The Canadian trade balance showed almost uninterrupted improvement throughout the period although, in real terms (that is, constant dollars), the increasing trade surplus was not as dramatic as in current dollars. The U.S. trade balance was in deficit throughout the 1950-76 period. In current dollars, the 1976 deficit was about 50 percent greater than that of 1950, whereas in constant dollars the 1976 deficit was only 63 percent of 1950s. Thus, in purchasing equivalents, the deficit declined. The dramatic and almost uninterrupted increase in the North American trade

---

[2]The physical balances are measured in actual physical volume traded and not, as is sometimes done, in roundwood equivalents. There is no presumption in the analysis that these physical units are interchangeable between commodity groups.

Table 4-4. North American Net Forest Products Trade for Selected Years in Millions of Current U.S. Dollars

Year	North American net trade	U.S. net trade[a]	Canadian net trade
1950	56	(917)	973
1956	267	(1,139)	1,406
1957	412	(976)	1,388
1958	351	(978)	1,329
1959	291	(1,140)	1,431
1960	465	(965)	1,430
1961-65	577	(1,013)	1,590
1966	753	(1,153)	1,906
1967	978	(952)	1,930
1968	1,127	(1,042)	2,169
1969	1,334	(1,115)	2,449
1970	1,905	(678)	2,583
1971	1,541	(1,202)	2,743
1972	1,752	(1,474)	3,226
1973	2,622	(1,353)	3,975
1974	4,584	(340)	4,924
1975	3,833	(446)	4,279
1976	4,212	(1,437)	5,649

Source: Food and Agricultural Organization, *Yearbook of Forest Products*, selected issues (Rome, FAO).

[a]Parenthetical numbers denote a deficit.

Table 4-5. North American Net Forest Products Trade for Selected Years in Millions of Constant 1976 U.S. Dollars

Year	North America net trade	U.S. net trade[a]	Canadian net trade
1950	140	(2,284)	2,424
1956	568	(2,423)	2,991
1957	745	(2,008)	2,753
1958	712	(1,978)	2,690
1959	379	(2,260)	2,639
1960	797	(1,991)	2,788
1961-65	1,078	(1,893)	2,971
1966	1,314	(2,012)	3,326
1967	1,669	(1,612)	3,281
1968	1,826	(1,689)	3,515
1969	1,997	(1,782)	3,779
1970	2,793	(994)	3,787
1971	2,104	(1,649)	3,753
1972	2,344	(1,975)	4,319
1973	3,319	(1,713)	5,032
1974	5,288	(391)	5,679
1975	4,036	(468)	4,504
1976	4,212	(1,437)	5,649

Source: Food and Agricultural Organization, Yearbook of Forest Products, selected issues (Rome, FAO).

[a]Parenthetical numbers denote a deficit.

surplus reflects a growing net outflow of forest products over the 1950-76 period. This is consistent with the findings of chapter 3, in which North America was found to be the principal intercontinental supplier of forest products.

Figure 4-1 presents the constant dollar trade balance trends. Table 4-6 presents that constant-dollar trade balance as disaggregated by major commodity groups for the three periods--1950, 1961-65, and 1976. Table 4-7 analyzes the changes that occurred in each of the commodity groups over the two subperiods (1950 to 1961-65 and 1961-65 to 1976) and estimates the contribution of the various commodity groups to the overall trade balance.

Overall, the trade balance for North American forest products achieved, in 1976, a surplus in excess of \$4.2 billion. Of this, over \$4 billion represents an increase in the surplus (in constant 1976 dollars) over the 1950 level (table 4-6). The greater part of the improved North American trade balance--76.9 percent, or \$3.1 billion--occurred during 1961-65 to 1976.

Canada's trade balance exhibits two striking features. First, Canada experienced a substantial surplus for each time period. Second, this surplus exhibited a dramatic increase of \$3.2 billion in constant 1976 dollars. Of this, a \$500 million surplus occurred in the 1950 to 1961-65 subperiod, and an additional \$2,700 million improvement occurred between 1961-65 and 1976.

The United States, on the other hand, experienced a trade deficit for each period. However, the U.S. deficit, in constant dollars, declined over the twenty-six-year period. Thus, despite its continuing

Figure 4-1. North American Forest Product Net Trade Balance in 1976 Constant Dollars

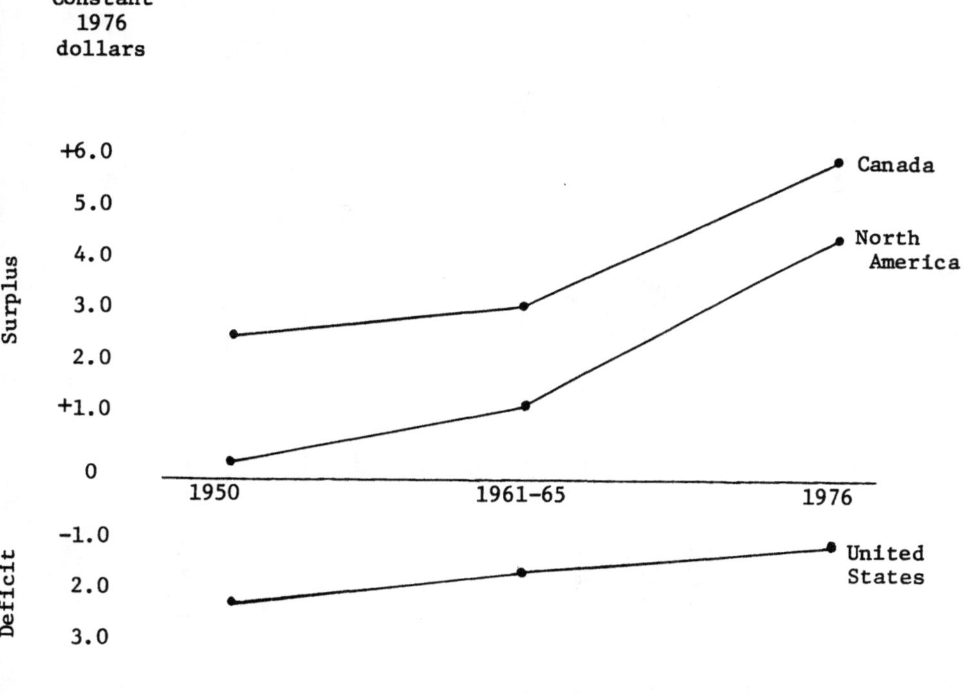

Source: Food and Agricultural Organization, Yearbook of Forest Products, selected issues (Rome, FAO).

Table 4-6. Net Trade Balance by Value of North American Forest Products for Selected Years (millions of 1976 dollars)

	1950			1961-65			1976		
Commodity	U.S.	Canada	North America	U.S.	Canada	North America	U.S.	Canada	North America
Industrial roundwood	(101.7)[a]	87.8	(13.9)	86.8	43.4	130.2	1,039.6	13.9	1,053.5
Sawnwood	(533.4)	730.2	175.8	(456.1)	757.1	300.9	(1,008.3)	1,458.6	450.3
Wood panels	(30.2)	24.3	(5.9)	(267.5)	68.9	(198.6)	(332.4)	14.7	(317.7)
Wood pulp	(568.8)	469.2	(99.6)	(312.2)	708.0	395.8	(401.8)	2,115.6	1,713.7
Paper and paperboard	(1,029.8)	1,111.6	81.6	(943.7)	1,393.5	449.8	(733.8)	2,045.9	1,312.1
Total	(2,263.9)	2,423.1	138.0	(1,892.6)	2,970.8	1,078.2	(1,436.8)	5,648.7	4,211.9

Note: Table developed from data in selected issues of the FAO's Yearbook of Forest Products.

[a]Parenthetical numbers denote a deficit.

Table 4-7. Changes in North America's Net Value Trade Balance for Selected Years

(millions of 1976 dollars)

Commodity	Changes between 1950 and 1961-65			Changes between 1961-65 and 1976			Changes between 1950-76		
	U.S.	Canada	North America	U.S.	Canada	North America	U.S.	Canada	North America
Industrial roundwood	189.5	(44.4)[a]	145.1	952.8	(29.4)	923.3	1,141.3	(73.8)	1,067.4
Sawnwood	97.2	26.9	124.1	(552.2)	701.5	149.3	(454.9)	728.4	273.4
Wood panels	(237.3)	44.6	(192.7)	(64.9)	(54.2)	(119.1)	(302.3)	(9.5)	(311.8)
Wood pulp	256.6	238.9	495.4	(89.6)	1,407.6	1,317.9	166.9	1,646.4	1,813.4
Paper and paperboard	86.1	281.9	368.0	209.9	652.4	862.3	295.9	934.3	1,230.3
Total	392.1	547.9	939.9	456.0	2,677.9	3,133.7	846.9	3,225.8	4,072.7
Contribution to trade balance change for each period (%)	41.7	58.3	100.0	14.5%	85.5%	100.0	20.8%	79.2%	100.0
Contribution of period to total trade balance change (%)			23.1%			76.9%			100.0

Note: Table developed from data in selected issues of the FAO's Yearbook of Forest Products.

[a]Parenthetical numbers denote a deficit.

trade deficit, the United States contributed to the overall _improvement_ in North America's trade surplus through a gradual but persistent reduction in the size of its forest products deficit (table 4-7). Of the $850 million improvement in the U.S. deficit over the twenty-six-year period, almost $400 million occurred from 1950 to 1961-65, and an additional improvement of $450 million occurred between 1961-65 and 1976.

The contributions of the United States and Canada to the increase in the North American surplus were 20.8 percent and 79.2 percent, respectively (table 4-7). While the U.S. contribution in the first half of the twenty-six-year period constituted a larger share of the total improvement than it did in the second half (41.6 percent as compared with only 14.5 percent), the absolute size of the U.S. improvement was slightly larger during the second half of the period. However, the Canadian trade balance improved dramatically between 1961-65 and 1976, thus resulting in a decline of the _relative_ contribution of the United States for this subperiod.

Table 4-7 shows that the largest source of improvement in North America's trade balance, occurring between 1950 and 1976 ($4.1 billion in constant 1976 dollars), came from the wood pulp sector ($1.8 billion). This was followed closely by paper and paperboard ($1.2 billion) and industrial roundwood ($1.1 billion). Sawnwood contributed $273 million to the improved surplus, whereas the wood panel balance deteriorated by $312 million. Thus, in four of the five major categories, the North American trade balance showed improvement between 1950 and 1976.

Of this improvement, Canada accounted for $3.2 billion (or 79.2 percent), with most of this attributable to wood pulp and paper and paperboard. Canada experienced a modest deterioration in its trade balance for industrial roundwood and wood panels. The U.S. contribution to the improvement in North America's trade balance was $847 million (or 20.8 percent), with a large improvement in industrial roundwood and a smaller one in paper and paperboard and wood pulp. However, a deterioration in the commodity trade balance was experienced for sawnwood and wood panels.

Table 4-8 and figures 4.2 and 4.3 present the U.S., Canadian, and North American trade balance in physical units for solid wood and fiber products. The finding of the physical trade balance examination is in basic agreement with that of the value trade balance. For North America, the solid wood and fiber balances showed exceptional improvement for the entire twenty-six-year period as well as for the two subperiods. For solid wood, the improvement was striking, with the physical trade surplus increasing from 1.7 million cubic meters in 1950 to 23.7 million cubic meters in 1976. For fiber, the North American trade balance went from a physical deficit of 169,000 metric tons in 1950 to a physical surplus of 9 million metric tons in 1976.

As in the value analysis, Canada maintained a trade surplus for each period, in both solid wood and fiber. Also, the Canadian surplus grew substantially for both subperiods between 1950 and 1976.

The U.S. physical balances exhibited changes similar to those experienced by its value balance, in that both solid wood and fiber had deficit balances in 1950 and subsequently showed improvement through both subperiods. The most striking feature of this trade was the dramatic improvement in the U.S. solid wood trade balance moving from a substantial deficit (11.9 million cubic meters) in 1950 to a modest surplus (2.3 million cubic meters) in 1976 with most of the improvement occurring in the

Table 4-8. Net Trade Balance by Volume of North American Forest Products for Selected Years

Commodity	1950 U.S.	1950 Canada	1950 North America	1961-65 U.S.	1961-65 Canada	1961-65 North America	1976 U.S.	1976 Canada	1976 North America
Industrial roundwood (1,000 m^3)	(4,696)[a]	4,518	(178)	109	2,643	2,752	18,054	478	18,532
Sawnwood (1,000 m^3)	(7,069)	9,004	1,935	(9,654)	13,930	4,276	(14,125)	21,152	7,027
Wood panels (1,000 m^3)	(125)	87	(38)	(1,100)	259	(841)	(1,620)	(248)	(1,868)
Total solid wood (1,000 m^3)	(11,890)	13,609	1,719	(10,645)	16,832	6,187	2,309	21,382	23,691
Wood pulp (1,000 metric tons)	(2,077)	1,629	(448)	(1,322)	2,971	1,649	(1,076)	6,048	4,972
Paper and paperboard (1,000 metric tons)	(4,262)	4,541	279	(4,211)	6,062	1,851	(3,511)	7,528	4,017
Total fiber products (1,000 metric tons)	(6,339)	6,170	(169)	(5,533)	9,033	3,500	(4,587)	13,576	8,989

Source: Food and Agricultural Organization, *Yearbook of Forest Products*, selected issues (Rome, FAO).

[a]Parenthetical numbers denote a deficit.

Figure 4-2. North American Forest Product Net Trade Balance (Solid Wood)

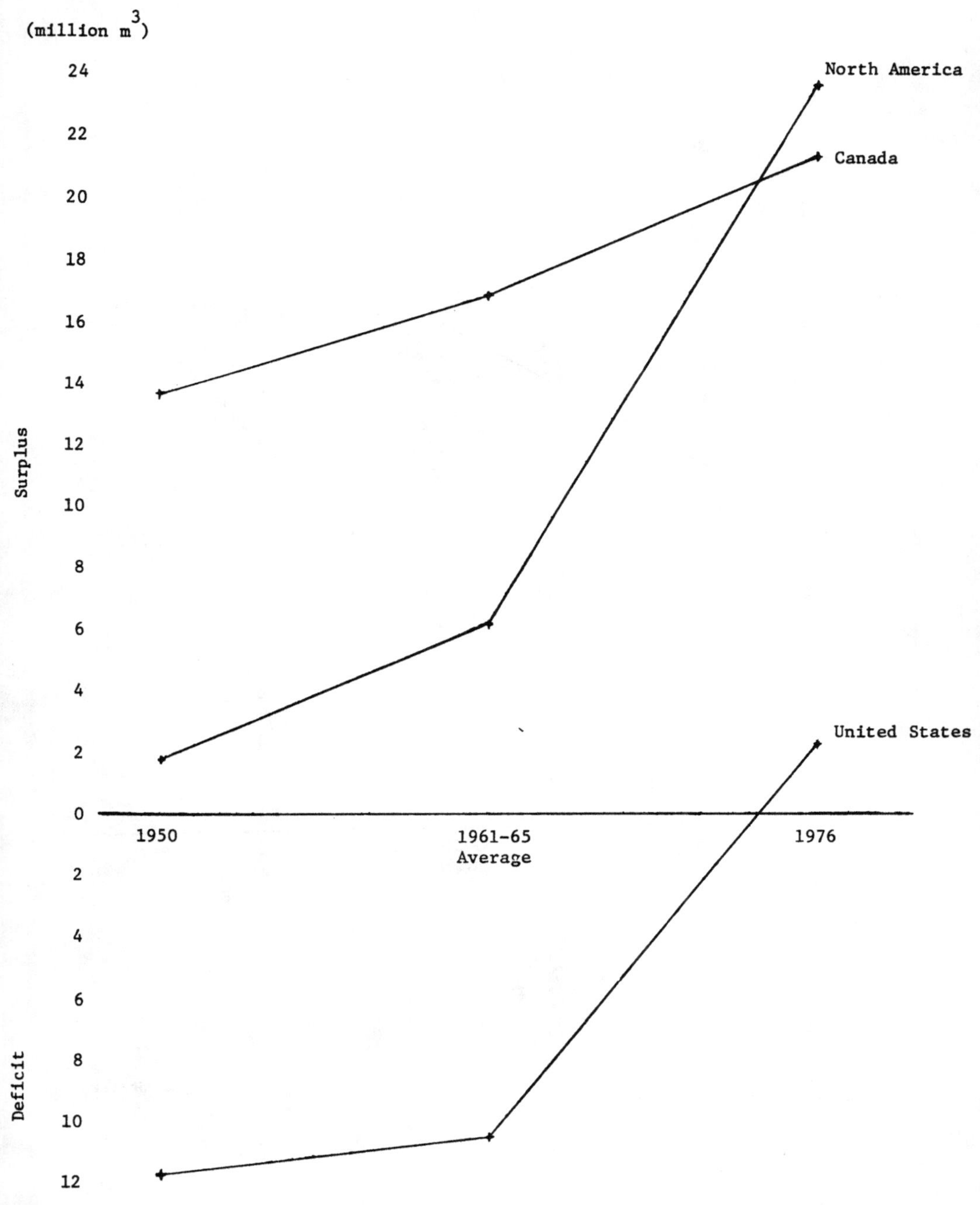

Figure 4-3. North American Forest Product Net Trade Balance (Fiber)

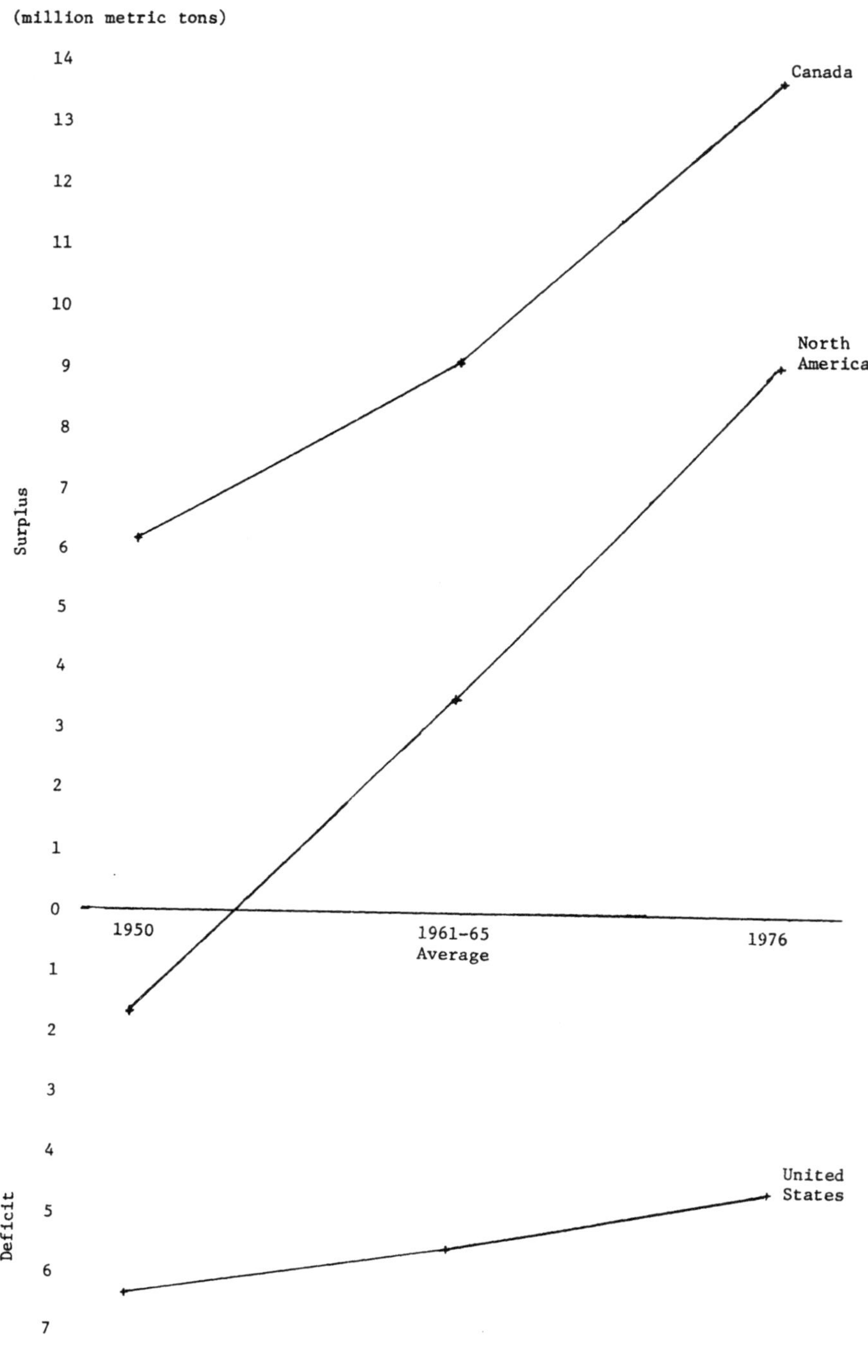

1961-65 to 1976 subperiod. Of the total improvement in the North American trade balance for solid wood (physical volume), the United States accounted for 14.2 million cubic meters (or 64.6 percent) of the total improvement.

Although the United States continued to experience substantial deficits in fiber, it improved modestly its fiber trade balance over the twenty-six-year period. During this time, North American trade balance improved by 9.2 million metric tons, and the United States accounted for 1.6 million metric tons (or 19.1 percent).

Commodity Performance

Tables 4-6 and 4-7 also present the North American (value) trade balance and its U.S. and Canadian components, with the changes between 1950, 1961-65, and 1976 disaggregated into five major commodity groups.

North America experienced trade balance improvements for both subperiods in each of the five commodity groups except wood panels. While the wood panel group experienced a substantial deterioration over the total twenty-six-year period and both subperiods, increased surpluses were experienced for the other four commodity groups. Great improvements in the trade balance were experienced by wood pulp ($1,813 million), paper and paperboard ($1,230 million), nonprocessed industrial roundwood ($1,067 million) and sawnwood ($273 million). All four commodities experienced trade balance improvements for both subperiods, with the larger absolute improvement experienced during 1961-65 to 1976. Similar trends are noted, focusing upon physical volume changes (see figures 4-2 and 4-5).

Over the period, Canada experienced a declining trade balance in unprocessed industrial roundwood ($74 million decline) and wood panels ($10 million decline). However, these declines were offset by trade balance improvements in sawnwood ($728 million), wood pulp ($1.3 billion) and paper and paperboard ($934 million), resulting in the overall trade balance improvement observed.

The United States exhibited trade balance deficits in all five commodities in 1950, and in all, except unprocessed industrial roundwood, in 1976. Three commodities--industrial roundwood, wood pulp, and paper and paperboard--experienced trade balance improvements of $1.1 billion, $167 million, and $296 million, respectively, between 1950 and 1976. Improvements were also observed for the physical balances, as reflected in figures 4-2 and 4-3. However, for wood pulp, 1961-65 was marked by a trade balance deterioration in contrast to the large improvement of the earlier subperiod.

Summary and Conclusions

As noted in Chapter 3, North America (the United States and Canada) possesses about 22.3 percent of the world's forest inventories and about 35.2 percent of its conifer resources. North America accounted for 34 percent of the world's 1976 industrial roundwood production and 39.6 percent of the world's industrial conifer roundwood production.

In this chapter we have examined the respective roles of the United States and Canada, not only as producers, but more important, as international traders of forest products. In total, North America has had an important and growing role as a net supplier of forest products to non-North American markets. Within this context, the roles of Canada and the United States have differed greatly. Canada has been a dominant net exporter of forest products to the United States and countries throughout the world. The United States, on the other hand, has been a persistent net importer, with the vast majority of her forest product imports originating in Canada. Overall, Canada's net exports substantially exceed the U.S. net imports, resulting in a large and growing net forest products flow outside of North America. Their behavior was generally consistent with the hypothesized "free trade paradigm" although some trade distortions appear to result from the Canadian restriction on unprocessed wood exports.

An analysis of North America's forest product trade balance reveals substantial improvement, whether the analysis is undertaken in physical terms or in constant dollar values. The trade balance showed a marked improvement between 1950 and 1961-65 and an even more striking one over the 1961-65 to 1976 period.

Although the United States experienced substantial forest product deficits both in 1950 and in 1976 and Canada experienced substantial trade balance surpluses, both countries made substantial contributions to the improvement in North America's overall balance. Of the total trade balance improvement of over $4 billion (1976 dollars) over the twenty-six-year period, the U.S. contribution was about $850 million.

Disaggregation of North American forest product flows reveals a positive and improving trade balance in four of five commodity groups--industrial roundwood, sawnwood, wood pulp, and paper and paperboard. Only one, wood panels, experienced a trade balance deficit and a declining trend.

Chapter 5

U.S. TRADE IN FOREST PRODUCTS

In chapter 4 we examined the North American trade balance in forest products, using the standard Food and Agricultural Organization (FAO) definition. The nature of U.S.-Canadian forest product trade interrelations were examined, and an assessment of their respective contributions to the overall North American balance were analyzed. In this chapter we will focus exclusively upon the United States. Its overall trade balance will be examined briefly, followed by a short commodity-by-commodity analysis involving the major forest commodities.

The U.S. Trade Balance

Value Measures

Table 5-1 presents the U.S. forest products trade balance (in current dollars) for selected years between 1950 and 1976. For each of these years the U.S. balance was in deficit and, as noted in Chapter 4, the balance exhibits considerable variability. One could hypothesize that the variability is related to fluctuations in the exchange rate, particularly after 1973, when the world's international monetary system was radically altered after the fixed rate of exchange was replaced by floating exchange rates. However, the testing of these hypotheses is beyond the scope of this report. As noted in chapter 4, the deficit (in current

Table 5-1. U.S. Forest Products Trade for Selected Years (millions of current U.S. dollars)

Year	Exports	Imports	Net trade[a]
1950	118	1,035	(917)
1956	321	1,460	(1,139)
1959	391	1,531	(1,140)
1964	707	1,427	(720)
1967	986	1,938	(952)
1970	1,621	2,299	(678)
1971	1,481	2,683	(1,202)
1972	1,817	3,291	(1,474)
1973	2,663	4,016	(1,353)
1974	3,537	3,877	(340)
1975	3,505	3,951	(446)
1976	3,994	5,431	(1,437)

Source: Food and Agricultural Organization, Yearbook of Forest Products, selected issues (Rome, FAO).

[a]Parenthetical numbers denote a deficit.

dollars unadjusted for inflation) shows a modest rise between 1950 and 1976. However, if the deficit is deflated by a common price deflator (in this case, the GNP deflator), it exhibits a modest decline in real terms between 1950 and 1976.

Physical Measure

Table 5-2 shows U.S. industrial forest product production and trade in physical volume (roundwood equivalents) through the same 1950-76 period. As in value terms, the physical volumes of forest product exports and imports have increased considerably, with export growth exhibiting more rapid growth. The trade deficit in physical wood volumes remained remarkably stable during that period, with the 1976 physical deficit only slightly less than that of 1950. Net imports, as a fraction of domestic production, remained in the 10 percent to 16 percent range, with a slight decline indicating that about 10 percent of U.S. forest consumption originates abroad. The lowest fraction observed was 7.6 percent in 1975.

Structure and Growth in the U.S. Forest Products Trade

Tables 5-3 through 5-8 present the levels of U.S. forest product exports and imports by major commodity group, their growth rates, and their contributions to the total value of exports and imports. A brief discussion of these changes is instructive. The principal forest product exports in value terms for 1976 are softwood logs (U.S. Census Schedule B, 242.2),[1] about 21 percent; softwood lumber (243.2), about 12 percent;

---

[1] For a discussion of the commodity groups used, see appendix A.

Table 5-2. U.S. Production and International Trade in Industrial Roundwood (million cubic feet, roundwood equivalents)

Year	Domestic production	Imports	Exports	Net imports	Net imports as % of domestic production
1950	8,525	1,520	140	1,380	16.2
1954	8,755	1,460	270	1,190	13.6
1959	9,390	1,700	355	1,345	14.3
1964	10,170	2,035	720	1,315	12.9
1970	11,120	2,420	1,355	1,065	9.6
1971	11,100	2,695	1,175	1,520	13.7
1972	11,430	3,040	1,380	1,660	14.5
1973	11,750	3,105	1,470	1,635	13.9
1974	11,280	2,690	1,515	1,175	10.4
1975	10,195	2,170	1,400	770	7.6
1976	11,535	2,735	1,445	1,290	11.2

Source: U.S. Forest Service, The Demand and Price Situation: 1976-77 Forest Service U.S.D.A., Miscellaneous Publication No. 1357 (Washington, D.C., 1977).

Table 5-3. U.S. Forest Product Exports, 1967-76
(million current dollars)

	U.S. Census Schedule B[a]													
Year	Pulpwood 242.1	Softwood logs 242.2	Hardwood logs 242.3	Softwood lumber 243.2	Hardwood lumber 243.3	Wood pulp[b] 251EX	Veneers 631.1	Plywood 631.2	Reconstituted wood 631.4	Wood chips 631.8320	News-print 641.1	Building board[c] 641.6	Paper and board[d] 641EX	Others	Total
1967	3.0	159.4	32.4	104.7	27.2	244.4	7.7	11.2	1.0	12.1	11.4	8.2	357.7	5.6	985.8
1968	2.4	237.7	41.5	126.1	21.3	265.2	11.6	11.3	1.1	26.6	17.1	8.7	422.6	3.9	1,197.0
1969	1.8	255.3	39.1	148.3	24.4	288.9	14.5	27.6	2.3	36.3	15.8	11.1	451.3	3.3	1,319.9
1970	1.9	322.5	35.8	159.4	29.4	468.4	13.3	16.2	2.2	36.6	18.7	10.2	488.1	4.7	1,607.4
1971	.5	264.3	29.7	142.4	32.8	355.9	14.9	15.4	4.5	41.1	20.7	11.9	527.0	3.5	1,465.1
1972	3.3	392.8	42.1	208.7	72.4	368.9	21.9	36.7	8.7	57.1	20.3	13.1	554.9	5.8	1,806.6
1973	2.1	760.3	53.2	421.7	50.3	434.5	35.2	70.8	15.6	85.7	16.9	25.0	677.3	7.0	2,655.6
1974	7.1	670.3	49.8	398.9	61.5	833.6	36.8	97.3	21.8	113.9	54.1	27.8	1,119.9	9.1	3,502.1
1975	6.4	683.7	42.7	342.4	58.5	908.1	42.9	136.3	17.2	115.4	52.8	25.3	1,012.1	13.5	3,457.4
1976	7.2	855.1	65.1	470.7	87.2	866.4	45.7	147.9	18.6	153.0	48.9	31.7	1,182.8	15.5	3,995.8

[a]See appendix A.

[b]Less wastepaper.

[c]Of wood pulp.

[d]Less newsprint and building boards.

Table 5-4. U.S. Forest Product Exports, 1967-76: Trade Structure and Percentage of Total Trade

	U.S. Census Schedule B[a]													
Year	Pulpwood 242.1	Softwood logs 242.2	Hardwood logs 242.3	Softwood lumber 243.2	Hardwood lumber 243.3	Wood pulp[b] 251EX	Veneers 631.1	Plywood 631.2	Reconstituted wood 631.4	Wood chips 631.8320	News-print 641.1	Building board[c] 641.6	Paper and board[d] 641EX	Others	Total
1967	0	16	3	11	3	25	1	1	0	1	1	1	36	1	100
1968	0	20	3	11	3	22	1	1	0	1	1	1	35	0	100
1969	0	19	3	11	2	22	1	2	0	2	1	1	35	0	100
1970	0	20	3	10	2	29	1	1	0	1	1	1	30	0	100
1971	0	18	2	10	2	24	1	1	0	2	2	1	37	0	100
1972	0	22	2	12	4	21	1	2	0	2	1	1	31	0	100
1973	0	28	2	16	2	16	1	3	1	2	1	1	26	0	100
1974	0	19	1	11	2	24	1	3	1	2	2	1	32	0	100
1975	0	20	1	10	2	26	1	4	1	2	2	1	29	0	100
1976	0	21	2	12	2	22	1	4	0	4	1	1	30	0	100

[a]See appendix A.

[b]Less wastepaper.

[c]Of wood pulp.

[d]Less newsprint and building boards.

Table 5-5. U.S. Forest Product Exports, 1967-76: Trade Trends and Percentage of 1967 Trade

U.S. Census Schedule B[a]

Year	Pulpwood 242.1	Softwood logs 242.2	Hardwood logs 242.3	Softwood lumber 243.2	Hardwood lumber 243.3	Wood pulp[b] 251EX	Veneers 631.1	Plywood 631.2	Reconstituted wood 631.4	Wood chips 631.8320	News-print 641.1	Building board[c] 641.6	Paper and board[d] 641EX	Others	Total
1967	100	100	100	100	100	100	100	100	100	100	100	100	100	100	100
1968	79	149	128	120	78	108	151	102	115	220	149	105	118	70	121
1969	60	150	121	142	90	118	189	249	239	301	139	134	126	59	134
1970	64	202	110	152	108	192	173	146	255	309	163	124	136	84	164
1971	16	166	92	136	120	145	193	139	522	340	181	144	147	63	149
1972	109	246	130	199	266	151	285	331	913	472	178	159	155	104	183
1973	70	477	164	403	185	178	458	639	1,646	709	148	303	189	126	273
1974	235	421	154	381	226	341	479	878	2,289	942	473	337	313	164	355
1975	213	429	132	327	215	372	559	1,230	1,809	954	461	306	283	242	351
1976	240	537	201	450	320	354	594	1,335	1,958	1,265	428	384	331	278	405

[a] See appendix A.

[b] Less wastepaper.

[c] Of wood pulp.

[d] Less newsprint and building boards.

Table 5-6. U.S. Forest Product Imports, 1967-76
(million current dollars)

	U.S. Census Schedule A[a]														
Year	Pulpwood 242.1	Softwood logs 242.2	Hardwood logs 242.3	Softwood lumber 243.2	Hardwood lumber 243.3	Wood pulp[b] 251EX	Veneers 631.1	Plywood 631.2	Reconstituted wood 631.4	Wood chips 631.8320	News-print 641.1	Building board[c] 641.6	Paper and board[d] 641EX	Others	Total
1967	16.6	1.7	5.6	327.4	61.4	358.3	42.7	142.3	.5	9.2	892.1	15.1	59.4	0.7	1,933.1
1968	16.6	1.4	6.0	494.3	63.0	390.3	50.6	217.7	.5	8.3	862.8	20.3	62.9	0.6	2,195.5
1969	14.3	3.0	4.8	545.9	77.2	458.9	47.0	250.2	1.9	5.1	938.6	22.2	77.5	0.6	2,447.3
1970	14.6	8.5	4.5	438.5	61.2	444.4	38.9	209.0	1.3	7.5	929.6	17.0	92.5	0.8	2,268.4
1971	10.9	3.9	3.5	681.9	64.6	425.6	46.4	259.6	2.2	12.0	987.8	20.6	93.4	0.7	2,613.4
1972	10.0	0.7	3.9	1,062.6	88.0	492.1	68.4	336.9	2.4	10.4	1,053.7	55.8	99.2	1.2	3,285.5
1973	8.0	0.6	3.6	1,430.3	120.5	654.9	83.3	391.5	2.4	13.9	1,185.0	44.4	152.7	0.7	4,091.8
1974	11.0	8.6	4.5	930.8	131.0	1,093.0	79.4	290.4	1.9	13.6	1,484.1	39.8	221.0	3.2	4,312.3
1975	9.2	12.4	3.4	730.7	67.7	1,006.6	50.9	262.0	4.2	14.9	1,418.1	14.6	145.1	1.5	3,741.3
1976	11.7	7.4	2.8	1,354.6	96.4	1,244.3	74.6	382.5	12.6	24.0	1,742.1	26.0	219.0	1.4	5,199.6

[a]See appendix A.

[b]Less wastepaper.

[c]Of wood pulp.

[d]Less newsprint and building boards.

Table 5-7. U.S. Forest Product Imports, 1967-76: Trade Structure and Percentage of Total Trade

U.S. Census Schedule A[a]

Year	Pulpwood 242.1	Softwood logs 242.2	Hardwood logs 242.3	Softwood lumber 243.2	Hardwood lumber 243.3	Wood pulp[b] 251EX	Veneers 631.1	Plywood 631.2	Reconstituted wood 631.4	Wood chips 631.8320	News-print 641.1	Building board[c] 641.6	Paper and board[d] 641EX	Others	Total
1967	1	0	0	17	3	19	2	8	0	0	46	1	3	0	100
1968	1	0	0	23	3	18	2	10	0	0	39	1	3	0	100
1969	1	0	0	22	3	19	3	10	0	0	38	1	3	0	100
1970	1	0	0	19	3	20	2	9	0	0	41	1	4	0	100
1971	0	0	0	26	2	16	2	10	0	0	38	1	4	0	100
1972	0	0	0	33	3	15	2	10	0	0	32	2	3	0	100
1973	0	0	0	35	3	16	2	10	0	0	29	1	4	0	100
1974	0	0	0	22	3	25	2	7	0	0	35	1	5	0	100
1975	0	0	0	20	3	27	1	7	0	0	38	0	4	0	100
1976	0	0	0	26	2	24	2	7	0	0	34	1	4	0	100

[a]See appendix A.

[b]Less wastepaper.

[c]Of wood pulp.

[d]Less newsprint and building boards.

Table 5-8. U.S. Forest Product Imports, 1967-76: Trade Trends and Percentage of 1967 Trade

U.S. Census Schedule A[a]

Year	Pulpwood 242.1	Softwood logs 242.2	Hardwood logs 242.3	Softwood lumber 243.2	Hardwood lumber 243.3	Wood pulp[b] 251EX	Veneers 631.1	Plywood 631.2	Reconstituted wood 631.4	Wood chips 631.8320	News-print 641.1	Building board[c] 641.6	Paper and board[d] 641EX	Others	Total
1967	100	100	100	100	100	100	100	100	100	100	100	100	100	100	100
1968	100	82	108	151	103	109	119	153	93	91	97	134	106	80	114
1969	86	181	86	167	126	128	110	178	353	55	105	146	130	77	127
1970	88	509	81	134	100	124	91	147	239	81	104	112	155	118	117
1971	66	233	62	208	105	119	109	182	408	130	111	136	157	97	135
1972	60	44	70	325	143	138	160	237	433	113	118	369	167	167	170
1973	48	33	65	437	196	183	195	275	432	151	133	294	257	95	212
1974	67	507	81	281	213	306	186	204	350	148	166	263	372	449	223
1975	55	737	62	220	110	282	119	184	774	162	159	97	244	214	193
1976	71	439	50	414	157	348	175	269	2,304	261	195	172	368	191	269

[a]See appendix A.

[b]Less wastepaper.

[c]Of wood pulp.

[d]Less newsprint and building boards.

wood pulp, excluding wastepaper (251), about 22 percent; and paper and paperboard, excluding newsprint (641EX), about 30 percent). These four groups constitute roughly 85 percent of the value of forest products exports during the period.

On the import side, three commodities make up 84 percent of the value of U.S. imports. These are softwood lumber (243.2), 26 percent; wood pulp (251EX), 24 percent; and newsprint (641.1), about 34 percent.

It is interesting to note that two of the major U.S. exports are also major imports--softwood lumber and wood pulp. Much of this trade is with Canada and represents imports from Canada based upon locational considerations and exports to non-North American countries.

Tables 5-5 and 5-8 indicate the growth rates of exports and imports in the major commodities. Three groups have experienced very rapid export growth. These are plywood (632.2), improved or reconstituted wood (631.4), and wood chips (631.8320). For plywood and chips particularly, their rapid growth has moved plywood and chips from an almost negligible fraction of U.S. exports to become major export commodities. By 1976, each constituted about 4 percent of total U.S. exports.

On the import side, one commodity--improved or reconstituted wood--(principally particleboard)--has exhibited extremely rapid growth. Imports of this commodity, however, still constitute only a negligible fraction of total U.S. imports of forest products. Other rapidly growing commodities include softwood lumber and wood pulp, which traditionally have been important U.S. forest product imports.

U.S. Trade in Major Forest Commodities

The following section will examine individually the U.S. post-World War II trade and production experience for a number of major forest commodities. The focus will be on the trends in production, net trade, and self-sufficiency (as indicated by the ratio of net trade to domestic production). Also, we will identify the important U.S. exporting and importing regions for each commodity and the major foreign trading partners. The data utilized for the discussion of the longer trends (1950-76) are drawn primarily from FAO data. U.S. Census data for the period 1967-76 are used for the analysis of U.S. international trade by region, and for identification of foreign trading partners. The tables utilized for the discussion are found in appendix A. Because different classification systems have been used, the FAO data utilizes the Standard Industrial Trade Classificaion (SITC) designation, whereas the Census data is designated by Schedules A and B (see appendix A).

Conifer Sawlogs

In recent years the United States has become the dominant exporter of conifer sawlogs. In 1950 the United States produced about 38 percent of the world's total production of softwood conifer sawlogs. By 1976, the fraction was reduced to about 31 percent. However, U.S. production in 1976 led that of other major producers, including the USSR (25 percent), Canada (13 percent), the Nordic countries (6.7 percent), western Europe (8 percent), and Japan (2.8 percent). (See table 5-9.)

In 1950 U.S. exports of conifer sawlogs totaled 130,000 cubic meters. By 1976, this figure had reach 14,295 cubic meters. The export volume of conifer sawlogs initially increased very gradually until

Table 5-9. U.S. Conifer Sawlog[a] Production and Trade for Selected Years

	1950	1954	1959	1964	1970	1976
Production (1,000 m^3)	127,072	139,533	150,351	150,804	156,468	176,575
Production as % of world total	37.6	35.5	32.1	28.6	26.2	30.6
Exports (1,000 m^3)	130	482	759	4,633	12,180	14,295
Exports as % of world total exports	7.8	30.9	23.9	47.0	50.0	50.5
Imports (1,000 m^3)	713	581	115	40	482	305
Imports as % of world total imports	26.7	22.0	3.1	0.4	2.0	1.1
Net trade (1,000 m^3) (Exports - imports)	(583)[b]	(99)	644	4,593	11,698	13,990
Net trade/production, %	(0.46)	(0.07)	0.43	3.0	7.5	7.9

Source: Food and Agriculture Organization, Yearbook of Forest Products, selected issues (Rome, FAO).

[a]SITC 247.1

[b]Parenthetical numbers denote a deficit.

the early 1960s, when the United States began actively exporting to the Japanese market. Sawlog exports exhibited substantial growth until about 1972. After 1972, sawlog exports fluctuated at a high level, with the 1976 level slightly exceeding 1973's previous high. In recent years, the United States has been the world's major exporter of conifer sawlogs, with the 1976 volume constituting over 50 percent of total world exports. Other major exporters include the USSR (with 32 percent), New Zealand, the German Federal Republic, and Canada. Japan was the major world importer of conifer sawlogs, accounting for about two-thirds of the world's total. Other important countries were Canada, Korea, Italy, and Sweden. Generally, imports by the European countries originated within Europe. The major U.S. market was Japan, with South Korea assuming an expanding role. Other major suppliers to the East Asian market included the USSR and New Zealand.

On the import side, U.S. conifer log imports dropped from a modest 713,000 cubic meters in 1950 to only 305,000 cubic meters in 1976. Essentially all of the imports originated in Canada.[1] The U.S. net trade balance in physical terms was reduced from an import deficit of 583,000 cubic meters in 1950 to an export surplus of 13,990 cubic meters in 1976. In 1976 the net trade surplus was 7.9 percent of domestic production, reflecting the self-sufficient position of the United States in conifer sawlogs.

In value terms, the U.S. export dominance is somewhat larger than is suggested by the volume figures, indicating a higher-valued product.

[1]It should be noted that the Canadian prohibition of unprocessed wood exports allows for some exceptions.

In 1976 U.S. exports were $855 million or about 58 percent of world exports, while imports were only $7.4 million, resulting in a net trade surplus of $847.6 million. This compares with the 1967 net trade surplus of $157.7 million based on exports of $159.4 million and imports of only $1.7 million.

Conifer sawlogs accounted for 21 percent of total export earnings from U.S. forest products in 1976; this was up from 16 percent in 1967 (table 5-4). During 1967-76, conifer sawlog imports never reached 1 percent of U.S. forest product import payments.

The dominant exporting region for conifer sawlogs has been the Pacific Northwest which, in 1976, accounted for 91 percent of U.S. exports. This area is also the dominant importer of conifer sawlogs, having generally accounted for over 90 percent of U.S. imports during 1967-76.

In both value and volume terms, Japan is the dominant recipient of U.S. sawlog exports, constituting 91 percent by value and 86 percent by volume of the 1976 U.S. export market. In values terms, Japan's portion has remained near 90 percent over 1967-76, whereas, Canada's has declined to 4 percent and South Korea's share has increased to 4 percent.

Overall, the most rapid growth in value was experienced in U.S. exports to Korea (1,339 percent), whereas Europe (778 percent) and Japan (542 percent) also experienced significant growth.

<u>Nonconifer Sawlogs</u>

For 1950-76, U.S. nonconifer sawlog production remained virtually unchanged. However, as a fraction of total worldwide production, the U.S. share declined from 34.4 percent in 1950 to 14.3 percent in 1976. During this period, U.S. exports of nonconifer sawlogs slowly increased and imports decreased. Therefore, during the latter part of the twenty-seven-year period, the United States achieved a modest trade surplus to replace its modest trade deficit of earlier years. In value terms, the 1976 export earnings were $65.1 million as compared with import expenditures of only $2.7 million. This resulted in a trade surplus of $62.4 million.

Nonconifer sawlogs constitute only a small fraction of total U.S. international payments and receipts for forest products. On the export side, the share of total international trade earnings attributed to nonconifer logs fell from 3 percent in 1967 to 2 percent in 1976, but for imports, nonconifer sawlogs were a negligible portion (table 5-7).

The U.S. region exporting the most nonconifer sawlogs in 1976 was the North Atlantic (51 percent) followed by the Great Lakes (22 percent) and the rapidly growing Gulf region. The Gulf region experienced dramatic growth over the ten-year period and accounted for 17 percent of U.S. nonconifer log exports in 1976 as compared with 2 percent in 1967. Seventy-eight percent of nonconifer sawlogs imports in 1976 entered the North Atlantic region.

In 1976 Europe received 72 percent of U.S. nonconifer log exports, with Germany receiving 44 percent. Canada (14 percent) and Japan (7 percent) also received significant volumes; however, their shares fell during 1967-76. The small volume of U.S. nonconifer imports in 1976 originated predominantly in Canada (56 percent) and Africa (23 percent). See table 5-10.

Conifer Sawnwood

Until World War II the United States had been a net exporter of conifer sawnwood (softwood lumber). However, by 1950 the volume of U.S. net imports (gross imports minus gross exports) of softwood lumber was

Table 5-10. U.S. Nonconifer Sawlog[a] Production and Trade for Selected Years

	1950	1954	1959	1964	1970	1976
Production (1,000 m^3)	35,457	39,251	33,389	35,824	34,551	32,285
Production as % of world total	34.4	31.9	23.3	20.1	16.4	14.3
Exports (1,000 m^3)	90	153	167	289	312	427
Exports as % of world total exports	2.7	2.8	1.6	1.5	0.8	1.0
Imports (1,000 m^3)	505	420	330	255	172	55
Imports as % of world total imports	18.4	8.5	3.3	1.3	0.5	0.1
Net trade (1,000 m^3) (Exports - imports)	(415)	(267)	(163)	34	140	372
Net trade/production, %	(1.2)	(0.7)	(0.5)	0.1	0.4	1.2

Source: Food and Agriculture Organization, *Yearbook of Forest Products*, selected issues (Rome, FAO).

[a] SITC 247.2

[b] Parenthetical numbers denote negative values.

already 9 percent of domestic production. The trend of increased dependence upon foreign supply sources, which have been almost exclusively Canadian, continued through 1976, when the volume of net imports equaled almost 20 percent of U.S. domestic production. During this twenty-seven-year period U.S. production of conifer sawnwood remained virtually unchanged, whereas domestic consumption (that is, domestic production plus net imports) grew gradually.

During this period U.S. exports grew both absolutely and as a percentage of the total world volume of softwood lumber exports (from 4.3 percent to 6.6 percent), while gross U.S. import volumes increased. U.S. softwood lumber imports constituted a substantial portion of total volume of world imports throughout the twenty-six-year period, being about one-third of the 1976 total (see table 5-11).

In 1976 the value (in current dollar terms) of conifer sawnwood imports was 413.9 percent above the 1967 level. However, the ten-year data series exhibit considerable volatility, depending importantly upon the U.S. housing cycle. The value of U.S. exports has shown a slightly higher growth rate than have imports during that period, reaching 450 percent of the 1967 level in 1976. Moreover, since this growth was on a much lower base the absolute gap between the value of imports and exports (trade deficit) has continued to grow fairly rapidly.

As a share of the total value earnings from U.S. forest product exports, conifer sawnwood trade was fairly stable for slight over 11 percent of the value in 1967 and only 12 percent in 1976. For imports however, the portion of the total value of forest product imports, accounted for by conifer sawnwood, increased from 17 percent in 1967 to 26 percent in 1976.

Table 5-11. U.S. Sawnwood[a] Conifer[b] Production and Trade for Selected Years

	1950	1954	1959	1964	1970	1976
Production (1,000 m^3)	72,293	69,106	69,160	69,110	64,971	72,782
Production as % of world total	36.5	31.4	26.8	23.9	20.8	22.3
Exports (1,000 m^3)	912	1,310	1,429	1,881	2,720	3,715
Exports as % of world total exports	4.3	4.7	4.5	4.2	5.5	6.6
Imports (1,000 m^3)	7,463	6,766	8,885	11,350	13,460	17,712
Imports as % of world total imports	34.8	25.2	27.9	26.2	27.5	33.0
Net trade (1,000 m^3) (Exports - imports)	(6,551)[b]	(5,456)	(7,456)	(9,469)	(10,740)	(13,997)
Net trade/production, %	(9.1)	(7.9)	(10.8)	(13.7)	(16.5)	(19.2)

Source: Food and Agriculture Organization, Yearbook of Forest Products, selected issues (Rome, FAO).

[a]SITC 243.2

[b]Parenthetical numbers denote negative values.

Regionally, the Pacific Northwest accounted for almost half the value of U.S. softwood lumber exports, increasing from 38 percent in 1967 to 48 percent in 1976. Other major exporting regions included the Great Lakes and Alaska, both accounting for 14 percent of exports in 1976. The South Pacific and North Central regions are the only ones exhibiting a marked change over the period. The South Pacific region fell from 15 percent in 1967 to 7 percent in 1976, while the North Central region's share increased from 2 to 4 percent.

The major recipients of U.S. softwood lumber exports by value in 1976 were Canada (24 percent), Japan (23 percent), Italy (13 percent), other European countries (18 percent), and Oceania (11 percent). The structure of U.S. lumber exports did not change markedly during that period except in increased exports to Japan (up to 23 percent from 17 percent) and a decline in the share of other European countries (down to 18 percent from 24 percent). The change in the trade pattern reflected the somewhat more rapid growth of lumber exports to Japan than to other purchasers. However, the trend was volatile, with the Japanese share reaching a peak of 32 percent in 1976 before falling to 23 percent in 1976.

Canada continued to be the dominant supplier of U.S. softwood imports, providing between 98 and 99 percent of the total during the entire 1967-76 period.

Nonconifer Sawnwood

The U.S. world position in both production and trade of hardwood lumber declined steadily between 1950 and 1976. Production declined from 17,403 cubic meters, or 36.6 percent of world production, in 1950 to 16,027 cubic meters, or 15.4 percent of world production in 1976. Also,

Table 5-12. U.S. Nonconifer Sawnwood[a] Production and Trade for Selected Years

	1950	1954	1959	1964	1970	1976
Production (1,000 m^3)	17,403	16,695	17,051	17,169	16,846	16,027
Production as % of world total	36.6	31.0	26.2	21.2	18.2	15.4
Exports (1,000 m^3)	149	185	295	301	279	491
Exports as % of world total exports	4.8	6.2	7.6	7.0	3.9	4.5
Imports (1,000 m^3)	667	492	766	714	792	673
Imports as % of world total imports	21.9	15.4	18.5	14.0	11.7	6.6
Net trade (1,000 m^3) (Exports - imports)	(518)[b]	(307)	(471)	(413)	(513)	(182)
Net trade/production, %	(3.0)	(1.8)	(2.8)	(2.4)	(3.0)	(1.1)

Source: Food and Agriculture Organization, Yearbook of Forest Products, selected issues (Rome, FAO).

[a]SITC 248.3

[b]Parenthetical numbers denote negative values.

whereas U.S. imports remained constant in volume, they declined as a percentage of total world imports, dropping from 21.9 percent to 6.6. percent between 1950 and 1976. Simultaneously, U.S. exports increased absolutely. The result was a significant volume decline in the U.S. wood deficit over the twenty-seven-year period, with the 1976 net deficit equaling 1.1 percent of domestic production (see table 5-12).

In 1976 the value of nonconifer sawnwood exports increased 320 percent over the 1967 value. Despite this absolute rise, however, exports' share of total forest product earnings decreased to 2 percent in 1976 from 3 percent in 1967. On the import side, gross imports in current dollars increased only 57 percent between 1967 and 1976, less than what could be expected to occur solely through price inflation. The share of forest products imports declined during this period from 3 percent to 2 percent of total expenditures. Imports of nonconifer sawnwood by the United States in 1976 were worth $96.5 million and exports amounted to $87.2 million resulting in a modest deficit of about $9.3 million.

The principal regions involved in hardwood lumber exports were the Great Lakes, with over 50 percent by value of U.S. exports, followed by the North Atlantic region, with about 25 percent. The North Atlantic, South Atlantic, and the Pacific Northwest exhibited rapid growth in the value of exports; however, only the North Atlantic had a relatively large export share. Exports in current prices from the Gulf region declined substantially during 1967-76, from 17 percent to 7 percent.

Imports of hardwood lumber were distributed over a number of regions. The Gulf (26 percent), North Atlantic (21 percent), Great Lakes (20 percent), Pacific Northwest (12 percent), South Atlantic (10 percent) and South

Pacific (9 percent) regions shared the gross import flows on current price terms. All these regions experienced increases in their share of imports except for the Great Lakes and North Atlantic, whose combined shares fell from 67 percent to 41 percent.

U.S. exports of hardwood lumber during 1967-76 went to Canada (57 percent in 1976) and to other European countries, excluding West Germany and the Netherlands (22 percent in 1976). These proportions remained stable, but both West Germany and the Netherlands dramatically increased their shares of U.S. hardwood lumber exports. The share for West Germany increased from 1 percent to 5 percent, and that for the Netherlands rose from zero to 8 percent.

U.S. imports, which declined in both real value and physical volume, were drawn from a number of regions. The dominant Canadian share declined from 52 percent to 29 percent during the ten-year period, whereas imports from Brazil increased from 5 percent to 15 percent and those from Malaysia increased from 6 percent to 12 percent.

Wood chips

Wood chips and particles are wastes associated with some types of wood-processing activities. Because wood chips can typically be transported more inexpensively than pulpwood, they have become a substitute for pulpwood in international trade. While pulpwood trade has increased only modestly, world trade in wood chips has mushroomed since the eariler 1960s. In 1976 world exports were twenty-one times higher than the average 1961-65 volumes. In 1976 U.S. wood chip exports equaled $153 million and imports amounted to $24 million. In 1967 these figures were $12.1 million and $9.2 million, respectively. (See tables 5-3 and 5-6.)

Table 5-13. U.S. Wood Chips[a] Trade for Selected Years

	1961-65	1967	1970	1976
World exports (1,000 m^3)	476	2,573	5,774	10,303
Exports (1,000 m^3)	34	1,084	3,510	5,723
Exports as % of world total exports	7.1	42.1	60.8	55.5
Imports (1,000 m^3)	382	1,195	881	672
Imports as % of world total imports	80.3	46.4	15.3	6.5
Net trade (1,000 m^3) (Exports - imports)	(348)[b]	(111)	2,629	5,051

Source: Food and Agriculture Organization, Yearbook of Forest Products, selected issues (Rome, FAO).

[a]SITC 246.02

[b]Parenthetical numbers denote negative values.

The U.S. international trade in wood chips has undergone very substantial changes in the past two decades. As with pulpwood, in the 1950s and early 1960s, the United States was the principal importer of wood chips, importing over 80 percent of total worldwide wood chip exports in 1961-65. By 1967, however, U.S. wood chip trade approached a balance, and by 1976, The United States was the dominant wood chip exporter, maintaining a large wood chip trade surplus, both by value and by volume. In 1976, wood chips accounted for 4 percent of total receipts for U.S. exports of wood products, reflecting the fact that wood chips experienced the most rapid growth of any major wood product.

U.S. wood chip exports were dominated in 1976 by the Pacific Northwest (85 percent), South Pacific (8 percent), and Alaskan (3 percent) regions, with the Pacific Northwest becoming somewhat less important in recent years. Wood chip imports in 1976 were dominated by the Pacific Northwest (77 percent) and the Great Lakes (19 percent) regions, with the Great Lakes experiencing substantial increases over 1967-76.

Japan was the primary recipient of U.S. wood chip exports for this period, receiving over 98 percent of the export value of U.S. wood chips in each year. All U.S. imports of wood chips for 1967-76 came from Canada.

Plywood

Like other wood panels, plywood is a product that came into general use after World War II. FAO statistics do not distinguish between hardwood and softwood plywood, although their production and trade patterns differ significantly. Confusion in distinguishing between hardwood and softwood plywood is compounded because the same piece of plywood may have both hardwood and softwood layers. Here we will treat plywood as

Table 5-14. U.S. Plywood[a] Production and Trade for Selected Years

	1950	1954	1959	1964	1970	1976
Production (1,000 m^3)	3,350	4,460	7,950	11,630	14,078	16,727
Production as % of world total	54.3	47.8	53.4	52.2	42.7	43.3
Exports (1,000 m^3)	3.1	5.9	44	18	115	478
Exports as % of world total exports	0.8	0.7	2.7	0.8	2.6	8.0
Imports (1,000 m^3)	37	256	785	815	1,771	1,936
Imports as % of world total exports	9.1	29.9	43.1	34.0	37.7	30.9
Net trade (1,000 m^3) (Exports - imports)	(33.9)	(250.1)	(741.0)	(797.0)	(1,656)	(1,458)
Net trade/production, %	(1.0)	(5.6)	(9.3)	(6.8)	(11.8)	(8.7)

Source: Food and Agriculture Organization, Yearbook of Forest Products, selected issues (Rome, FAO).

[a]SITC 631.2

[b]Parenthetical numbers denote a negative value.

homogeneous, but subsequently will examine briefly the distinction between hardwood and softwood (see table 5-14).

In 1950 the United States produced 54.3 percent of the world's plywood, or 3.3. million cubic meters. By 1976, this had increased more than five times that volume--to 17.7 million cubic meters--although the U.S. portion of the world's total had declined to 43.3 percent. In 1976 the U.S. portion of total world plywood exports had risen to 8 percent from less than 1 percent in 1950. U.S. imports of plywood have shown a strongly rising trend through the twenty-seven-year period, reaching almost 2 million cubic meters in 1976. As a result, there has been a deficit trade balance for plywood throughout most of the postwar period, with net imports in 1976 of about 1.5 million cubic meters. The net trade/production ratio was a minus 8.7 percent in 1976, indicating a fairly high degree of dependence upon foreign supplies.

In value terms, U.S. exports of plywood increased from $11.1 million to $147.9 million between 1967 and 1976, an increase of more than thirteen times the 1967 level. However, imports have gone up from $141.3 million to $382.5 million, thus increasing the U.S. import deficit from $131.2 million to $234.6 million. When adjusted for inflation, however, the trade deficit experienced a small decline in real terms. As a fraction of U.S. forest product export earnings, plywood exports increased their share from about 1 percent to about 4 percent between 1967 and 1976. Imports of plywood continued to account for about 7 percent of total forest product import payments in 1976, after experiencing a high of 10 percent in the late 1960s and early 1970s.

The regional shares of total U.S. plywood exports showed a high degree of volatility over the period 1967-76. In the latter part of the period, the dominant regional exporters were the Pacific Northwest (63 percent) and the Great Lakes (17 percent). The Great Lakes in particular increased its share toward the end of the period, whereas the South Atlantic and the South Pacific regions experienced substantial declines over the ten-year period.

For plywood imports, the North Atlantic (28 percent in 1976), South Atlantic (20 percent), Gulf (24 percent), South Pacific (14 percent), and Pacific Northwest (9 percent) accounted for 96 percent of U.S. plywood imports in 1976. The South Atlantic in particular showed substantial growth over the ten-year period, and the South Pacific, Pacific Northwest, and the Great Lakes have experienced sizable declines.

Bureau of Census data present U.S. exports and imports of plywood, disaggregated by hardwood and softwood. Unfortunately, about 7 percent of the 1976 imports, by value, are not specified as to wood type. However, of the imports specified in 1976, less than one-half of 1 percent were hardwood (see table 5-15). This percentage is almost identical to that for 1967.

In 1976 unspecified plywood exports constituted slightly over 4 percent of total U.S. plywood exports, and hardwood plywood was also approximately 4 percent by value. This compares with 1967, when about 9 percent of U.S. plywood exports were hardwood and about 12.3 percent were unspecified. Thus, despite fairly rapid growth in U.S. hardwood plywood exports, softwood plywood exports are increasing their relative dominance.

Despite fairly high duties (20 percent _ad valorem_ tax) on U.S.-Canadian plywood trade Canada received the major portion of U.S. plywood exports in recent years; for example, her 1976 share was 28 percent by value.

Table 5-15. U.S. Exports of Hardwood and Softwood Plywood for Selected Years
(quantity in thousand square feet, value in current dollars)

Year	Hardwood		Softwood		Not specified	
	Quantity	Value	Quantity	Value	Quantity	Value
1950[a]	365	97,998	3,279	501,233	172	88,193
1954[a]	431	183,776	6,682	838,207	222[b]	65,264[b]
1959[a]	1,951	562,122	71,864	7,366,912	1,298[b]	260,357[b]
1964[a]	2,022	790,150	28,233	3,483,952	802[b]	238,379[b]
1967	6,656	967,964	84,922	8,914,472	2,518	1,379,713
1968	10,004	1,349,654	64,165	8,521,988	5,829	1,636,500
1969	9,209	1,451,171	194,268	23,633,261	9,073	2,695,789
1970	8,505	1,220,261	113,781	13,270,525	59,461	1,877,580
1971	10,175	1,360,566	99,028	13,135,804	8,080	1,490,743
1972	21,478	2,474,283	220,439	31,548,318	10,018	3,016,424
1973	33,417	4,423,670	462,427	62,079,343	15,449	5,733,422
1974	40,334	7,058,024	541,984	82,597,783	40,050	9,574,457
1975	48,837	8,151,357	790,896	124,778,480	31,435	8,474,079
1976	59,687	10,144,955	718,060	128,486,225	33,864	10,564,940

Note: Domestically produced merchandise only.

Source: U.S. Bureau of the Census, U.S. Exports, Schedule B Commodity by Country, Report FT-410, Annual.

[a]Data are not perfectly comparable with other years because of changes in commodity classifications and estimation methods.

[b]Includes some veneer and composite board.

Table 5-16. U.S. Imports of Hardwood and Softwood Plywood for Selected Years
(quantity in thousand square feet, value in current dollars)

Year	Hardwood		Softwood		Not specified	
	Quantity	Value	Quantity	Value	Quantity	Value
1950[a]	51,273	6,042,334	95	6,528	11,894	617,685
1954[a]	110,149	13,612,184	650[b]	52,606[b]	323,673	19,357,591
1959[a]	136,838	14,573,520	2,400[b]	108,895[b]	1,190,987	84,092,982
1964[a]	1,922,651	120,865,599	4,853	573,295	-	-
1967	2,509,761	140,609,024	3,099	348,348	14,848	1,440,092
1968	3,807,005	214,116,337	10,554	915,604	36,091	2,924,373
1969	4,242,559	243,473,396	14,418	1,388,269	74,512	5,681,177
1970	4,115,076	203,959,745	2,330	253,068	79,281	5,110,765
1971	5,047,503	252,873,310	3,480	226,836	152,777	9,242,195
1972	6,228,373	323,901,433	5,948	488,901	229,597	14,199,036
1973	4,896,734	368,568,787	8,535	1,526,771	259,999	20,864,857
1974	3,244,776	275,641,541	4,155	605,128	148,735	14,073,224
1975	3,756,222	245,595,728	6,640	715,504	198,984	15,794,146
1976	4,513,028	351,832,693	11,929	1,631,453	317,092	29,304,768

Note: Data for 1950-64 represent imports of merchandise for consumption; and data for 1967-76 represent general imports.

Sources: U.S. Bureau of the Census, U.S. General Imports, Report FT-135 Annual, 1967-76; U.S. imports of Merchandise for Consumption, Report FT-125, Annual 1964 and Report FT-110, Annual 1950-59.

[a]Data are not perfectly comparable with other years because of changes in commodity classifications and estimation methods. Dashes denote no commodities in the classification.

[b]Includes some alder plywood.

Europe accounted for 63 percent of the U.S. value of exports in 1976, with the principal recipients being Denmark (19 percent), the United Kingdom (17 percent), and the Netherlands (9 percent). See table 5-16.

Ninety-three percent of U.S. imports of plywood originated primarily in Asia with the 1976 shares of value being Korea (49 percent), Japan (15 percent), Philippines (8 percent), and other Asian countries (22 percent). Undoubtedly, the imports originating in Asia were dominantly hardwood plywood. Although Asia was the dominant supplier throughout the period, the countries involved have changed their roles, with Korea dramatically increasing its share and the Japanese and Philippines experiencing significant declines.

Veneer Sheets

During the 1950-76 period, U.S. exports of veneer sheets rose both in absolute volume terms (from 76,000 to 267,000 cubic meters) and also as a percentage of total world veneer exports (6.3 percent to 22.4 percent).

U.S. veneer imports also increased for each of the years examined. The import volume increased from 80,000 to 441,000 cubic meters between 1950 and 1976. As a percentage of total world veneer imports, however, the U.S. share declined, particularly after 1964. For the selected years examined, the United States experienced a deficit throughout 1950-76. However, the 1976 deficit was only about half that of 1970 (see table 5-17).

In value terms, veneer sheets constitute only a small portion of total U.S. trade in forest products, with both exports and imports in 1976 amounting to about 1 percent of total export and import values of forest products. During 1967-76, exports rose by a factor of about 6 while imports increased by a factor of only 1.75.

Table 5-17. U.S. Veneer Sheet[a] Trade for Selected Years

	1950	1954	1959	1964	1970	1976
Exports (1,000 m^3)	7.6	9.6	15	32	73	267
Exports as % of world total exports	6.3	5.8	4.4	5.7	8.8	22.4
Imports (1,000 m^3)	80	129	235	384	415	441
Imports as % of world total imports	51.6	52.6	54.7	58.0	43.0	31.9
Net trade (1,000 m^3) (Exports - imports)	(72.4)[b]	(119.4)	(220)	(352)	(342)	(174)

Source: Food and Agriculture Organization, Yearbook of Forest Products, selected issues (Rome, FAO).

[a]SITC 634.1

[b]Parenthetical numbers denote negative value.

Regionally, the dominant exporters are the North Atlantic (54 percent), the Pacific Northwest (18 percent), and the Great Lakes (13 percent).

The structure of exports changed radically between 1967 and 1976, with the North Atlantic's share rising rapidly and that of the Pacific Northwest and the Great Lakes falling dramatically. This change reflects the rapid increase of exports from the North Atlantic region rather than an absolute decline in exports by other regions.

Recipients of U.S. veneer exports changed markedly between 1967 and 1976. The European countries dramatically increased their share, by value, of U.S. veneer exports to 66 percent in 1976, from 24 percent in 1967, while the Canadian share declined. West Germany's 1976 share was 37 percent, surpassing Canada's 32 percent.

In contrast to the export situation, countries supplying veneer to the United States have maintained quite stable shares, with Canada providing slightly over 60 percent in both 1967 and 1976. Other major suppliers in 1976 were Brazil (7 percent), the Philippines (16 percent), and the other Asian countries (6 percent). Africa's share appreciably declined between 1967 and 1976, but Brazil experienced major growth.

As with plywood, veneer sheets can be either hardwood or softwood. Although we do not have regional or intercountry flows disaggregated by wood type, we do have disaggregated national figures on veneer imports and exports classified by wood type. In 1976 hardwood imports constituted about 87.3 percent of U.S. imports by value, compared with 97 percent in 1967, whereas exports of hardwood veneer accounted for about three-fourths of U.S. veneer exports in both 1967 and 1976 (see tables 5-18 and 5-19).

Table 5-18. U.S. Exports of Hardwood and Softwood Veneer for Selected Years
(quantity in thousand square feet, value in current dollars)

Year	Hardwood		Softwood		Not specified	
	Quantity	Value	Quantity	Value	Quantity	Value
1950[a]	-	-	-	-	34,517	1,210,420
1954[a]	-	-	-	-	43,293	1,396,406
1959[a]	-	-	-	-	66,115	2,876,485
1964[a]	-	-	-	-	144,345	7,115,597
1967	105,794	5,870,132	87,044	1,812,706	-	-
1968	173,552	8,479,694	132,733	3,118,774	-	-
1969	194,191	10,326,851	166,361	4,173,676	-	-
1970	183,827	9,736,996	143,293	3,614,476	-	-
1971	172,633	9,978,882	398,814	5,010,395	-	-
1972	204,338	12,710,466	287,426	9,206,759	-	-
1973	346,026	22,693,823	314,459	12,590,780	-	-
1974	380,766	25,754,884	218,626	11,269,913	-	-
1975	389,811	27,229,977	346,407	15,013,170	-	-
1976	855,353	34,354,774	262,434	11,359,921	-	-

Note: Domestic merchandise only

Source: U.S. Bureau of the Census, U.S. Exports, Schedule B Commodity by Country, Report FT-410, Annual.

[a]Data are not perfectly comparable with other years because of changes in commodity classifications and estimation methods. Values of "not specified" for 1950-64 are totals for all veneers. Dashes denote no commodities in the classification.

Table 5-19. U.S. Imports of Hardwood and Softwood Veneer for Selected Years
(quantity in thousand square feet, value in current dollars)

Year	Hardwood		Softwood		Not specified	
	Quantity	Value	Quantity	Value	Quantity	Value
1950[a]	161,852	3,382,450	-	-	200,078	1,938,560
1954[a]	312,215	8,591,544	-	-	271,990	3,835,060
1959[a]	496,500	17,341,654	-	-	567,481	11,767,031
1964[a]	1,407,300	35,471,368	28,464	362,516	301,003	6,596,319
1967	1,223,514	58,819,991	194,222	1,790,303	573,785	9,689,856
1968	2,175,198	48,866,658	161,360	2,034,999	-	-
1969	1,866,784	44,671,119	198,883	3,233,078	-	-
1970	1,615,119	36,066,402	270,826	3,066,216	-	-
1971	2,034,410	43,825,422	266,848	3,381,128	-	-
1972	2,788,299	63,702,786	365,441	5,755,867	-	-
1973	2,583,277	76,452,493	384,669	7,518,420	-	-
1974	1,351,310	43,751,782	317,700	6,608,300	-	-
1975	1,142,280	44,145,431	350,891	6,927,460	-	-
1976	1,596,070	62,973,688	397,883	10,387,034	-	-

Note: Data for 1950-64 represent imports of merchandise for consumption; data for 1967-76 represent general imports.

Sources: U.S. Bureau of the Census, U.S. General Imports, Report FT-135 Annual, 1967-76; U.S. Imports of Merchandise for Consumption, Report FT-125, Annual, 1964, and Report FT-110, Annual, 1950-59.

[a]Data are not perfectly comparable with other years because of changes in commodity classifications and estimation methods. Dashes denote no commodities in the classification.

Reconstituted Wood (Including Particle Board)

Since reconstituted wood and particle board are fairly new products,[1] we do not have as much experience with them as we have had with some of the more traditional forest products. However, U.S. production of particle board increased by a factor of almost ten between 1959 and 1976. Nevertheless, the U.S. fraction of world production fell from 24.7 percent to 15.4 percent, reflecting the rapid expansion of production worldwide. The U.S. portion of international trade in particle board has always been small; for example, in 1976 the United States exported only 3.6 percent of the world total and imported 2.4 percent of the total volume of world imports. Overall, the U.S. physical trade in particle board has been fairly well balanced, with 1976 net exports equaling 1.1 percent of U.S. production. In 1976 the U.S. exports of particle board were valued at $18.6 million and imports at $12.7 million, resulting in a modest trade surplus. See table 5-20.

Reconstituted wood (including particle board) was a negligible fraction of U.S. earnings from and expenditures on the forest products trade during 1967-76. However, the growth rate of both exports and imports of this commodity was higher than that for any of the commodities we examined for the period 1967-76, with both export and import values increasing by a factor of about 20 during the period.

The major exporting regions are the North Central, the Pacific Northwest, and the Great Lakes,which together accounted for 87 percent of the value of exports in 1976.

---

[1]The FAO data are limited to particle board (SITC 631.42), while the U.S. Census data refer to reconstituted wood (SITC 631.4). See appendix B.

Table 5-20. U.S. Particle Board[a] Production and Trade for Selected Years

	1959[b]	1964	1970	1976
Production (1,000 m^3)	553	1,120	3,127	5,318
Production as % of world total	24.7	14.7	16.3	15.4
Exports (1,000 m^3)	0	0	21	169
Exports as % of world total exports	0	0	1.0	3.6
Imports (1,000 m^3)	7	3	12	108
Imports as % of world total imports	5.6	0.4	0.6	2.4
Net trade (1,000 m^3) (Exports - imports)	(7)[c]	(3)	9	61
Net trade/production, %	(1.3)	(0.3)	0.3	1.1

Source: Food and Agriculture Organization, Yearbook of Forest Products, selected issues (Rome, FAO).

[a]SITC 631.42

[b]Data did not become available until 1959.

[c]Parenthetical numbers denote negative value.

On the import side, the North Atlantic region has been the major importing region of reconstituted wood for most of the past ten years. Between 1973 and 1976, however, the Great Lakes region was the largest regional importer. Nonetheless, given the relatively small volumes of imports involved, this shift may not be very significant.

Canada was the major recipient of U.S. reconstituted wood exports between 1967 and 1976 and accounted for 85 percent of the value of U.S. 1976 exports. Likewise, Canada was the source of 90 percent of the value of 1976 reconstituted wood imports by the United States. Canada's share during the period increased as the European share declined, dropping from 66 percent in 1967 to 10 percent in 1976.

Pulpwood

U.S. pulpwood production in 1976 had increased to 213 percent above its 1950 level. Throughout this period, the United States maintained nearly 30 percent of its share of total worldwide production. U.S. international trade in pulpwood has been modest, but during this time the volume of U.S. pulpwood exports rose gradually and the volume of its pulpwood imports declined. As the U.S. physical trade balance deficit progressively diminished, there was a substantial reduction of the U.S. dependence on foreign suppliers. In value terms, the U.S. pulpwood deficit in 1976 amounted to only $4.5 million, with exports valued at $7.2 million and imports at $11.7 million. This compares with a deficit of $13.6 million in 1967 (see table 5-20).

The Great Lakes region has tended to be the dominant pulpwood exporter, although regional flows exhibit a great deal of variability. The North Atlantic, South Atlantic, and Great Lakes are the major importers, with the North Atlantic being the largest in recent years.

Table 5-21. U.S. Pulpwood[a] Production and Trade for Selected Years

	1950	1954	1959	1964	1970	1976
Production (1,000 m^3)	48,796	56,698	68,846	81,137	109,598	104,217
Production as % of world total	32.6	30.4	28.8	35.4	34.9	31.3
Exports (1,000 m^3)	69	105	254	163	201	372
Exports as % of world total exports	0.9	1.2	2.8	1.2	1.1	2.1
Imports (1,000 m^3)	3,616	4,087	2,653	1,537	1,065	621
Imports as % of world total imports	52.8	43.8	27.8	12.3	5.8	3.9
Net trade (1,000 m^3) (Exports - imports)	(3,547)[b]	(3,982)	(2,399)	(1,374)	(864)	(249)
Net trade /production, %	(7.3)	(7.0)	(3.5)	(1.7)	(0.8)	(0.2)

Source: Food and Agriculture Organization, *Yearbook of Forest Products*, selected issues (Rome, FAO).

[a]SITC 246.01

[b]Parenthetical numbers denote negative values.

The major recipient of U.S. pulpwood exports has been Canada. U.S. pulpwood imports were dominated again by Canada followed by the Bahamas.

Wood Pulp

During 1950-76 total U.S. physical wood pulp production increased about 300 percent from 13.5 billion to 41.1 million metric tons. The U.S. fraction of total world production remained relatively constant, falling from 40 percent to 36.7 percent. U.S. exports, however, increased dramatically both in total volume (from 87,000 to 2.3 million metric tons) and as a fraction of world pulpwood exports (1.5 percent to 13.5 percent). Simultaneously, U.S. imports increased absolutely from 60 percent in physical terms, while falling as a percentage of the world total (from 37.7 percent in 1950 to 20.5 percent in 1976). Over the twenty-six years, there was a reduction in the U.S. trade balance of almost 50 percent in the volume of net imports (from about 3 million metric tons to 1.1. million metric tons). U.S. net imports as a fraction of domestic production decreased from 15.4 percent in 1950 to 2.6 percent in 1976, indicating a dramatic reduction in U.S. dependence upon foregin wood pulp sources (see table 5-22).

In value terms, the 1976 wood pulp imports amounted to about $1.2 billion and exports came to $866 million, resulting in a deficit of $387 million. This compares with the 1967 wood pulp deficit of $114 million. Therefore, even adjusting the deficit inflation (about 70 percent), the deficit increased between 1967 and 1976. In volume terms, net imports declined slightly between 1967 and 1976 (from 1.179 to 1.33 million metric tons). During this ten year period wood pulp exports constituted between 16 and 27 percent of the total value of U.S. forest products, while wood

Table 5-22. U.S. Wood Pulp[a] Production and Trade for Selected Years

	1950	1954	1959	1964	1970	1976
Production (1,000 metric tons)	13,472	16,562	22,066	27,761	37,318	41,130
Production as % of world total	40.0	39.1	40.2	37.6	36.5	36.7
Exports (1,000 metric tons)	87	400	592	1,434	2,808	2,283
Exports as % of world total exports	1.5	5.8	7.0	11.5	16.6	13.5
Imports (1,000 metric tons)	2,159	1,861	2,206	2,661	3,167	3,359
Imports as % of world total imports	37.7	27.0	25.9	21.5	19.0	20.5
Net trade (1,000 metric tons) (Exports - imports)	(2,072)[b]	(1,461)	(1,614)	(1,227)	(359)	(1,076)
Net trade/production, %	(15.4)	(8.8)	(7.3)	(4.4)	(1.0)	(2.6)

Source: Food and Agriculture Organization, Yearbook of Forest Products, selected issues (Rome, FAO).

[a]SITC 251

[b]Parenthetical numbers denote negative values.

pulp imports comprised a similar share of total forest product imports, reaching 24 percent in 1976. Although the United States had experienced a generally improving trade balance in wood pulp since 1950, this improvement appears to have reversed itself after 1970.

Four regions--the Great Lakes, Pacific Northwest, North Central, and North Atlantic--accounted for 97 percent of U.S. imports of wood pulp in 1976, with the Great Lakes and the Pacific Northwest having increased shares in 1967-76. During this time, the share of the North Atlantic region declined. Eighty-four percent of the 1976 U.S. export value of wood pulp came from the South Atlantic, the Gulf, Pacific Northwest, and South Pacific regions, with the Gulf and South Atlantic regions having dominance. The Great Lakes and South Pacific regions expanded their shares significantly in the ten-year period.

Europe, with a 56 percent share in 1976, was the major recipient of U.S. wood pulp exports, followed by Japan with a 17 percent share.

U.S. wood pulp imports came primarily from Canada and accounted for 97 percent of the value of 1976 imports. During 1967-76 the share originating in Sweden declined, falling from 5 percent to 1 percent despite the fact that the 1976 value of Swedish wood pulp exports was almost identical to their 1967 value.

Newsprint

In constrast to most of the forest products examined, U.S. production of newsprint as a proportion of total world production has risen significantly during 1950-76, increasing from 10.6 percent to 15 percent. During this period, U.S. total production in volume terms increased by a factor of more than three. Despite this substantial increase in domestic output, the United States still had imports equal to 188.3 percent of its domestic production in 1976. However, in 1950, U.S. imports had been equal to 475 percent of domestic production; thus, the data show a substantial decline in U.S. dependence upon foreign newsprint sources over the twenty-six year period. The U.S. share of total world imports of newsprint has declined markedly from almost 80 percent in 1950 to 54 percent in 1976 although, during this period, the absolute level of net imports of the United States rose from 4,412 metric tons to 5,305 metric tons (see table 5-23).

Between 1967 and 1976 the value and volume of newsprint exports rose. In 1976 U.S. earnings from newsprint exports were $48.9 million, while U.S. payments for newsprint imports were over $1.7 billion. Thus the deficit on newsprint approached $1.6 million in 1976. This compares with the 1967 export level of $11.4 million and an import level of $892 million for a deficit of about $880 million. Adjusted for inflation, the newsprint deficit has increased modestly.

Newsprint exports constitute a small fraction (about 1 percent) of total earnings from exports of forest products. In contrast, newsprint import payments accounted for 34 percent of the total forest product import payments of the United States in 1976, and newsprint was the dominant forest product import in value terms. However, this share was down from 1967, when newsprint payments constituted 46 percent of total payments.

The principal regions involved in newsprint exports are the North Atlantic, Gulf, and Pacific Northwest regions, which together accounted for 90 percent of the value of U.S. newsprint exports in 1976. In 1967-76 the Great Lakes region increased its share of U.S. newsprint export earnings from 1 percent to 4 percent. The major importing regions are the Great Lakes and the North Atlantic regions. In 1976 they made up 75 percent of the value of U.S. newsprint imports.

Table 5-23. U.S. Newsprint[a] Production and Trade for Selected Years

	1950	1954	1959	1964	1970	1976
Production (1,000 metric tons)	919	1,081	1,745	2,040	3,035	3,153
Production as % of world total	10.6	10.4	13.4	12.6	14.1	15.0
Exports (1,000 metric tons)	40	127	109	107	130	159
Exports as % of world total exports	0.7	2.0	1.6	1.3	1.2	1.7
Imports (1,000 metric tons)	4,412	4,529	4,774	5,402	6,019	5,305
Imports as % of world total imports	79.6	73.7	67.4	63.3	56.8	54.0
Net trade (1,000 metric tons) (Exports - imports)	(4,372)[b]	(4,402)	(4,665)	(5,295)	(5,889)	(5,146)
Net trade/production, %	(475.7)	(407.2)	(267.3)	(259.6)	(194.0)	(163.2)

<u>Source</u>: Food and Agriculture Organization, <u>Yearbook of Forest Products</u>, selected issues (Rome, FAO).

[a]SITC 641.1

[b]Parenthetical numbers denote negative values.

U.S. exports of newsprint, although modest in total value, are broadly distributed internationally; for example, in 1976, 34 percent went to Mexico, 25 percent to Asia, 12 percent to Venezuela, 11 percent to Europe, and 3 percent to Africa. The most notable changes over the 1967-76 period were the increases in the shares going to Europe and Africa.

U.S. imports of newsprint come almost exclusively from Canada, with the Canadian share of import value rising during 1967-76 to 99 percent from 96 percent. Simultaneously, the Finnish share dropped from 4 percent to zero.

Paper and Paperboard

U.S. production of paper and paperboard (excluding newsprint) declined in physical terms from 59.2 percent of world production in 1950 to 39.6 percent in 1976. However, whereas U.S. production as a share of the world total declined, total U.S. production in 1976 was 250 percent of that of 1950. Also, U.S. gross exports as a fraction of total world exports rose to 15.9 percent of the world's total exports from 9.3 percent in 1950. Simultaneously, U.S. imports, while increasing absolutely, declined from 5.2 percent of the world's total in 1950 to 2.7 percent in 1976. The result of these trends has been the rapid increase in the U.S. trade surplus, both volume and value, in paper and paperboard. The volume surplus increased from 88,000 metric tons in 1950 to 2.3 million metric tons, while 1976 export earnings approached $1.2 billion and net foreign exchange receipts approached $1 billion (see table 5-24).

While the trade surplus in paper and paperboard has risen rapidly, its share of total export earnings from the forest products sector declined

Table 5-24. U.S. Paper and Paperboard[a] Production and Trade for Selected Years

	1950	1954	1959	1964	1970	1976
Production (1,000 metric tons)	20,087	21,735	27,375	33,738	43,082	50,886
Production as % of world total	59.2	59.2	48.9	44.5	40.4	39.6
Exports (1,000 metric tons)	190	397	613	1,243	2,301	2,754
Exports as % of world total exports	9.3	14.0	15.5	18.1	18.0	15.9
Imports (1,000 metric tons)	102	149	264	168	371	421
Imports as % of world total imports	5.2	5.6	6.8	2.5	3.0	2.7
Net trade (1,000 metric tons) (Exports - imports)	88	248	349	1,075	1,930	2,333
Net trade/production, %	0.4	1.1	1.3	3.2	4.5	4.6

Source: Food and Agriculture Organization, Yearbook of Forest Products, selected issues (Rome, FAO).

[a] SITC 641 less newsprint (SITC 641.1)

somewhat, from about 35 percent in the late 1960s and early 1970s to 30 percent in 1976.

Also, in recent years imports of paper and paperboard have grown somewhat more rapidly than the total value of forest products imports. However, in 1976, imports of this commodity constituted only 4 percent of the total value.

The major exporting regions for paper and paperboard are the North Atlantic, the Gulf, Great Lakes, South Atlantic, and the Pacific Northwest. All these regions experienced nearly average growth except the Great Lakes, which has grown somewhat more rapidly.

In 1976 more than 86 percent of paper and paperboard imports went to the North Atlantic and Great Lakes regions. The regional shares of these imports, that is, the percentage attributable to the various regions, remained relatively stable over the period 1967-76.

U.S. paper and paperboard exports were distributed broadly, with Canada receiving 27 percent; Europe, 27 percent; Asia, 12 percent; and others, 34 percent of the 1976 export value. In recent years Canada's share of paper and paperboard imports from the United States has increased more rapidly than that of the other major importing regions.

U.S. paper and paperboard imports by value in 1976 were dominated by Canada (64 percent), with Europe maintaining a significant 29 percent share. Sweden's share dropped substantially (from 10 percent to 4 percent) between 1967 and 1976, whereas minor suppliers experienced fairly rapid increases in their initially small shares.

Building Board (Fiberboard)

Building board production grew rapidly between 1950 and 1964, with U.S. physical production increasing in 1964 to over 300 percent of the 1950 level. After the mid-1960s, however, U.S. production leveled off, and the U.S. share of total world production subsequently declined gradually over the entire period, falling from 54 percent in 1950 to 38.4 percent in 1976. U.S. exports and imports grew somewhat faster than U.S. production and, in 1976, were almost the same fraction of the world total as they had been in 1950. Throughout most of the period the United States experienced a small deficit by both physical and value measures (see table 5-25).

Throughout 1967-76 building board's share of total forest products' expenditures and receipts from international trade remained relatively stable at about 1 percent.

The major recipients of U.S. building board exports in 1976, by value, were Canada (68 percent), Europe (11 percent), Asia (10 percent), and Central America (7 percent). Canada's share increased rapidly between 1967 and 1976. U.S. imports of building board in 1976 were from various regions including Brazil (33 percent), Canada (24 percent), and the Nordic countries (15 percent). Over 1967-76, Brazil's share expanded dramatically, while the Nordic countries' share declined.

The U.S. region exporting the greatest amount of building board in 1976 was the Great Lakes, with 46 percent of the total export value of building board. Next came the North Atlantic (14 percent), Pacific Northwest (11 percent), and the North Central regions (10 percent). The major importing regions were the North Atlantic (22 percent), the Gulf (22 percent), South Pacific (19 percent), and the South Atlantic regions (10 percent).

Table 5-25. U.S. Fiberboard[a] Production and Trade for Selected Years

	1950	1954	1959	1964	1970	1976
Production (1,000 m^3)	1,771	2,187	2,768	5,415	5,825	5,822
Production as % of world total	54.0	47.8	42.6	43.8	40.9	38.4
Exports (1,000 m^3)	34	37	29	81	95	179
Exports as % of world total exports	8.1	5.5	2.3	4.6	4.6	8.2
Imports (1,000 m^3)	40	27	32	244	283	228
Imports as % of world total imports	10.2	4.7	3.2	14.2	13.6	11.5
Net trade (1,000 m^3) (Exports - imports)	(6)[b]	10	(3)	(163)	(188)	(49)
Net trade/production, %	(0.4)	0.4	(0.1)	(3.0)	(3.2)	(0.8)

Note: 1950-59 data converted from metric tons to cubic meters, with 1.6 metric tons = 1 cubic meter.

Source: Food and Agriculture Organization, *Yearbook of Forest Products*, selected issues (Rome, FAO).

[a]SITC 641.6

[b]Parenthetical numbers denote negative values.

Summary and Conclusions

The discussion of U.S. forest products trade by commodity suggests something of the structure of U.S. forest products and changes occurring in that structure between 1950 and 1976. Looking first at the forest resource in its more primary form--sawlogs, pulpwood, and wood chips--the United States experienced a large relative increase in its net outflows (exports) for these primary resources over the period. Conifer sawlogs contributed a substantial net outflow for the United States, as did wood chips. For sawlogs, the net outflows equaled about 8 percent of domestic production in 1976. This contrasts with the 1950 situation, at which time trade in conifer sawlogs were essentially balanced, pulpwood trade was in a large deficit, and wood chips had not yet become an important commodity. By 1967, the volume of wood chips was about 750,000 short tons and this increased, by 1976, to almost 4 million short tons. Thus, the combination of pulpwood and chips was in substantial trade surplus in 1976, with the cubic meter equivalents equaling about two-thirds of the total volume exported in logs. Therefore, for both unprocessed solid wood and unprocessed wood fiber, the United States had a substantial export surplus in both physical volumes and also in values.

In addition to holding the position of a major net exporter of raw wood resources, the United States was a dominant producer and net exporter of paper and paperboard (exclusive of newsprint). Although the U.S. share of total world production of paper and paperboard fell from almost 60 percent in 1950 to just under 40 percent in 1976, total U.S. production increased by 150 percent. The U.S. trade position improved from a very small trade surplus in 1950 to a substantial trade surplus in 1976, at which time the United States was exporting 4.6 percent of its domestic production and achieving a net trade surplus in paper and paperboard of over $1 billion.

Additionally, the United States was an important net exporter of softwood based panels. Although the United States had a substantial trade deficit in plywood and veneers, disaggregation reveals that it was a fairly large exporter of softwood plywood and a large importer of hardwood plywood. Examination of the other wood panel products revealed a similar export surplus for the softwood products and a deficit for the hardwood. Of course, the large deficit in hardwood plywood dominated, to create a net deficit in wood panel products.

In contrast to unprocessed solid wood and paper and paperboard, the United States was a substantial net importer of pulp and newsprint. U.S. imports and exports of pulp were both large, with a trade deficit dominating. Also, U.S. imports of newsprint were far in excess of domestic production, resulting in a large deficit in newsprint trade. In the sawnwood category, the United States was both a large exporter and importer, with total sawnwood imports well above total exports.

To generalize, the United States was a major net exporter of raw forest products (logs and chips) and also a major exporter of the more highly processed forest products, such as paper and paperboard. It was a major net importer of intermediately processed forest products, such as sawnwood, pulp, and newsprint. In addition, the United States was a net aggregate importer of hardwood products, for example, hardwood panels, while it exported similar softwood products.

Chapter 6

INTERNATIONAL FOREST PRODUCTS TRADE OF MAJOR U.S. REGIONS

The previous chapters have provided a review of the world's forest resources and an overview of U.S. trade in forest products in the context of (1) North American (that is, U.S. and Canadian) trade with the rest of the world, and (2) aggregate U.S. trade with Canada and other major trading partners. Along with this latter perspective, we have indicated the roles played by various regions in the U.S. trade of particular forest products. The thrust of this chapter will be a closer analysis of the international trade of the individual U.S. regions, with emphasis on comparative contributions of the regions and on the commodity structure of trade.

Definitions and the Data

The regional trade figures presented in this chapter and in the appendixes must be interpreted with some care because of the manner in which regions have been defined and the trade flows have been recorded. A region, as defined here, does not represent a contiguous land area, but only a set of geographic points, each of which serves as an entry and exit point for internationally traded goods. The points represented are official U.S. customs districts (see appendix A, table A-6).

Although the trade regions are defined only by customs districts, there is a rough correspondence between the configuration of trade

Figure 6-1. Comparison of U.S. Customs Districts and International Trade Regions with U.S. Forest Service Resource Sections.

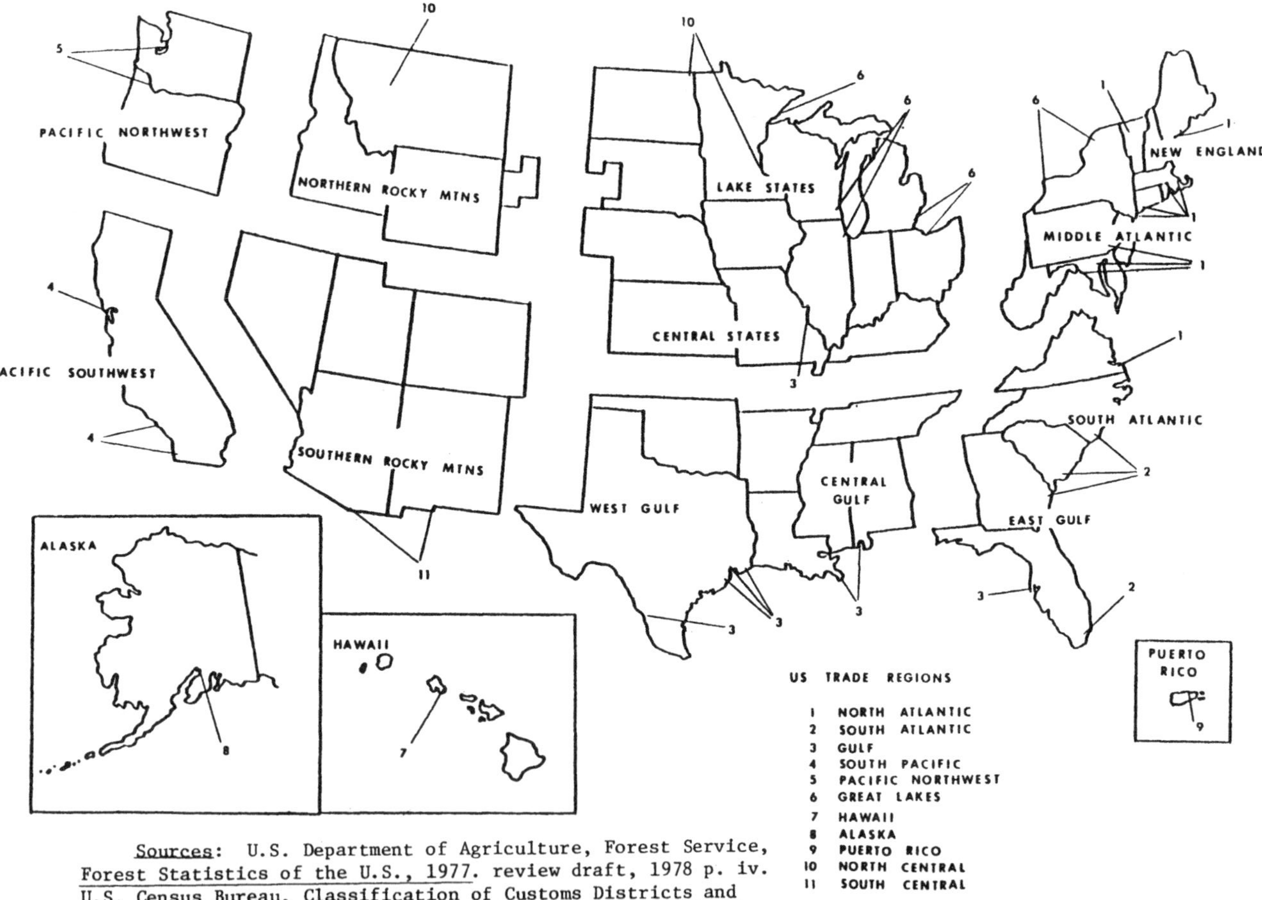

Sources: U.S. Department of Agriculture, Forest Service, Forest Statistics of the U.S., 1977. review draft, 1978 p. iv. U.S. Census Bureau, Classification of Customs Districts and

regions and the sectional and regional breakdown used by the U.S. Forest Service in compiling its forest resource data. In figure 6-1, the customs districts constituting trade regions are superimposed, by trade region numbers, on a map showing the Forest Service forest resource sections. Trade region 1 (North Atlantic) corresponds to the New England Middle Atlantic resource sections, plus the state of Virginia. Trade region 2 (South Atlantic) corresponds to the South Atlantic and East Gulf resource sections, exclusive of Virginia and Puerto Rico (trade region 9). The West Gulf and Central Gulf resource sections match up with trade region 3 (Gulf). Trade region 4 (South Pacific) corresponds to the Forest Service's Pacific Southwest section, exclusive of Hawaii, which constitute trade region 7. The Pacific Northwest resource section includes Washington and Oregon, which make up trade region 5 (Pacific Northwest) and Alaska, which is trade region 8. Trade region 6 (Great Lakes) matches up roughly with the Lake States and Central States resource sections, excluding North and South Dakota. Note also that two of the Great Lakes region's customs districts are located in the Middle Atlantic resource section. Trade regions 10 (North Central) and 11 (South Central) correspond, respectively, to the Northern Rocky Mountains (plus the Dakotas) and the Southern Rocky Mountains resource sections.

The comparison of trade regions with Forest Service resource sections was made primarily to give a general impression of the geographic location and extent of the trade regions. However, this comparison also suggests an investigation of the relationship between a region's forest resource endowment and its trade in forest products. In general, we would expect resource-rich regions to be net exporters, and heavily populated resource-poor regions to be net importers of forest products. A region's net trade position is determined by both its international trade and its trade with other domestic regions. The data provided, however, relate only to U.S. regions' international forest products trade and thus can be used only to determine a region's net international trade position. Thus, two important elements of interregional flows cannot be examined with the data developed. First, because the regions consist only of international entry and exit points, the international trade data do not necessarily indicate regions of production or consumption. For example, goods entering the country at Seattle, and recorded as an import to the Pacific Northwest might well be transported through the Pacific Northwest region to be consumed in some other region. Second, a region may be a large regional net exporter and yet this need not be reflected in the international data. This would be the case if, for example, a region like the Gulf had large net regional outflows that were predominantly to other U.S. regions. In addition, a region may be a large net exporter and yet this feature need not be reflected in its international trade data if the exports went predominantly to other U.S. regions.

Bearing these caveats in mind, one can still obtain useful insights from the investigation of the regional characteristics of a large country's international trade flows. For the United States, some interesting questions concerning the comparative roles of the various major U.S. forest products producing regions require, in part, knowledge of regional international trade patterns for their interpretation.

Important questions relate to the comparative roles of the Pacific Northwest and the U.S. South (South Atlantic and Gulf regions) in forest product production and trade both domestically and internationally.

It is generally believed that U.S. forest product production is gradually shifting to the South. However, a preliminary look at the data reveal that removals of growing stock from commercial timberlands over the 1970-76 period increased 2.1 percent in the Pacific Northwest and only 1 percent in the South. This apparent contradiction between belief and the data disappears, however, when the changing composition of removals nationwide is analyzed. Nationally, hardwood removals declined 12.3 percent over the 1970-76 period while softwood removals increased 9.2 percent, resulting in an overall increase in removals of 2.3 percent. The South exhibited a corresponding change in its composition of production with softwood removals increasing 18.7 percent while hardwood removals declined. This increase in removals allowed the South to account for over 80 percent of the nationwide increase in softwood removals over the 1970-76 period. In contrast, the Pacific Northwest increase in softwood removals was the same as its overall increase, 2.1 percent.

Table 6-1. Annual Removals of Growing Stock and Commercial Timberland (million cubic feet)

	South		Pacific Northwest	
	1970	1976	1970	1976
Softwood	3,768	4,471	3,163	3,209
Hardwood	2,733	2,100	87	109
Total	6,501	6,571	3,250	3,318

Source: Forest Service Statistics of the United States, 1977, Review Draft, USDA. (Washington, D.C., Forest Service, 1978) pp. 79-80.

U.S. exports of forest products are predominantly softwood products. Of the four commodities that are identified in chapter 5 as accounting for about 85 percent of U.S. forest product exports, two are exclusively softwood, softwood logs and softwood lumber. The other two, wood pulp and paper and paperboard, can use either softwood or hardwood inputs, however, U.S. exports of these products contain predominantly softwood inputs. In addition, over the 1967-76 period both wood chips and plywood exports increased dramatically. Again, although these products also can be either hardwood or softwood, the U.S. product exports have utilized predominantly softwood inputs.

A key question is the extent to which such a shift will affect both domestic and international forest product production and trade flows, particularly in the affected regions. We cannot hope in this brief chapter to thoroughly investigate this question. However, the pattern of regional international forest products trade and its changes over the 1967-76 period are examined and are suggestive of some hypotheses.

Regional Exports

Exports by region are reported in tables 6-2, 6-3, and 6-4. Exports in five of the ten regions grew faster than the overall national growth (table 6-3). The most dramatic growth was experienced by the North Central region (979 percent), followed by the Pacific Northwest (503 percent), the Great Lakes (462 percent), South Central (451 percent) and the South Pacific (450 percent). Of these regions, both the North Central and the South Central exported only a very small percentage of the U.S. total, thus the most dramatic increased in exports for a major exporting region was experienced by the Pacific Northwest. It is interesting to note that the South Atlantic and Gulf regions have experienced slower than average growth in their forest products exports. The two region share of

Table 6-2. U.S. Forest Product Exports by Region, 1967-76

(million current dollars)

Year	North Atlantic	South Atlantic	Gulf	South Pacific	Pacific Northwest	Great Lakes	Alaska	North Central	South Central	Others	Total
1967	166.8	145.1	167.5	46.7	312.8	92.3	42.0	6.0	3.5	3.2	985.9
1968	191.4	162.9	213.5	68.2	401.3	99.4	44.7	6.7	4.7	4.2	1197.0
1969	190.2	153.4	226.2	81.0	479.6	116.5	56.6	8.9	3.1	4.5	1319.9
1970	217.4	221.7	309.2	96.7	569.3	107.1	68.7	7.2	6.0	4.1	1607.4
1971	218.0	197.8	287.2	74.5	475.7	128.3	62.0	9.3	7.7	4.4	1465.0
1972	260.1	191.5	290.3	106.4	681.1	168.2	82.2	14.1	8.6	4.0	1806.6
1973	131.4	241.8	305.3	157.1	1254.7	213.6	131.1	22.3	9.5	6.9	2473.6
1974	499.3	418.5	554.4	226.0	1301.1	287.7	154.7	35.8	10.3	14.3	3502.1
1975	433.0	440.5	569.2	191.1	1266.1	352.6	133.5	51.8	13.4	6.2	3457.4
1976	499.3	492.9	567.7	210.2	1574.8	426.7	144.1	58.6	15.5	6.0	3995.8

Source: Resources for the Future compilation of U.S. Census Bureau data.

Table 6-3. U.S. Forest Product Exports by Region, 1967-76. Trade Structure: Percentage of Total Trade

Year	North Atlantic	South Atlantic	Gulf	South Pacific	Pacific Northwest	Great Lakes	Alaska	North Central	South Central	Others	Total
1967	17	15	17	5	32	9	4	1	0	0	100
1968	16	14	18	6	34	8	4	1	0	0	100
1969	14	12	17	6	36	9	4	1	0	0	100
1970	14	14	19	6	35	7	4	0	0	0	100
1971	15	14	20	5	32	9	4	1	1	0	100
1972	15	11	16	6	38	9	5	1	0	0	100
1973	5	10	12	6	51	9	5	1	0	0	100
1974	14	12	16	6	37	8	4	1	0	0	100
1975	13	13	16	6	37	10	4	1	0	0	100
1976	12	12	14	5	39	11	4	1	0	0	100

Source: Calculated from table 6-2.

Note: Totals may not equal 100.0 due to rounding.

Table 6-4. U.S. Forest Products Exports by Region, 1967-76. Trade Trends: Percentage of 1967 Trade

Year	North Atlantic	South Altantic	Gulf	South Pacific	Pacific Northwest	Great Lakes	Alaska	North Central	South Central	Others	Total
1967	100	100	100	100	100	100	100	100	100	100	100
1968	115	112	127	146	128	108	106	111	135	133	121
1969	114	106	135	173	153	126	135	148	88	140	134
1970	130	153	185	207	182	116	163	121	173	130	163
1971	131	136	171	160	152	139	147	156	222	139	149
1972	156	132	173	228	218	182	195	237	249	127	183
1973	79	167	182	336	401	232	311	372	273	217	251
1974	299	288	331	484	416	312	368	598	301	450	355
1975	260	304	340	409	405	382	317	866	385	195	351
1976	299	340	330	450	503	462	342	979	451	189	405

Source: Calculated from Table 6-2.

total exports fell from 32 percent in 1967 to 26 percent in 1976, indicating that despite the rapid overall increase in softwood removals, the South's share of total forest product exports fell. This data suggest that much of the South's increase in production was used both to meet increased domestic softwood requirements and to displace the softwood production of other domestic regions. The large increase in the Pacific Northwest's share of exports, from 32 percent to 39 percent, suggests that this region increasingly relied on offshore markets.

Regional Imports

Tables 6-5, 6-6, and 6-7 report U.S. forest product imports by region. The Great Lakes region was the dominant importer, accounting for about 45 percent of U.S. forest product imports. The North Atlantic was a distant second, at about 20 percent; and the Pacific Northwest was the third leading importing region, with 14 percent of the value of 1976 U.S. imports. A common characteristic of these three regions was their common border with Canada, the principal source of U.S. forest product imports. For these three regions, the share of both the Great Lakes and particularly the Pacific Northwest rose over the 1967-76 period, while that of the North Atlantic declined rather substantially. The other regions' shares remained essentially stable.

Regional Commodity Exports

In chapter 5 we discussed the major regions involved in importing and exporting the various forest products as part of our broader discussion of U.S. trade in the various commodities. In this section, we focus upon the comparative roles of the various regions in the export of some major forest products. We have identified four commodities--softwood logs,

Table 6-5. U.S. Forest Product Imports by Region, 1967-76

(million current dollars)

Year	North Atlantic	South Atlantic	Gulf	South Pacific	Pacific Northwest	Great Lakes	Alaska	North Central	South Central	Others	Total
1967	477.4	34.3	96.2	130.1	162.4	837.1	0.4	145.0	1.0	48.4	1932.3
1968	554.2	50.4	125.9	134.4	197.7	936.5	0.4	172.9	0.8	22.3	2195.5
1969	571.2	57.4	126.4	154.9	239.6	1073.3	0.8	193.2	1.4	29.1	2447.3
1970	527.4	57.8	121.9	133.6	194.0	1028.3	1.4	175.0	1.0	28.0	2268.4
1971	576.7	75.5	149.0	116.3	263.8	1157.7	1.2	236.6	0.8	35.7	2613.4
1972	714.6	108.3	199.0	152.8	368.9	1369.3	1.1	325.3	2.1	43.9	3285.5
1973	885.1	132.6	239.8	201.5	438.1	1695.1	0.9	385.2	2.9	50.5	4031.8
1974	904.3	111.9	225.6	207.7	476.7	1952.3	1.5	375.7	0.8	46.0	4302.6
1975	743.3	74.2	159.3	172.1	475.2	1772.1	2.0	297.2	0.6	35.7	3731.8
1976	1005.4	124.0	228.8	241.0	711.4	2387.1	1.9	454.9	1.3	43.7	5199.5

Source: Resources for the Future compilation of U.S. Census Bureau data.

Table 6-6. U.S. Forest Product Imports by Region, 1967-76. Trade Structure: Percentage of Total Trade

Year	North Atlantic	South Atlantic	Gulf	South Pacific	Pacific Northwest	Great Lakes	Alaska	North Central	South Central	Others	Total
1967	25	2	5	7	8	43	0	8	0	3	100
1968	25	2	6	6	9	43	0	8	0	1	100
1969	23	2	5	6	10	44	0	8	0	1	100
1970	23	3	5	6	9	45	0	8	0	1	100
1971	22	3	6	4	10	44	0	9	0	1	100
1972	22	3	6	5	11	42	0	10	0	1	100
1973	22	3	6	5	11	42	0	10	0	1	100
1974	21	3	5	5	11	45	0	9	0	1	100
1975	20	2	4	5	13	47	0	8	0	1	100
1976	19	2	4	5	14	46	0	9	0	1	100

Source: Calculated from Table 6-5.

Table 6-7. U.S. Forest Product Imports by Region, 1967-76. Trade Trends: Percentage of 1967 Trade

Year	North Atlantic	South Atlantic	Gulf	South Pacific	Pacific Northwest	Great Lakes	Alaska	North Central	South Central	Others	Total
1967	100	100	100	100	100	100	100	100	100	100	100
1968	116	147	130	103	122	112	112	119	84	46	114
1969	120	167	130	119	148	128	214	133	135	60	127
1970	110	169	126	103	119	123	380	121	95	58	117
1971	121	220	154	89	162	138	319	163	80	74	135
1972	150	316	205	117	227	164	312	224	213	91	170
1973	185	386	247	155	270	203	256	266	284	104	209
1974	189	326	233	160	293	233	420	259	84	95	223
1975	156	216	164	132	293	212	555	205	58	74	193
1976	211	361	236	185	438	285	533	314	128	90	269

Source: Calculated from table 6-5.

softwood lumber, wood pulp, and paper and paperboard (excluding newsprint) that accounted for about 85 percent of U.S. forest product exports. Also, as noted earlier, in recent years, wood chips and plywood export earnings share increased very rapidly as both commodities achieved status as major U.S. exports. In this section we compare the trade performance of the major U.S. regions, particularly the Pacific Northwest region and the South, with respect to these six commodities.

Looking first at conifer logs and softwood lumber, solid wood products in which the Pacific Northwest has been the major regional exporter, we observe that the export dominance of the Pacific Northwest for both products increased over the 1967-76 period while the South's modest share of lumber diminished and her negligible share continued for sawlogs.

Table 6-8. Regional Percentage Share of U.S. Conifer Sawlog Exports for 1967 and 1976

Region	1967	1976
Pacific Northwest	88	91
South	0	0
North Atlantic	7	3
South Pacific	3	4
Alaska	2	1
Total	100	100

Source: Resources for the Future compilation of U.S. Census Bureau data.

Table 6-9. Regional Percentage Share of U.S. Softwood Lumber Exports for 1967 and 1976

Region	1967	1976
Pacific Northwest	38	48
South	14	10
Great Lakes	14	14
Alaska	12	14
South Pacific	15	7
Total	93	93

<u>Source</u>: See table 6-8.

For fiber products--paper and paperboard and wood pulp--the South's dominant export share for paper and paperboard declined while its wood pulp share increased modestly. The Pacific Northwest, an important but certainly not dominant exporter of both products, experienced a slight decline in its paper and paperboard share and a somewhat larger decline in its wood pulp share.

Table 6-10. Regional Percentage Share of U.S. Paper and Paperboard Exports for 1967 and 1976

Region	1967	1976
Pacific Northwest	12	11
South	42	39
Great Lakes	12	19
North Atlantic	29	26
Total	95	95

<u>Source</u>: See table 6-8.

Table 6-11. Regional Percentage Share of U.S. Wood Pulp Exports for 1967 and 1976

Region	1967	1976
Pacific Northwest	27	20
South	51	54
Alaska	11	8
South Pacific	4	10
Total	93	93

Source: See table 6-7.

For the rapidly expanding exports of wood chips and plywood, the export performance of the Pacific Northwest again exceeded that of the South. While the Pacific Northwest region's share of wood chips fell somewhat, the Pacific Northwest region continued to account for almost 90 percent of total wood chip exports. For plywood, the Pacific Northwest substantially increased its already overwhelming exported share. By contrast, the South's share of wood chips remained very small (1 percent) while its share of plywood exports declined precipitiously.

Table 6-12. Regional Percentage Share of U.S. Wood Chip Exports for 1967 and 1976.

Region	1967	1976
Pacific Northwest	99	85
South	0	1
South Pacific	0	8
Great Lakes	1	0
Total	100	94

Source: See table 6-7.

Table 6-13. Regional Percentage Share of U.S. Plywood Exports for 1967 and 1976

Region	1967	1976
Pacific Northwest	53	63
South	18	8
South Pacific	13	3
North Central	3	6
Great Lakes	8	17
North Atlantic	4	1
Total	99	98

Source: See table 6-7.

Other regions that exhibited share changes of note for these six commodities include the Great Lakes region. Its share of both plywood and paper and paperboard increased substantially over the period as did its share of total forest product exports. The South Pacific also increased significantly its share of both wood chips and wood pulp while its plywood and softwood lumber shares declined. Overall, the South Pacific region's share of total forest product remained constant. Another region that experienced substantial changes was the North Atlantic region. This region's share of total exports declined as did its share of conifer sawlogs, paper and paperboard, and plywood.

Conclusions

Despite the relative shift in softwood forest product production away from the Pacific Northwest region and to the South, the Pacific Northwest region increased its dominance in forest product exports, which are predominantly softwood, even as relative production shifted to the South. Simultaneously, the South's share of forest products exports fell. This finding is consistent with the hypothesis that the South's increased softwood output found outlets in the domestic market, and perhaps displaced Pacific Northwest products, while simultaneously the Pacific Northwest relied increasingly on foreign markets.

APPENDIXES

Appendix A

DATA BASE DOCUMENTATION

The purpose of this appendix is to document the sources and compilation of the regional data base used in the Resources for the Future (RFF) study Post-World War II Trends in U.S. Forest Products Trade. It is hoped that this documentation will provide users of the data with a thorough understanding of how the data base was compiled, thereby minimizing its misuse and facilitating any manipulation or updating which may become necessary for specific uses.

In the first section of this report, the objectives of the data collection process, as initially formulated by RFF researchers, are outlined. Then the primary data sources and the logic behind the choice of sources are described. The next sections are devoted to a discussion of the problems which become apparent during our efforts to compile a consistent data base, and the reasoning which was used to try to resolve those problems. In the final section, the data base is summarized by a brief description of the data compilation process.

Objectives

A primary objective of this study was to focus on U.S. regional trade with the rest of the world. This is a fair departure from previous studies, which have generally looked at trade between the rest of the world and the United States as a whole, ignoring the differences in the trade situations of the various regions.

One of the desired attributes of the data base, then, was that it be disaggregated to meaningful U.S. regions. In the context of forest products, a "meaningful" regional breakdown would, at the minimum, treat the South and the Pacific Northwest individually; it would be desirable to further break out Alaska and the Pacific Northwest individually; it would be desirable to further break down the Alaska, Northeast, and Great Lakes regions; improvements in analytical ability resulting from further disaggregation must, however, be traded off against the reduction in manageability of the data base because of its increasing size.

To study the changes in trade flows which have occurred over time, it is desirable to have time series which are long enough to observe significant trends. While a ten-year series does not offer enough observations fro most econometric work, it does afford the analyst some indication of how trade has changed over time. The period 1967-76 fulfilled this objective, while also satisfying the data availability and manageability constraints.

In order to identify the important participants in U.S. forest products trade, it is necessary to identify the country of orgin in the case of U.S. imports, and the country of destination for U.S. exports.

Observations of both quantity and value of imports and exports provide the analyst with much more information than would either quantity or value alone. Both quantity and value data were desired for the Trends study.

The commodities of interest include those which derive a significant percentage of their value-added from the wood resource. Traditionally, this commodity list includes wood building materials, such as lumber, plywood, particleboard, and so forth, and paper and paper goods such as newsprint, printing paper, paperboard, and so forth but excludes such products as wood furniture and prefabricated building components. For the Trends project, the commodity classes listed in table A-1 are of interest, but primary attention is focused on these commodities noted by superscript a.

Table A-1. Commodity Codes and Descriptions to be Used in the Study of U.S. International Forest Products Trade (Schedule A: Imports, Schedule B: Export)

Schedule A	Schedule B	Commodity description
241.1	241.1	Fuel wood and wood waste
241.2	241.2	Wood charcoal
242.1[a]	242.1	Pulpwood
242.2[a]	242.2	Logs and bolts--softwood, in the rough or roughly squared, quartered, or halved, except pulpwood
242.3[a]	242.3	Logs and bolts--hardwood, in the rough or roughly squared, quartered, or halved, except pulpwood
242.8	242.8	Poles, piling, posts, and other wood in the rough, n.e.c.[b]
243.1	243.1	Railway ties--wood, except bridge and switch ties
243.2[a]	243.2	Lumber--softwood, rough-sawed, or surface-worked
243.3[a]	243.3	Lumber--hardwood, rough-sawed, or surface-worked
244.0	244.0	Cork--natural, raw and waste
251[ac]	251	Pulps and waste paper
251.1		Paper--waste and scrap; paperboard products for remanufacture; and flax and hemp fibers for paper making
	251.1	Paper--waste and old paper
251.2		Wood pulp--mechanical
	251.3	Wood pulp--mechanical and semi-chemical, including screenings and wood pulp, n.e.c
251.4		Pulp, n.e.c., including mixtures and screenings
	251.5	Pulp, except wood pulp
251.6	251.6	Wood pulp--chemical, dissolving grades

Table A-1 (cont.)

Schedule		
A	B	Commodity description
251.7	251.7	Wood pulp--sulphate
251.8	251.8	Wood pulp--sulphite
631.1[a]	631.1	Veneers--wood
631.2[a]	631.2	Plywood, veneer and cellular wood panels
631.4[a]	631.4	Wood--improved or reconstituted
a, d	631.8320	Pulpwood in chip form
631.8	631.8	Wood--simply shaped or worked, n.e.c.
632.1	632.1	Wood packing boxes, cases, crates, and containers, except cooperage products
632.3	632.3	Cooperage products
632.4		Flooring, hardwood (except in strips and planks), and wood doors
	632.4	Builders wood work and pre-fab buildings
632.7	632.7	Manufactures of wood for domestic or decorative use
632.8	632.8	Articles manufactured of wood, n.e.c.
633.0	633.0	Cork manufactures, including agglomerated cork and manufactures
641[a, c]	641	Paper and paperboard
641.1[a]	641.1	Standard newsprint paper
641.2	641.2	Printing and writing paper, n.e.c.
641.3	641.3	Kraft paper and paperboard
641.4	641.4	Cigarette paper, in bulk, rolls, sheets
641.5	641.5	Paper and paperboard, machine-made

Table A-1 (cont.)

Schedule		Commodity description
A	B	
641.6[a]	641.6	Building boards of wood pulp or of vegetable fiber
641.7	641.7	Papers--handmade
641.9	641.9	Paper and paperboard in rolls, or sheets, n.e.c.
642.1	642.1	Paper bags, paperboard boxes, containers
642.2	642.2	Paper stationery for correspondence, n.e.c.
642.3	642.3	Paper stationery, except correspondence
642.9	642.9	Articles of paper pulp, paper, or paperboard, n.e.c.

Source: U.S. Bureau of the Census, U.S. Foreign Trade Statistics Classifications and Cross-Classifications, 1974 (Washington, D.C. 1975). The classification schemes used for the Trends study data were in effect until 1978, when the Census Bureau overhauled its foreign trade commodity classification schedules.

[a]Trade study will focus on these commodities regionally, with less attention given to others at the national level.

[b]The abbreviation means "not elsewhere classified."

[c]The commodity 251.0 referred to in the international trade study includes all 251 commodities except 251.1, waste and scrap paper.

[d]Pulpwood in chip form is not classified as a separate commodity under Schedule A even at the seven-digit level. Import data for pulpwood in chip form does appear under a unique seven-digit TSUSA classification, 200.1500.

[e]The commodity 641.0 referred to in the international trade study includes all 641 commodities except 641.1 newsprint, and 641.6, building board. Both 641.1 and 641.6 are tabulated individually for the study.

It should be noted that some small differences in classifications occur because FAO data utilized in this study use the Standard International Trade Classification (SITC) designation, whereas the Census data is designated by Schedule A and B.

To summarize, the objective of the data collection process was to identify, for the period 1967-76, the quantity and value of forest product imports and exports by U.S. region of destination/origin by foreign country of origin/destination.

Data Collection Plan

Very nearly all U.S. trade in forest products with non-North American countries is transported by water. U.S. trade with Canada and Mexico is mostly overland, by rail or truck. The recognition of these characteristics heavily influenced how data were collected. A computer tape containing all of the U.S. waterborne trade data was available and became the primary source of this data. Data on trade with Canada and Mexico and on nonwaterborne trade with other countries were available from other sources and were used to supplement the waterborne trade data in order to complete the data base.

Data Source Descriptions

The primary source of U.S. international trade data consists of several unpublished tabulations of the U.S. Census Bureau. One data set was supplied to RFF by the U.S. Maritime Administration, and will be referred to in this report as "MARAD data." Other data were obtained through the U.S. Department of Commerce's Trade Statistics Reference Bureau, and will be referred to in this report as "Commerce data." The MARAD data identify all U.S. waterborne trade, whereas the Commerce data identify U.S. exports to, and imports from, Canada and Mexico, by all modes of transportation for all commodities, as well as U.S. exports to non-North American countries for selected commodities.

The original sources of the MARAD data are the Census Bureau's import tabulation SA 305 and export tabulation SA 705. The SA 305 is described: "Waterborne General Imports--Type of vessel service by customs district and port of unlading by foreign port of lading by Schedule A subgroup by country of origin." The SA 705 is described: "Waterborne Exports of Domestic and Foreign Merchandise--Type of vessel service by customs district and port of lading by foreign port of unlading by Schedule B subgroup by country of destination" (1). Both SA 305 and SA 705 are annual Census Bureau compilations.

The Maritime Administration has performed some aggregation to give RFF a version which contains waterborne General Imports and Exports of Domestic and Foreign merchandise, by nine U.S. coastal districts, by foreign country of origin/destination, and by year, for the period 1967-76.

The original sources of the Commerce data are the Census Bureau's import tabulation IA 263 (part 1) and export tabulations EA 631 and EA 622. The IA 236 is described: "Continent by country of origin by TSUSA schedule by TSUSA commodity by unit control by customs district and method of transportation." The EA 631 is described: "Continent by country of destination by Schedule B commodity by customs district of exportation and method of transportation." The EA 622 is described: "Schedule B commodity by country of destination by customs district of exportation and method of transportation" (2). The IA 236, EA 631, and EA 622 are all annual Census Bureau tabulations.

Conversions and aggregations have been performed at RFF to give the Commerce data the same regional and commodity configurations as the MarAd data. The conversion and aggregation methods are detailed below.

Comparability and Coverage of Data

There are two types of problems involved with using the international trade data base. The first type involves a lack of complete comparability between the data sources, caused largely by differences in definitions and in the types of quantity units and values used. The second type results from having uneven coverage of trade flows between the two sources. Below, each problem is described, the seriousness of the problem is discussed, and any appropriate remedies to the problem are explained in detail.

Comparability

Quantity Units. The Census Bureau reports quantities of goods traded in two ways: (1) waterborne and airborne quantities are recorded as "shipping weights," (2) trade "by all modes of transport" (including waterborne and airborne) is recorded in the Tariff Schedule of the United States (TSUSA)[1] (imports) and Schedule B (exports) units. Trade which occurs by rail or truck is not recorded as a separate category, as are waterborne and airborne flows, but is included in the "all modes of transport" figure. Because two different measures of quantity are used, there is no way to directly compare "total" trade quantities with waterborne trade quantities without making a conversion between unit measures. This type of conversion

[1]For the commodities under consideration here, TSUSA quantity units are identical to units used in Schedule A, Statistical Classification of Commodities Imported into the United States.

was necessary, in order to compare, for example, the total quantity of goods traded with Western Europe with the total of goods traded with Canada; trade with Western Europe (nearly all waterborne) is reported by MarAd as shipping weight, while trade with Canada (predominantly truck and rail) is reported in TSUSA and Schedule B units.

Unit conversion factors for forest products have been developed by the Food and Agriculture Organization (FAO) of the United Nations for use in its annual Yearbook of Forest Products (3). For every commodity except pulpwood (242.1) and softwood lumber (243.2), the FAO conversion factor has been combined with a standard weight conversion factor to derive the conversion factors shown in table A-2. The calculation of the conversion factor for softwood logs is shown as footnote a in table A-2. Separate conversion factors for imports and exports of hardwood logs have been derived to reflect the fact that many of the hardwood logs imported by the United States are of tropical origin, which are generally not as dense as the temperate zone hardwoods which constitute U.S. exports. The FAO's conversion factor for "Tropical nonconiferous" sawlogs was used in deriving the import conversion factor shown in table A-2, and the export conversion factor was derived, using FAO's figure for "other nonconiferous" sawlogs.

For pulpwood, the FAO's weight/volume conversion factor was used, but the volume/product measures given by the FAO was not appropriate. The FAO factor of 128 cubic feet per cord of pulpwood is nominally correct, given that a cord of wood is a stack measuring 4 ft x 4 ft x 8 ft. But that stack includes air, bark, and wood, so that use of 128 cubic feet per cord would overestimate the actual amount of pulpwood. A rule of thumb is that a cord of pulpwood contains 85 cubic feet of wood (4). This figure was

Table A-2. Factors for Converting Quantity Measures from/to Shipping Weights to/from TSUSA and Schedule B Units

Commodity code	Description	TSUSA and Schedule B units	Conversion factor
241.1	Pulpwood	cords	3,582 lb/cord
242.2	Softwood logs	MBF	6,992 lb/MBF[a]
242.3	Hardwood logs	MBF	7,990 lb/MBF (exports) 7,290 lb/MBF (imports)
243.2	Softwood lumber	MBF	2,862 lb/MBF (exports) 2,352 lb/MBF (imports)
243.3	Hardwood lumber	MBF	3,643 lb/MBF
631.1	Veneer	MSF	1,464 lb/MSF
631.2	Plywood	MSF	1,268 lb/MSF
631.4	Reconstituted wood	SFT	2,537 lb/MSF

[a]As an example, this factor was obtained by the following calculation:

MBF = 4.53 cubic meters (FAO factor)

1 cubic meter = 700 kilograms (FAO factor)

1 kilogram = 2.205 lb (standard factor)

MBF = 4.53 x 700 x 2.205 = 6,992 lb

used in combination with the FAO's weight/volume factor to derive the pulpwood conversion factor shown in table A-2.

The need for separate conversion factors for imports and exports of softwood lumber arises because of the way a board-foot of lumber is commonly defined and the consequent real difference in weight between a board-foot of lumber imported for U.S. domestic use and a board-foot of lumber exported from the United States. According to the Standard Grading Rules for Canadian lumber, "A board foot is the quantity of lumber contained in or derived by drying, dressing, or working from a piece of rough green lumber 1 inch thick, 1 foot wide and 1 foot long or its equivalent in thicker, wider, narrower or longer lumber" (5).

The real dimensions of a board-foot of softwood lumber used in the United States are less than the nominal 1" x 12" x 12", but lumber exported from the United States is usually cut to the full nominal dimensions. Imported lumber has generally been cut in the same real dimensions as domestically sawn lumber, before it enters the country. Therefore, a board-foot of export lumber contains more wood, and hence weighs more, than a board-foot of import lumber. The FAO volume/product conversion factor is based on a board-foot cut to the full nominal dimensions. Therefore, it was possible to derive the conversion factor for export lumber using FAO conversion factors in the same manner as conversion factors for other forest products had been derived. For domestic lumber, a common rule of thumb is that 1,000 board-feet contain 68.5 cubic feet of lumber.[2] This figure was

---

[2]The precise number of board-feet per cubic foot of lumber depends on the dimensions of the boards. Hartman and coauthors (ref. 5, p. 21) present calculations of the cubic foot contents of lumber for a range of nominal lumber sizes.

used in compilation with the FAO's weight/volume factor to derive the softwood lumber import conversion factor in table A-2.

The FAO conversion factors for veneer and plywood are given in surface measures based on a one-eighth-inch thickness. Because U.S. production and trade statistics are normally reported on a three-eighth-inch basis (ref. 4, p. 26), the FAO factors for veneer and plywood were multiplied by three and used in the same manner as the FAO factors for other products. Thus, 1,000 square feet (one-eighth-inch thickness) equals .295 cubic meters, and 1,000 square feet (three-eighth-inch thickness) equals .885 cubic meters.

In compiling the Commerce data, commodities at the seven-digit TSUSA or Schedule B levels were aggregated to the four-digit Schedule A or B levels. The only problem encountered in this process was in aggregating the seven-digit TSUSA commodities to the four-digit Schedule A commodity 243.2, imports of softwood lumber.

The value of imports of softwood lumber is reported in dollars for all of the seven-digit commodities, but quantities are reported in several units. Nearly all of the commodities are reported in 1,000 board-feet (MBF); commodity quantities which are reported in units other than MBF are as follows:

TSUSA code	Commodity description	Quantity Units
200.6500	Laths of wood	Thousands (m)
200.7520	Wood fence pickets assembled	Not reported
202.4720 202.4800	Wood siding, western red cedar	Thousand square feet
202.4740 202.5000	Wood siding, except western red cedar	Thousand square feet

Because the nominal width for wood siding is 1 inch, 1,000 square feet is equivalent to 1,000 board-feet. Therefore, the wood siding was aggregated directly with the softwood lumber commodities which were reported in board-feet.

The remaining commodities which were not reported in board-feet--wood laths and fence pickets (TSUSA 200.6500 and 200.7520)--were imported in quantities which were very small relative to the total imports of softwood lumber.

Table A-3 reveals that, in an average year during 1967-76, wood laths and fence pickets imported from Mexico represented only about one-half of 1 percent of all softwood lumber imports from Mexico. Table A-4 shows a similar situation for laths and pickets imported from Canada.

Because the amounts of wood lath and fence picket imports were so small, there was no attempt made to derive board-foot quantities for the two commodities. The value of those imports was aggregated with the rest of the softwood lumber commodities, but the volumes were essentially ignored.

Commodity Classification. Two commodity classification schemes were used in the trade study data, but it was necessary to make only minor adjustments to instill comparability between the different schemes. The MARAD data were reported in terms of Schedule A: Statistical Classification of Commodities Imported into the United States, and Schedule B: Statistical Classification of Domestic and Foreign Commodities Exported from the United States (6). The Commerce data were reported in terms of the Tariff Schedules of the United States Annotated (TSUSA) statistical classification system (7), and Schedule B. TSUSA codes have been converted to Schedule A

Table A-3. Value of Imports from Mexico of Wood Laths and Fence Pickets, as a Percentage of the Total Value of Softwood Lumber Imports from Mexico

(current dollars)

Year	Total value	Value laths and pickets	% of total
1967	529,510	1,421	0.3
1968	565,512	0	0
1969	1,038,004	3,565	0.3
1970	985,716	0	0
1971	737,644	0	0
1972	2,028,430	0	0
1973	2,758,873	4,285	0.2
1974	607,085	4,352	0.7
1975	203,865	5,329	2.6
1976	533,910	3,600	0.7

Source: Calculated from data given in U.S. Bureau of the Census, Import Tabulation IA 236 (part 1), "Content by Country of Origin by TSUSA Schedule by TSUSA Number by Unit Control by Customs District and Method of Transportation," 1967-76.

Table A-4. Value of Imports from Canada of Wood Laths and Fence Pickets, as a Percentage of the Total Value of Softwood Lumber Imports from Canada

(current dollars)

Year	Total value	Value laths and pickets	% of total
1967	322,157,617	2,811,585	0.9
1968	488,059,831	3,158,090	0.6
1969	537,094,584	3,391,983	0.6
1970	431,158,421	5,503,133	1.3
1971	673,943,936	6,792,659	1.0
1972	1,048,225,688	5,856,227	0.6
1973	1,345,457,393	7,417,366	0.6
1974	906,034,297	7,767,357	0.9
1975	712,274,687	6,789,436	1.0
1976	1,346,043,014	8,193,289	0.6

Source: Calculated from data given in U.S. Burea of the Census, Import Tabulation IA 236 (part 1), "Content by Country of Origin by TSUSA Schedule by TSUSA Number by Unit Control by Customs District and Method of Transportation," 1967-76.

codes using a U.S. Census Bureau concordance showing to which Schedule A codes each TSUSA commodity corresponds (8). After making this conversion, all trade data can be considered to be classified according to Schedules A and B.

Over the period of the study, 1967-76, there have been some minor changes incorporated into Schedules A and B which make the early years, as reported, not comparable to the later years. Changes which affect the principal commodities of interest (a) can be corrected quite easily.

The first type of change relates to revisions in the commodity classification scheme. Effective January 1, 1968, for Schedule A, and January 1, 1970, for Schedule B, the following changes were made: Commodity 241.0, "Fuel wood, wood charcoal, and wood waste" was split into 241.1, "Fuel wood, and wood waste" and 241.2, "Wood charcoal." Effective January 1, 1970, for Schedule B, 641.8 "Paper and paperboard, machine-made, simply finished, in rolls, or sheets, n.e.c." was abolished, and part of it was allocated to 641.2, "Printing and writing paper, n.e.c.," 641.5, "Paper and paperboard, machine made" and 641.7 "Papers-handmade."

The 241 commodities are not of major concern, and no attempt was made to adjust the data. The 641 commodities are of interest only at the three-digit level, and since the commodities remained in the 641 group, no adjustment was necessary.

The second type of change that occurred relates to the quantity measures used. Effective January 1, 1970, 631.1, "Veneers--wood" and 631.2, "Plywood veneer and cellular wood panels," which had previously been reported in square feet, are now reported in 1,000 square feet. The only adjustment necessary, then, is to divide the 1967-69 data by 1,000. Also effective January 1, 1970, 642.9, "Articles of paper pulp, paper, or paperboard," which had previously been reported in thousands of articles is now reported in pounds. This commodity is not of special interest, and no adjustment was made.

Methods of Valuation. The Bureau of Census trade valuations are of two types: Customs valuations and free alongside ship (f.a.s.) valuations.

Customs valuations are made for the purpose of levying duties on goods entering the United States. There is no set formula for calculating the Customs value, and it may be based on the foreign market value, export value, constructed value, American selling price, or some other value. Generally, the Customs value reflects a market value in the foreign country.

F.a.s. values, be they for imports or exports, reflect the purchase price and "all charges incurred in placing the merchandise alongside the carrier at the port of exportation in the country of exportation" (6, p. 3).

For the MARAD and Commerce data, all export values are reported as f.a.s. values. All import values except the 1976 values reported by Commerce are Customs values. The 1976 Commerce import values are reported as f.a.s. values. However, since both Customs and f.a.s. values reflect values at the foreign port, it was decided that no adjustment would be made to the 1976 Commerce values.

U.S. Regional Configuration. The study data are designed to focus on eleven U.S. regions of origin/destination, although two of the regions, Hawaii and Puerto Rico, are negligible in the U.S. forest products trade situation. This eleven-region scheme has been dictated by the way in which the statistics were reported by the Maritime Administration, but it is felt that the regional breakdown fits reasonably well with the regional pattern of forest products production and trade in the United States.

Although both of the primary sources have been built up from the same basic data, the Commerce data set is more disaggregate regionally than the MARAD data set.

The MARAD data were reported by nine coastal districts, while the Commerce data were reported by forty-six customs districts. The U.S. Department of Commerce has prescribed a scheme for allocating customs districts among coastal districts. According to this scheme, each customs district is either (1) allocated 100 percent to a coastal district, or (2) split between two coastal districts, or (3) not allocated to any coastal district at all.

This study has, with a few exceptions, followed the Department of Commerce scheme. The majority of the customs districts has been allocated directly to a coastal district. However, the Department of Commerce scheme involves a six-coastal-district country, in which Alaska, Hawaii, and Puerto Rico customs districts have been included in the North Pacific, South Pacific, and South Atlantic coastal districts, respectively. For this study, these three customs districts have been broken into three additional coastal districts.

Four customs districts have been split between coastal districts in the Department of Commerce scheme. Each customs district contains several ports, and the district has been divided according to ports. For customs district 18 (Tampa, Florida), ports 18-01 (Tampa), 18-07 (Boca Grande), and 18-14 (Saint Petersburg) have been included in coastal district 3 (Gulf). The forest products trade data reveal that customs district 18 is insignificant relative to the total U.S. trade in forest products with either Canada and Mexico. Therefore, the allocation of the entire customs district to one or the other of the coastal districts should not distort the trade picture, while making the task of allocating customs districts much simpler. Since port 18-01 is the largest in customs district 18, the entire customs district will be allocated to coastal district 3. A similar rationale is used to allocate customs district 52 (Miami, Florida) to coastal district 2 (South Atlantic) and to allocate customs district 41 (Cleveland, Ohio) to coastal district 6 (Great Lakes).

Customs district 34 (Pembina, North Dakota) is a special case. The district encompasses twenty-three ports, and all but two of the ports have not been allocated to any coastal district in the Department of Commerce scheme. Ports 34 through 23 (Warroad, Minnesota) and 34 through 24 (Baudette, Minnesota), both of which provide ports for water-tranpsorted goods, have been included in coastal district 6 (Great Lakes). District 34 is important to U.S.-Canadian forest products trade, but only for goods transported overland. Waterborne trade through district 34 is insignificant. Therefore, the entire customs district will be allocated to a new coastal district 10, North Central. Similarly, customs districts 33 (Great Falls, Montana) and 35 (Minneapolis, Minnesota), neither of which has been assigned to a coastal district by Commerce, are also allocated to the North Central district.

Customs districts 24 (El Paso, Texas) and 26 (Nogales, Arizona) have not been allocated by Commerce. For this study, they will form a new district 11, South Central.

The allocation scheme of the Commerce Department is shown in table A-5, and the final customs district-coastal district allocation for the Trends study is shown in table A-6.

Data Coverage

Re-Exports. A re-export of an "export of foreign merchandise" occurs when goods imported into acountry are subsequently exported in essentially the same condition as when they were imported. Since, in this case, foreign merchandise is simply passing through a country, inclusion of these goods in trade statistics inflates both imports and exports.

At the time goods are formally imported in a country, it is assumed that the merchandise will be consumed in the importing country. There is no way for customs officials to know if the goods will be re-exported at some future date. Therefore, all goods--either for consumption or for re-export--are recorded as imports once they enter the country, whether directly from a foreign country or out of bonding.

Goods are never officially classified as re-exports until they are actually exported. At that time, the shipper declares whether the goods are of foreign or domestic origin. This is the basis for the Census Bureau's dual classifications: exports of domestic merchandise and exports of foreign merchandise.

An inconsistency arises in the data base because of the different coverage of re-exports given by the two data sources. The MARAD data includes exports of foreign as well as domestic merchandise, while the Commerce Data covers exports of domestic merchandise only.

Table A-5. Department of Commerce Allocation of U.S. Coastal Districts

Coastal district			Coastal district		
Code no.	Description	Customs districts and ports included	Code no.	Description	Customs districts and ports included
1	North Atlantic	Districts 01 through 06 District 10 District 11 District 13 District 14 District 54			District 45 Port 52-02 District 53
			4	South Pacific	District 25 District 27 District 28 District 32
2	South Atlantic	Districts 15 through 17 Ports 18-03, 18-05, and 18-16 District 49 District 52 (except 52-02)	5	North Pacific	Districts 29 through 31
			6	Great Lakes	District 07 District 09 Ports 34-23 and 34-24
3	Gulf	Ports 18-01, 18-07, and 18-14 Districts 19 through 23 Ports 41-13 and 41-15			Districts 36 through 39 District 41 (except ports 41-15)

Source: U.S. Department of Commerce, Classification of Customs Districts and Ports, U.S. Foreign Trade Schedule D, Bureau of the Census (Washington, D.C. 1976).

Table A-6. Trends Study Allocation

Region	Description	Customs district	Description
1	North Atlantic	1	Portland, Maine
		2	Saint Albans, Vt.
		4	Boston, Mass.
		5	Providence, R.I.
		6	Bridgeport, Conn.
		10	New York, N.Y.
		11	Philadelphia, Pa.
		13	Baltimore, Md.
		14	Norfolk, Va.
		54	Washington, D.C.
2	South Atlantic	15	Wilmington, N.C.
		16	Charleston, S.C.
		17	Savannah, Ga.
		52	Miami, Fla.
3	Gulf	18	Tampa, Fla.
		19	Mobile, Ala.
		20	New Orleans, La.
		21	Port Arthur, Tex.
		22	Galveston, Tex.
		23	Laredo, Tex.
		45	Saint Louis, Mo.
		53	Houston, Tex.
4	South Pacific	25	San Diego, Calif.
		27	Los Angeles, Calif.
		28	San Francisco, Calif.
5	North Pacific	29	Portland, Oreg.
		30	Seattle, Wash.
6	Great Lakes	7	Ogdensburg, N.Y.
		9	Buffalo, N.Y.
		36	Duluth, Minn.
		37	Milwaukee, Wisc.
		38	Detroit, Mich.
		39	Chicago, Ill.
		41	Cleveland, Ohio
7	Hawaii	32	Honolulu, Hawaii
8	Alaska	31	Anchorage, Alaska

Table A-6 (cont.)

Region	Description	Customs district	Description
9	Puerto Rico	49	San Juan, P.R.
10	North Central	33	Great Falls, Mont.
		34	Pembina, N.D.
		35	Minneapolis, Minn.
11	South Central	24	El Paso, Tex.
		26	Nogales, Ariz.

The degree to which U.S. import statistics are inflated by including merchandise which is later re-exported can be observed by comparing U.S. exports of foreign merchandise with U.S. imports. Such comparisons are made in table A-7 for 1976. For all but a few commodities, relatively insignificant amounts of U.S. imports are re-exported. Some commodities for which re-exports make up a significant percentage of imports, such as softwood logs (242.2), and hardwood logs (242.3), are commodities of which, on balance, the United States is a substantial net exporter. The export of foreign merchandise, then, is not very significant to the net U.S. trade, and no attempt has been made to correct the import data.

Similarly, the degree to which re-exports inflate U.S. export statistics can be observed in table A-8, again for 1976. As mentioned, for most commodities, re-exports are not significant, but where re-exports do make up a significant percentage of total exports, the U.S. trade balance is in the other direction, that is, the United States is a net importer of the commodity.

As mentioned above, the MARAD data includes exports of foreign merchandise, while the Commerce data includes only exports of domestic merchandise. Tables A-7 and A-8 indicate that the magnitude of re-exports is relatively insignificant, and it is felt that the loss in precision from the inclusion or exclusion of re-exports does not seriously bias the data base.

Transshipments. As regards the actual physical movement of goods, transshipments are identical to re-exports. That is, foreign merchandise simply passes through a country en route to another foreign country. The statistical difference between re-exports and transshipments stems from the different customs treatments given the two kinds of shipments.

Table A-7. Re-exports as a Percentage of U.S. Imports, by Commodity, 1976

Commodity code	Description	Exports of foreign merchandise ($)	U.S. general Imports ($)	Re-exports/ imports (%)
242.1	Pulpwood	-	10,199,586	0
242.2	Softwood logs	3,182,395	7,189,633	44.3
242.3	Hardwood logs	60,185	2,714,623	2.2
243.2	Softwood lumber	889,159	1,240,272,402	0.1
243.3	Hardwood lumber	2,235,961	96,091,358	2.3
251	Pulp	4,559,625	1,215,384,031	0.4
631.1	Veneer	375,151	73,360,722	0.5
631.2	Plywood	851,948	382,769,414	0.2
631.4	Reconstituted wood	53,452	12,018,292	0.4
641.1	Newsprint	12,517,312	1,741,549,798	0.7
641.6	Fiberboard	81,907	25,861,818	0.3
641	Paper and paperboard	14,246,206	1,991,787,336	0.7

Source: Data for "Exports of foreign merchandise ($)" are from Bureau of the Census, U.S. Exports/Schedule B Commodity Groupings by World Areas, Report FT 450, Annual (Washington, D.C., GPO). Data for "U.S. General Imports ($)" are from Bureau of the Census, U.S. General Imports/World Areas by Schedule A Commodity Groupings, Report FT 155, Annual.

Table A-8. Re-exports as a Percentage of U.S. Exports, by Commodity, 1976

Commodity code	Description	Exports of foreign merchandise ($)	Total exports (domestic and foreign ($)	Re-exports/ total exports (%)
242.1	Pulpwood	0	7,230,364	0
242.2	Softwood logs	3,182,395	855,130,523	0.4
242.3	Hardwood logs	60,185	65,884,051	0.1
243.2	Softwood lumber	889,159	471,521,871	0.2
243.3	Hardwood lumber	2,235,961	89,450,677	2.5
251	Pulp	4,559,625	1,001,345,117	0.5
631.1	Veneer	375,151	46,089,846	0.8
631.2	Plywood	851,948	150,048,068	0.6
631.4	Reconstituted wood	53,452	18,709,452	0.3
641.1	Newsprint	12,517,312	53,754,593	23.3
631.6	Fiberboard	81,907	32,102,470	0.3
641	Paper and paperboard	14,246,206	1,305,497,827	1.1

Source: Data for "Exports of foreign merchandse ($)" and "Total exports" columns are from Bureau of the Census, U.S. Exports/Schedule B Commodity Groupings by World Areas, Report FT 450, Annual (Washington, D.C., GPO).

When merchandise enters the U.S. customs area, it can be handled in one of two ways: (1) the merchandise can clear customs, including payment of duties. The merchandise is then officially imported and is entered as such in U.S. import statistics; or (2) the merchandise can be entered in bond and placed in a warehouse or transferred to another customs port. The merchandise is not officially imported until it is withdrawn from the warehouse for consumption or until it clears customs at the second port. If the merchandise is withdrawn from the warehouse for re-export or is re-exported from the second port, it is neither entered as an import nor an export, and is called a transshipment. The essential difference, then, between a re-export and a transshipment is that a re-export has been officially imported before export while a transshipment is exported out of bonding, and was never considered to have been imported.

Transshipments of foreign goods through the United States are thought to be commonplace, although movement of these goods is not reported in Census trade statistics, and is not of concern for this study. The transshipment of U.S. merchandise through Canada is important, however, because it represents a flow to non-North American countries which is not picked up by the MARAD data. For example, goods entering Canada from the United States destined for France, would appear in Census Bureau statistics as an <u>export to France</u> by the mode of transport with which it left the United States. Thus, while not physically possible, it is statistically possible to have a U.S. export to Europe by rail or truck! Likewise, imports are recorded at the port at which they cleared customs by the mode of transport with which they entered that port. Therefore, U.S. imports by rail or truck from Europe could have been transshipped through Canada or Mexico, or alternatively might have been shipped by rail or truck in bond from the U.S. port of entry to the U.S. port of clearance.

Since U.S. nonwaterborne trade with non-North American countries is not included in the MARAD data, it was necessary to further supplement the MARAD data with Commerce data on nonwaterborne trade. However, since transshipments do not occur to a significant extent in all commodities, nonwaterborne trade data was gathered selectively for a few products.

Table A-9 shows U.S. exports by land by commodity as a percentage of total exports of the commodity for various world areas and for the entire world, for 1976. Only two commodities had significant transshipments relative to the U.S. total exports of the commodity: newsprint (641.1) and hardwood lumber (243.3). The United States is a substantial net importer of both commodities, so transshipments are not as significant relative to net trade flows.

Other than newsprint and hardwood lumber, no commodity is important relative to total U.S. exports, but transshipments are important in the U.S. export trade of reconstituted wood (631.4) with western European countries. The United States is also a substantial net importer of reconstituted wood.

Table A-10 shows the 1976 U.S. imports by land by commodity as a percentage of total imports of the commodity, by world areas. Transshipments were not significant for any of the commodities in U.S. world trade. For those commodities in which there were significant transshipments of goods from certain foreign areas, for example, softwood logs from Europe, the absolute amount of trade was small; moreover, the United States is a very substantial net exporter of softwood sawlogs.

Table A-9. U.S. Exports by Rail or Truck to Non-North American World Areas, as a Percentage of Exports by All Modes of Transport, by Commodity, 1976

Commodity code	Description	Western Europe	Asia	Oceania	Africa	World
242.1	Pulpwood	0	0	0	0	0.8
242.2	Softwood logs	0.4	0	6.2	0	0
242.3	Hardwood logs	1.6	0	0	0	1.1
243.2	Softwood lumber	4.1	0.3	0.3	0.9	1.4
243.3	Hardwood lumber	30.3	1.0	0	0	10.5
631.1	Veneer	0	2.1	0	0	0
631.2	Plywood	1.4	0.8	0.9	0	0.9
631.4	Reconstituted wood	12.5	0	0	0	0.1
641.1	Newsprint	63.9	0	0	89.8	12.3
641.6	Fiberboard	2.7	0	7.0	0	0.4
251	Pulp	1.2	3.2	0	9.1	1.7
641	Paper and paperboard	3.3	0.6	2.0	3.1	1.4

Source: Bureau of the Census, U.S. Exports, FT 450, Annual 1976.

Table A-10. U.S. Imports by Rail or Truck from Non-North American World Areas, as a Percentage of Imports by All Modes of Transport, by Commodity, 1976

Commodity code	Description	Percentage of total imports to				
		Western Europe	Asia	Oceania	Africa	World
242.1	Pulpwood	0	0	0	0	0
242.2	Softwood logs	73.8	0	0	0	0
242.3	Hardwood logs	0	0	0	0	0
243.2	Softwood lumber	32.6	2.9	0	100	0
243.3	Hardwood lumber	13.6	0.4	0	0	0.2
251	Pulp	0	2.9	0	0.5	0
631.1	Veneer	0.9	0.1	0	11.8	0.2
631.2	Plywood	0.8	0	0	0	0
631.4	Reconstituted wood	1.1	0	0	0	0.1
641.1	Newsprint	1.6	0	0	0	0
641.6	Fiberboard	0	0	0	0	0
641	Paper and paperboard	2.1	0.3	0	0	0.1

Source: Calculated from data given in Bureau of the Census, *U.S. General Imports/Schedule A Commodity Groupings by World Areas*, Report FT 150, Annual 1976 (Washington, D.C., GPO).

Based on the information given in tables A-8 and A-9, and similar information from other years, it was decided to collect transshipments data for exports of softwood logs (243.2), hardwood logs (243.3), and newsprint (641.1). No transshipments data were collected for imports.

The U.S. Department of Commerce has tabulated transshipments of U.S. merchandise through Canada for 1974 (8) and 1976 (9). The value of goods exported overland from the United States is the residual of the "total" value less the value of goods transported by water and air. The derivation of quantities shipped overland is not as straightforward. Because the total quantity shipped overland is reported in Schedule B units and the water and air transported quantities are reported as shipping weights, some relationship between quantity units and the shipping weights must be established. The Commerce Department has assumed that the weight/value relationship for goods transported by land is identical to the relationship for goods transported by water, and used this relationship to estimate the weight of goods transported overland.

Symbolically, the weight of goods transported overland is determined by:

$$W_L = V_L \times Ww/_{Vw}$$

where $V_L = V_T - V_w - V_A$

V_L = value of goods transported overland

V_T = value of goods transported by all modes

V_W = value of goods transported by water

V_A = value of goods transported by air

W_L = weight of goods transported overland

W_w = weight of goods transported by water

The weights estimated by the Department of Commerce were used in this study after being converted to Schedule B quantity units using the conversion factors of table A-2.

Transshipments for 1967-73 and 1975 were derived from the Census Bureau's export tabulation EA 622. The value of overland shipments is again a residual from the total value. The quantity of overland shipments was imputed, using the same weight/value relationship assumption as did the Commerce Department. However, because the desired end figure was a measurement in Schedule B units rather than in weight units, the calculations differed somewhat from the above. Symbolically, the quantity of goods transported overland is determined by:

$$Q_L = Q_T - \frac{Q_T}{V_T} (V_w + V_A)$$

where Q_L = quantity of goods transported overland

Q_T = quantity of goods transported by all modes

V_T, V_w, V_A defined as above

It is felt that this method, used with data at the seven-digit commodity level, gives more precise results than would the use of the general conversion factor, because variations in commodity type are implicity considered in the calculation.

Airborne Trade. Since many forest products have a relatively low value to weight ratio, economics dictates against transporting these goods by air. Table A-11 shows the percentage share that airborne imports and exports constituted of the total imports and exports of 1976. It is apparent that air transport of merchandise to and from non-North American

Table A-11. U.S. Airborne Trade with Non-North American Countries as a Percentage of Trade by All Modes of Transport, by Commodity, 1976

		Imports			Domestic exports		
Commodity code	Description	Airborne ($)	Total ($)	Airborne (%)	Airborne ($)	Total ($)	Airborne (%)
242.1	Pulpwood	400	10,199,586	0	0	7,230,364	0
242.2	Softwood logs	0	7,189,633	0	866	851,948,128	0
242.3	Hardwood logs	0	2,714,623	0	15,482	65,823,866	0
243.2	Softwood lumber	24,518	1,240,272,402	0	52,807	470,632,712	0
243.3	Hardwood lumber	81,668	96,091,358	0.1	128,545	87,214,716	0.1
251	Pulp	12,757	1,215,384,031	0	19,507	996,785,492	0
631.1	Veneer	43,622	73,360,722	0.1	16,334	45,714,695	0
631.2	Plywood	18,140	382,769,414	0	164,194	149,196,120	0.1
631.4	Reconstituted wood	12,797	12,018,292	0.1	42,680	18,656,000	0.2
641.1	Newsprint	0	1,741,549,798	0	108,217	41,237,281	0.3
641.6	Fiberboard	2,946	25,861,818	0	228,938	32,020,563	0.7
641	Paper and paperboard	3,949,670	1,991,787,336	0.2	27,119,206	291,251,621	2.1

Source: Data for "Imports" are taken from Bureau of the Census, General Imports, FT 155, Annual 1976. Data for "Domestic exports" are taken from Bureau of the Census, U.S. Exports, FT 450, Annual 1976.

countries was only minor compared with other modes of transport, as only one commodity, paper and paperboard (641), had more than 1 percent of its total value transported by air. When imports and exports are combined, paper and paperboard transported by air make up less than 1 percent of the paper and paperboard transported by all modes.

Since air transport plays such a relatively insignificant role in the movement of forest products, it was decided that omission of airborne trade data would not seriously bias the data base; therefore, no airborne statistics were collected for trade with non-North American countries.

Compilation of Data Base. The above material has given an indication of the aggregation and conversion problems and methods used to resolve those problems. Below, the general procedure in compiling the final data base is outlined, and in the process the components of the data base are described in summary fashion.

The manipulations which were performed on each component of the data base are:

1. MARAD data--all U.S. waterborne trade data were converted from shipping weight units to TSUSA and Schedule B quantity units.
2. Canadian and Mexican imports and exports to or from the United States by all modes of transport were tabulated from Commerce sources in TSUSA and Schedule B units. Aggregation from the seven-digit commodity level to the four-digit level, and conversion to an eleven-region country were performed. The data were then substituted for the Canadian and Mexican waterborne data provided by MARAD.
3. Data on U.S. imports and exports of "Pulpwood in chip form" (Schedule B, 631.8320) to all countries, by all modes of transport

were tabulated from Commerce sources. Conversion to an eleven region country was performed. This data supplemented the other Commerce and MARAD data, in which pulpwood chips were not included as a commodity.

4. Transshipments data were derived from Commerce sources for Schedule B commodities softwood lumber (243.2); hardwood lumber (243.3); and reconstituted wood (631.4); as detailed above (see page 175). Aggregation from the seven-digit commodity level to the four-digit level, and conversion to an eleven-region country were performed. This data supplemented the other Commerce and MARAD data, in which exports by land to non-North American countries were not included.

In summary, the final data base consists of the quantity and value of U.S. imports and exports, for thirteen Schedule A and B commodities by U.S. region or origin/destination, by foreign country of destination/origin, for the years 1967-76.

References

1. U.S. Bureau of the Census, Guide to Foreign Trade Statistics, 1975 (Washington, D.C. 1975).

2. U.S. Bureau of the Census, Guide to Foreign Trade Statistics.

3. Food and Agricultural Organization, 1972 Yearbook of Forest Products (Rome, FAO, 1974).

4. Hartman, David A., William A. Atkinson, Ben S. Bryant, and Richard C. Woodfin, Jr., Conversion Factors for the Pacific Northwest Forest Industry (Seattle, Institute of Forest Products, College of Forest Resources, University of Washington, n.d.).

5. National Lumber Grades Authority. Standard Grading Rules for Canadian Lumber (Vancouver, British Columbia).

6. Bureau of the Census, U.S. Foreign Trade Statistics Classifications and Cross-Classifications, 1974 (Washington, D.C. 1975).

7. U.S. International Trade Commission, Tariff Schedules of the United States Annotated (1976), ITC Publication 749 (Washington, D.C. GPO 1975).

8. U.S. Department of Commerce, Maritime Administration, unpublished materials.

9. U.S. Department of Commerce, U.S. Exports Transshipped Via Canadian Ports, Maritime Administration, Office of Maritime Technology Market Analysis Program (Washington, D.C., GPO, 1978).

Appendix B

THE COMPOSITION OF RECONSTITUTED WOOD TRADE

Reconstituted wood (631.4) is a four-digit, Schedules A and B commodity group composed of two seven digit commodities, 631.4100 and 631.4200. The former includes "improved wood, densified and/or impregnated with resin or resin-like materials." This describes various types of compressed, impregnated, and laminated boards that are _not_ considered particle board. The second commodity, 631.4200, is simply wood particle board. Table B-1 shows the breakdown, by seven-digit commodity, of 1976 U.S. imports and exports of reconstituted wood. Imports in 1976 were composed largely of particle board (99 percent of total volume and 89 percent of total value), and exports were approximately 75 percent particle board (77 percent of total volume and 72 percent of total value). The United States was a net exporter of both seven digit commodities, exporting about 50 percent more (by value) than it imported of the aggregate 631.4.

Table B-1. 1976 U.S. Trade in Reconstituted Wood

Schedules A-B	Imports		Exports	
	Quantity (lb)	Value ($)	Quantity (SFT)	Value ($)
631.4100	1,433,518	1,308,925	27,686,023	5,255,141
631.4200	153,005,916	10,709,367	95,620,767	13,400,859
Total 631.4	154,439,434	12,018,292	123,306,790	18,656,000

Source: U.S. Bureau of the Census, _U.S. Exports_, Report FT-410. _U.S. General Imports_, Report FT-135 (Washington, D.C., GPO, 1976).

Appendix C

U.S. TRADE OF FOREST PRODUCTS BY REGION OF ORIGIN AND DESTINATION

The tables in this appendix detail U.S. international trade in specific forest products for 1967-76, by origins and destinations (both U.S. regions and foreign countries or areas). The discussion in chapter 5 is based on these tables.

At the head of each table is given the type of flow (import or export) and the commodity name and Schedule A or B code. Four kinds of tables are presented: (1) the value of trade, (2) the physical quantity of trade, (3) the country or region structure of the trade value, and (4) the time trends in the value of trade. The source of the data in tables of the first two types is the Resources for the Future compilation of U.S. Census Bureau data. Information on this compilation can be found in appendix A. The "structure" and "trends" tables were calculated directly from data given in the trade value table.

EXPORTS
SOFTWOOD LOGS (2422)
UNITED STATES

VALUE--MILLION CURRENT DOLLARS

YEAR	CANADA	EUROPE	JAPAN	KOR REP	OTH ASIA	OTHERS	TOTAL
1967	13.094	0.493	143.279	2.451	0.000	0.042	159.360
1968	15.841	0.428	214.286	6.879	0.139	0.116	237.688
1969	16.619	0.677	234.273	3.429	0.004	0.267	255.270
1970	14.737	0.615	300.830	5.380	0.726	0.244	322.531
1971	18.722	0.615	235.233	9.162	0.023	0.505	264.259
1972	30.431	0.888	354.602	6.239	0.011	0.581	392.751
1973	28.581	1.310	707.463	21.350	0.034	1.575	760.313
1974	32.187	3.049	602.806	31.043	0.313	0.938	670.333
1975	30.492	4.154	629.472	17.257	0.808	1.546	683.730
1976	37.725	3.839	776.512	32.835	2.990	1.188	855.090

QUANTITY--MILLION BF

YEAR	CANADA	EUROPE	JAPAN	KOR REP	OTH ASIA	OTHERS	TOTAL
1967	260.662	1.432	1847.501	36.981	0.000	0.147	2146.722
1968	280.525	2.266	2617.896	91.651	1.255	1.570	2995.164
1969	276.869	7.428	2537.672	40.620	0.011	2.091	2864.690
1970	248.850	3.201	3046.754	61.146	8.759	2.484	3371.194
1971	300.476	3.067	2441.904	86.900	0.334	4.587	2837.268
1972	462.919	2.201	3220.377	62.829	0.041	3.140	3751.406
1973	353.380	2.715	3302.042	142.993	0.050	5.032	3806.212
1974	279.334	4.738	2701.907	196.224	1.205	1.697	3185.104
1975	245.179	7.121	2860.069	128.046	2.725	2.793	3245.934
1976	319.821	5.307	3400.584	217.538	4.085	2.309	3949.644

EXPORTS
SOFTWOOD LOGS (2422)
UNITED STATES

TRADE STRUCTURE-- % OF TOTAL TRADE (VALUE BASIS)

YEAR	CANADA	EUROPE	JAPAN	KOR REP	OTH ASIA	OTHERS	TOTAL
1967	8.	0.	90.	2.	0.	0.	100.
1968	7.	0.	90.	3.	0.	0.	100.
1969	7.	0.	92.	1.	0.	0.	100.
1970	5.	0.	93.	2.	0.	0.	100.
1971	7.	0.	89.	3.	0.	0.	100.
1972	8.	0.	90.	2.	0.	0.	100.
1973	4.	0.	93.	3.	0.	0.	100.
1974	5.	0.	90.	5.	0.	0.	100.
1975	4.	1.	92.	3.	0.	0.	100.
1976	4.	0.	91.	4.	0.	0.	100.

TRADE TRENDS-- % OF 1967 TRADE (VALUE BASIS)

YEAR	CANADA	EUROPE	JAPAN	KOR REP	OTH ASIA	OTHERS	TOTAL
1967	100.	100.	100.	100.	******	100.	100.
1968	121.	87.	150.	281.	******	280.	149.
1969	127.	137.	164.	140.	******	644.	160.
1970	113.	125.	210.	219.	******	587.	202.
1971	143.	125.	164.	374.	******	1215.	166.
1972	232.	190.	247.	254.	******	1398.	246.
1973	218.	265.	494.	871.	******	3790.	477.
1974	246.	616.	421.	1266.	******	2256.	421.
1975	233.	842.	439.	704.	******	3721.	429.
1976	288.	778.	542.	1339.	******	2859.	537.

EXPORTS
SOFTWOOD LOGS (2422)

VALUE--MILLION CURRENT DOLLARS

YEAR	N ATLANTC	S ATLANTC	GULF	S PACIFIC	PACIFC NW	GRT LAKES	ALASKA	N CENTRAL	S CENTRAL	OTHERS	TOTAL
1967	10.968	0.021	0.459	4.468	139.956	0.486	2.980	0.021	0.000	0.000	159.360
1968	11.435	0.044	0.414	19.038	202.651	0.644	3.336	0.126	0.000	0.000	237.688
1969	12.786	0.008	0.020	19.271	219.050	0.322	3.793	0.019	0.000	0.000	255.270
1970	12.648	0.021	0.248	21.285	281.592	0.253	6.471	0.011	0.000	0.001	322.531
1971	13.606	0.072	0.211	11.206	234.190	0.000	4.965	0.000	0.009	0.001	264.259
1972	15.772	0.140	0.048	10.268	357.808	0.266	8.388	0.057	0.003	0.001	392.751
1973	18.408	0.695	0.106	23.897	698.730	0.595	17.773	0.089	0.019	0.000	760.313
1974	17.354	2.573	0.167	24.455	616.239	0.948	8.422	0.146	0.029	0.000	670.333
1975	21.589	2.660	0.634	25.220	623.002	1.283	9.079	0.247	0.002	0.013	683.730
1976	28.645	2.095	1.908	35.765	778.394	2.130	5.884	0.241	0.019	0.010	855.090

QUANTITY--MILLION BF

YEAR	N ATLANTC	S ATLANTC	GULF	S PACIFIC	PACIFC NW	GRT LAKES	ALASKA	N CENTRAL	S CENTRAL	OTHERS	TOTAL
1967	205.380	0.058	4.777	55.214	1836.459	8.564	35.874	0.397	0.000	0.000	2146.722
1968	216.090	0.174	5.121	277.779	2432.112	13.550	48.820	1.516	0.002	0.000	2995.164
1969	226.739	0.038	0.067	253.134	2337.101	4.738	42.544	0.228	0.000	0.000	2864.690
1970	221.503	0.107	1.332	245.326	2845.849	1.121	55.857	0.096	0.000	0.002	3371.194
1971	242.079	0.327	1.599	136.765	2405.463	0.000	50.777	0.000	0.272	0.008	2837.289
1972	281.690	0.524	0.225	95.023	3298.042	1.965	70.548	3.272	0.013	0.003	3751.406
1973	277.594	1.779	0.342	144.441	3299.143	3.315	78.867	0.523	0.208	0.000	3806.212
1974	200.105	4.109	0.627	106.862	2826.060	3.778	42.801	0.626	0.137	0.000	3185.104
1975	178.070	4.392	0.946	119.927	2895.906	5.273	40.374	1.027	0.004	0.014	3245.934
1976	257.924	3.213	5.782	146.784	3492.500	10.085	30.204	1.062	0.079	0.007	3949.644

EXPORTS
SOFTWOOD LOGS (2422)

TRADE STRUCTURE-- % OF TOTAL TRADE (VALUE BASIS)

YEAR	N ATLANTC	S ATLANTC	GULF	S PACIFIC	PACIFC NW	GRT LAKES	ALASKA	N CENTRAL	S CENTRAL	OTHERS	TOTAL
1967	7.	0.	0.	3.	88.	0.	2.	0.	0.	0.	100.
1968	5.	0.	0.	8.	85.	0.	1.	0.	0.	0.	100.
1969	5.	0.	0.	8.	86.	0.	1.	0.	0.	0.	100.
1970	4.	0.	0.	7.	87.	0.	2.	0.	0.	0.	100.
1971	5.	0.	0.	4.	89.	0.	2.	0.	0.	0.	100.
1972	4.	0.	0.	3.	91.	0.	2.	0.	0.	0.	100.
1973	2.	0.	0.	3.	92.	0.	2.	0.	0.	0.	100.
1974	3.	0.	0.	4.	92.	0.	1.	0.	0.	0.	100.
1975	3.	0.	0.	4.	91.	0.	1.	0.	0.	0.	100.
1976	3.	0.	0.	4.	91.	0.	1.	0.	0.	0.	100.

TRADE TRENDS-- % OF 1967 TRADE (VALUE BASIS)

YEAR	N ATLANTC	S ATLANTC	GULF	S PACIFIC	PACIFC NW	GRT LAKES	ALASKA	N CENTRAL	S CENTRAL	OTHERS	TOTAL
1967	100.	100.	100.	100.	100.	100.	100.	100.	******	******	100.
1968	104.	214.	93.	426.	145.	133.	112.	588.	******	******	149.
1969	117.	41.	4.	431.	157.	66.	127.	91.	******	******	160.
1970	115.	103.	54.	476.	201.	52.	217.	50.	******	******	202.
1971	124.	349.	46.	251.	167.	0.	167.	0.	******	******	166.
1972	144.	683.	10.	230.	256.	55.	281.	268.	******	******	246.
1973	168.	3384.	23.	535.	499.	122.	596.	417.	******	******	477.
1974	158.	12532.	36.	547.	440.	195.	283.	684.	******	******	421.
1975	197.	12956.	138.	564.	445.	264.	305.	1156.	******	******	429.
1976	261.	10206.	416.	800.	556.	438.	197.	1125.	******	******	537.

EXPORTS
HARDWOOD LOGS (2423)
UNITED STATES

VALUE--MILLION CURRENT DOLLARS

YEAR	CANADA	W GERMNY	OTH EURO	JAPAN	OTH ASIA	OTHERS	TOTAL
1967	7.205	10.277	10.869	3.957	0.064	0.025	32.396
1968	6.162	15.259	15.267	4.676	0.053	0.063	41.480
1969	5.660	12.672	11.131	9.532	0.065	0.041	39.101
1970	5.150	13.174	10.497	5.020	0.049	1.883	35.774
1971	5.634	11.062	7.822	3.797	0.022	1.385	29.722
1972	7.001	13.802	10.664	6.661	0.138	3.811	42.077
1973	9.407	17.805	14.525	5.823	1.498	4.126	53.184
1974	9.194	15.595	14.656	7.525	0.714	2.076	49.760
1975	5.858	18.181	13.614	2.142	0.318	2.595	42.708
1976	8.947	28.381	18.528	4.823	0.900	3.518	65.096

QUANTITY--MILLION BF

YEAR	CANADA	W GERMNY	OTH EURO	JAPAN	OTH ASIA	OTHERS	TOTAL
1967	75.147	10.042	13.984	4.512	0.137	0.056	103.877
1968	62.339	13.620	16.556	9.216	0.062	0.103	101.896
1969	47.686	12.555	12.763	14.472	0.093	0.262	87.851
1970	42.976	13.198	12.463	6.490	0.109	2.000	77.236
1971	39.465	10.502	8.997	4.100	0.049	2.050	65.063
1972	56.272	14.070	10.843	6.495	0.137	3.744	91.560
1973	64.454	20.645	16.942	8.506	1.221	5.350	117.119
1974	52.966	18.227	16.939	40.049	1.078	3.249	132.507
1975	32.493	16.349	14.391	9.042	0.510	2.131	74.916
1976	42.807	25.929	22.220	5.420	0.836	4.321	101.532

EXPORTS
HARDWOOD LOGS (2423)
UNITED STATES

TRADE STRUCTURE-- % OF TOTAL TRADE (VALUE BASIS)

YEAR	CANADA	W GERMNY	OTH EURO	JAPAN	OTH ASIA	OTHERS	TOTAL
1967	22.	32.	34.	12.	0.	0.	100.
1968	15.	37.	37.	11.	0.	0.	100.
1969	14.	32.	28.	24.	0.	0.	100.
1970	14.	37.	29.	14.	0.	5.	100.
1971	19.	37.	26.	13.	0.	5.	100.
1972	17.	33.	25.	16.	0.	9.	100.
1973	18.	33.	27.	11.	3.	8.	100.
1974	18.	31.	29.	15.	1.	4.	100.
1975	14.	43.	32.	5.	1.	6.	100.
1976	14.	44.	28.	7.	1.	5.	100.

TRADE TRENDS-- % OF 1967 TRADE (VALUE BASIS)

YEAR	CANADA	W GERMNY	OTH EURO	JAPAN	OTH ASIA	OTHERS	TOTAL
1967	100.	100.	100.	100.	100.	100.	100.
1968	86.	148.	140.	118.	83.	256.	128.
1969	79.	123.	102.	241.	101.	167.	121.
1970	71.	128.	97.	127.	77.	7611.	110.
1971	78.	108.	72.	96.	35.	5598.	92.
1972	97.	134.	98.	168.	216.	15402.	130.
1973	131.	173.	134.	147.	2342.	16677.	164.
1974	128.	152.	135.	190.	1116.	8392.	154.
1975	81.	177.	125.	54.	498.	10487.	132.
1976	124.	276.	170.	122.	1407.	14216.	201.

EXPORTS
HARDWOOD LOGS (2423)

VALUE--MILLION CURRENT DOLLARS

YEAR	N ATLANTIC	S ATLANTC	GULF	S PACIFIC	PACIFC NW	GRT LAKES	ALASKA	N CENTRAL	S CENTRAL	OTHERS	TOTAL
1967	19.676	1.826	0.563	3.352	0.115	6.847	0.008	0.009	0.000	0.000	32.396
1968	23.317	3.221	1.770	4.074	0.254	8.796	0.027	0.020	0.000	0.000	41.480
1969	18.824	1.854	1.166	6.893	0.888	9.425	0.016	0.035	0.000	0.000	39.101
1970	17.926	3.249	1.106	3.294	1.000	9.196	0.000	0.003	0.000	0.000	35.774
1971	14.028	1.759	1.412	1.912	0.859	9.510	0.208	0.032	0.002	0.000	29.722
1972	20.647	4.138	1.964	2.333	1.070	11.889	0.000	0.024	0.000	0.012	42.077
1973	27.933	4.233	4.004	3.049	1.976	11.559	0.236	0.121	0.001	0.072	53.184
1974	27.595	1.641	2.967	1.884	5.341	10.066	0.060	0.044	0.003	0.159	49.760
1975	23.054	2.102	4.070	0.884	1.033	11.442	0.000	0.071	0.012	0.040	42.708
1976	32.977	3.338	11.134	1.617	1.715	14.011	0.000	0.160	0.028	0.116	65.096

QUANTITY--MILLION BF

YEAR	N ATLANTIC	S ATLANTC	GULF	S PACIFIC	PACIFC NW	GRT LAKES	ALASKA	N CENTRAL	S CENTRAL	OTHERS	TOTAL
1967	86.995	1.403	1.150	3.643	0.180	10.360	0.115	0.031	0.000	0.000	103.877
1968	77.905	2.634	2.363	7.042	1.424	9.896	0.477	0.146	0.006	0.000	101.896
1969	60.113	1.930	1.605	10.718	2.044	11.043	0.309	0.090	0.000	0.000	87.851
1970	54.545	3.346	2.936	3.637	2.589	10.056	0.000	0.075	0.000	0.000	77.236
1971	43.451	2.126	3.009	1.984	1.230	9.681	2.593	0.080	0.010	0.000	65.063
1972	59.999	4.122	2.055	2.000	5.618	17.650	0.000	0.090	0.000	0.026	91.560
1973	76.572	5.739	4.640	2.148	6.235	19.320	1.955	0.524	0.002	0.083	117.118
1974	66.959	2.928	3.618	1.576	36.670	17.950	2.470	0.165	0.001	0.170	132.507
1975	43.625	1.976	3.969	5.009	4.394	16.622	0.000	0.266	0.021	0.034	74.916
1976	54.309	3.480	11.138	1.278	4.379	26.231	0.000	0.501	0.149	0.068	101.532

EXPORTS
HARDWOOD LOGS (2423)

TRADE STRUCTURE-- % OF TOTAL TRADE (VALUE BASIS)

YEAR	N ATLANTIC	S ATLANTC	GULF	S PACIFIC	PACIFC NW	GRT LAKES	ALASKA	N CENTRAL	S CENTRAL	OTHERS	TOTAL
1967	61.	6.	2.	10.	0.	21.	0.	0.	0.	0.	100.
1968	56.	8.	4.	10.	1.	21.	0.	0.	0.	0.	100.
1969	48.	5.	3.	18.	2.	24.	0.	0.	0.	0.	100.
1970	50.	6.	3.	9.	3.	26.	0.	0.	0.	0.	100.
1971	47.	6.	5.	6.	3.	32.	1.	0.	0.	0.	100.
1972	49.	10.	5.	6.	3.	28.	0.	0.	0.	0.	100.
1973	53.	8.	8.	6.	4.	22.	0.	0.	0.	0.	100.
1974	55.	3.	6.	4.	11.	20.	0.	0.	0.	0.	100.
1975	54.	5.	10.	2.	2.	27.	0.	0.	0.	0.	100.
1976	51.	5.	17.	2.	3.	22.	0.	0.	0.	0.	100.

TRADE TRENDS-- % OF 1967 TRADE (VALUE BASIS)

YEAR	N ATLANTIC	S ATLANTC	GULF	S PACIFIC	PACIFC NW	GRT LAKES	ALASKA	N CENTRAL	S CENTRAL	OTHERS	TOTAL
1967	100.	100.	100.	100.	100.	100.	100.	100.	******	******	100.
1968	119.	176.	315.	122.	221.	128.	330.	210.	******	******	128.
1969	96.	101.	207.	206.	772.	138.	190.	377.	******	******	121.
1970	91.	178.	197.	98.	870.	134.	0.	29.	******	******	110.
1971	71.	56.	251.	57.	747.	139.	2504.	343.	******	******	92.
1972	105.	227.	349.	70.	930.	174.	0.	254.	******	******	130.
1973	142.	232.	712.	91.	1718.	169.	2841.	1301.	******	******	164.
1974	140.	90.	527.	56.	4644.	147.	725.	472.	******	******	154.
1975	117.	115.	723.	26.	898.	167.	0.	761.	******	******	132.
1976	168.	183.	1979.	48.	1491.	205.	0.	1724.	******	******	201.

EXPORTS
SOFTWOOD LUMBER (2432)
UNITED STATES

VALUE--MILLION CURRENT DOLLARS

YEAR	CANADA	ITALY	OTH EURO	JAPAN	OCEANIA	OTHERS	TOTAL
1967	23.038	13.581	25.411	17.610	11.388	13.636	104.664
1968	26.247	18.951	29.950	21.375	14.023	15.517	126.063
1969	29.072	29.867	28.655	27.595	16.112	16.960	148.263
1970	25.393	28.198	35.298	35.461	16.487	18.533	159.370
1971	30.249	20.291	27.791	32.422	15.866	15.783	142.402
1972	45.824	30.615	38.438	58.210	20.002	15.581	208.670
1973	68.368	70.862	93.326	114.947	48.850	25.278	421.630
1974	82.272	53.786	55.768	126.366	46.021	34.717	398.930
1975	82.086	40.567	46.573	108.776	32.563	31.825	342.391
1976	114.551	60.457	82.838	110.027	55.320	47.423	470.616

QUANTITY--MILLION BF

YEAR	CANADA	ITALY	OTH EURO	JAPAN	OCEANIA	OTHERS	TOTAL
1967	201.642	100.776	182.689	340.343	118.123	128.670	1072.242
1968	204.335	137.112	203.483	367.657	126.583	129.073	1168.242
1969	195.917	150.830	156.637	387.302	132.531	125.993	1149.210
1970	197.247	135.329	184.650	500.193	118.389	139.890	1275.698
1971	201.264	114.050	145.802	350.786	115.355	111.599	1038.856
1972	282.668	132.922	171.000	513.384	131.401	99.489	1330.864
1973	491.359	180.026	313.277	768.320	196.321	126.211	2075.513
1974	350.256	130.886	199.341	780.618	171.951	146.844	1779.896
1975	367.108	118.313	136.698	729.511	141.062	149.524	1642.216
1976	418.702	146.908	221.642	621.337	210.321	182.406	1801.316

EXPORTS
SOFTWOOD LUMBER (2432)
UNITED STATES

TRADE STRUCTURE-- % OF TOTAL TRADE (VALUE BASIS)

YEAR	CANADA	ITALY	OTH EURO	JAPAN	OCEANIA	OTHERS	TOTAL
1967	22.	13.	24.	17.	11.	13.	100.
1968	21.	15.	24.	17.	11.	12.	100.
1969	20.	20.	19.	19.	11.	11.	100.
1970	16.	18.	22.	22.	10.	12.	100.
1971	21.	14.	20.	23.	11.	11.	100.
1972	22.	15.	18.	28.	10.	7.	100.
1973	16.	17.	22.	27.	12.	6.	100.
1974	21.	13.	14.	32.	12.	9.	100.
1975	24.	12.	14.	32.	10.	9.	100.
1976	24.	13.	18.	23.	12.	10.	100.

TRADE TRENDS-- % OF 1967 TRADE (VALUE BASIS)

YEAR	CANADA	ITALY	OTH EURO	JAPAN	OCEANIA	OTHERS	TOTAL
1967	100.	100.	100.	100.	100.	100.	100.
1968	114.	140.	118.	121.	123.	114.	120.
1969	126.	220.	113.	157.	141.	124.	142.
1970	110.	208.	139.	201.	145.	136.	152.
1971	131.	149.	109.	184.	139.	116.	136.
1972	199.	225.	151.	331.	176.	114.	199.
1973	297.	522.	367.	653.	429.	185.	403.
1974	357.	396.	219.	718.	404.	255.	381.
1975	356.	299.	183.	618.	286.	233.	327.
1976	497.	445.	326.	625.	486.	348.	450.

EXPORTS
SOFTWOOD LUMBER (2432)

VALUE--MILLION CURRENT DOLLARS

YEAR	N ATLANTC	S ATLANTC	GULF	S PACIFIC	PACIFC NW	GRT LAKES	ALASKA	N CENTRAL	S CENTRAL	OTHERS	TOTAL
1967	4.389	6.278	8.460	15.489	39.437	15.018	13.002	1.770	0.740	0.080	104.664
1968	5.336	6.678	8.646	17.332	49.549	17.098	17.435	2.153	1.730	0.106	126.063
1969	5.066	7.546	7.848	16.853	62.911	19.451	24.722	2.575	1.156	0.132	148.263
1970	5.480	7.017	10.299	15.954	70.778	16.369	29.828	1.759	1.730	0.157	159.370
1971	6.362	5.916	9.246	8.617	59.104	19.693	24.507	2.231	1.663	0.232	137.570
1972	10.142	6.002	15.201	20.696	63.998	29.390	38.109	3.420	1.351	0.362	208.670
1973	10.464	8.213	16.433	36.003	216.506	42.550	82.173	6.113	2.489	0.685	421.630
1974	17.000	9.442	12.599	23.200	200.857	45.583	79.884	7.615	1.826	0.922	398.930
1975	7.830	9.308	12.933	21.941	161.457	46.643	67.779	11.262	2.502	0.735	342.391
1976	14.393	18.041	25.931	32.149	224.451	68.142	67.056	16.624	3.422	0.408	470.616

QUANTITY--MILLION BF

YEAR	N ATLANTC	S ATLANTC	GULF	S PACIFIC	PACIFC NW	GRT LAKES	ALASKA	N CENTRAL	S CENTRAL	OTHERS	TOTAL
1967	37.096	43.404	59.144	140.361	390.621	116.120	261.382	15.063	8.490	0.561	1072.242
1968	43.849	43.725	65.724	129.809	415.155	120.478	319.745	15.691	13.532	0.565	1168.242
1969	42.471	44.083	48.490	107.283	394.395	121.394	364.102	16.876	9.607	0.506	1149.210
1970	44.359	43.639	57.924	100.821	434.710	117.545	447.577	14.191	14.260	0.671	1275.698
1971	40.541	33.385	45.888	83.448	382.437	117.546	305.992	15.765	13.187	0.668	1038.856
1972	48.178	31.466	56.171	104.028	450.364	165.785	443.489	20.946	8.831	1.605	1330.864
1973	54.117	33.063	57.319	133.410	939.796	212.731	571.330	55.439	15.148	3.159	2075.513
1974	75.464	32.125	43.328	92.736	796.182	176.042	518.119	35.800	7.681	2.397	1779.896
1975	32.341	32.563	45.109	90.507	741.309	189.900	445.517	54.797	9.351	0.824	1642.216
1976	52.959	55.911	78.852	107.527	900.053	227.537	401.894	64.003	11.967	0.613	1801.316

EXPORTS
SOFTWOOD LUMBER (2432)

TRADE STRUCTURE-- % OF TOTAL TRADE (VALUE BASIS)

YEAR	N ATLANTIC	S ATLANTC	GULF	S PACIFIC	PACIFC NW	GRT LAKES	ALASKA	N CENTRAL	S CENTRAL	OTHERS	TOTAL
1967	4.	6.	6.	15.	38.	14.	12.	2.	1.	0.	100.
1968	4.	5.	7.	14.	39.	14.	14.	2.	1.	0.	100.
1969	3.	5.	5.	11.	42.	13.	17.	2.	1.	0.	100.
1970	3.	4.	6.	10.	44.	10.	19.	1.	1.	0.	100.
1971	5.	4.	7.	6.	43.	14.	16.	2.	1.	0.	100.
1972	5.	3.	7.	10.	40.	14.	18.	2.	1.	0.	100.
1973	2.	2.	4.	9.	51.	10.	19.	1.	1.	0.	100.
1974	4.	2.	3.	6.	50.	11.	20.	2.	0.	0.	100.
1975	2.	3.	4.	6.	47.	14.	20.	3.	1.	0.	100.
1976	3.	4.	6.	7.	48.	14.	14.	4.	1.	0.	100.

TRADE TRENDS-- % OF 1967 TRADE (VALUE BASIS)

YEAR	N ATLANTIC	S ATLANTC	GULF	S PACIFIC	PACIFC NW	GRT LAKES	ALASKA	N CENTRAL	S CENTRAL	OTHERS	TOTAL
1967	100.	100.	100.	100.	100.	100.	100.	100.	100.	100.	100.
1968	122.	106.	102.	112.	126.	114.	134.	122.	234.	131.	120.
1969	115.	120.	93.	109.	160.	130.	190.	145.	156.	165.	142.
1970	125.	112.	122.	103.	179.	109.	229.	99.	234.	196.	152.
1971	145.	94.	109.	56.	150.	131.	188.	126.	225.	288.	131.
1972	231.	96.	180.	134.	213.	196.	293.	193.	182.	450.	199.
1973	236.	131.	194.	232.	549.	283.	632.	245.	336.	852.	403.
1974	387.	150.	149.	150.	509.	304.	614.	430.	247.	1147.	381.
1975	178.	148.	153.	142.	409.	311.	571.	636.	336.	914.	327.
1976	328.	287.	307.	208.	569.	454.	516.	939.	462.	507.	450.

EXPORTS
HARDWOOD LUMBER (2433)
UNITED STATES

VALUE--MILLION CURRENT DOLLARS

YEAR	CANADA	NETHLNDS	W GERMNY	OTH EURO	ASIA	OTHERS	TOTAL
1967	17.390	0.077	0.325	5.681	2.251	1.504	27.227
1968	12.484	0.108	0.382	5.007	1.991	1.298	21.269
1969	14.131	0.183	0.389	4.957	3.126	1.639	24.425
1970	11.292	0.180	0.470	6.937	7.722	2.762	29.364
1971	12.866	0.053	0.482	6.510	10.818	2.046	32.775
1972	23.133	0.194	0.526	7.653	37.890	2.990	72.386
1973	27.938	0.750	1.299	12.935	2.995	4.363	50.280
1974	32.852	4.213	1.757	15.997	2.129	4.557	61.505
1975	32.127	4.270	2.521	14.081	2.565	2.916	58.481
1976	49.519	6.551	4.376	19.095	3.593	4.095	87.228

QUANTITY--MILLION BF

YEAR	CANADA	NETHLNDS	W GERMNY	OTH EURO	ASIA	OTHERS	TOTAL
1967	125.761	0.321	1.161	23.437	7.284	6.466	164.450
1968	80.793	0.427	1.574	17.693	4.777	5.865	111.129
1969	82.690	0.775	1.276	15.759	7.416	7.923	115.839
1970	60.440	0.612	1.612	20.818	20.336	13.053	116.870
1971	80.743	0.207	1.663	18.939	29.395	9.697	140.644
1972	121.021	0.626	2.687	21.651	53.711	13.198	212.893
1973	172.889	1.681	3.047	29.585	5.984	17.619	230.804
1974	121.299	7.807	3.065	28.118	4.141	14.574	179.004
1975	226.097	6.763	4.427	22.359	5.252	6.997	271.895
1976	168.734	10.100	7.235	29.762	7.778	9.566	233.176

EXPORTS
HARDWOOD LUMBER (2433)
UNITED STATES

TRADE STRUCTURE-- % OF TOTAL TRADE (VALUE BASIS)

YEAR	CANADA	NETHLNDS	W GERMNY	OTH EURO	ASIA	OTHERS	TOTAL
1967	64.	0.	1.	21.	8.	6.	100.
1968	59.	1.	2.	24.	9.	6.	100.
1969	58.	1.	2.	20.	13.	7.	100.
1970	36.	1.	2.	24.	26.	9.	100.
1971	39.	0.	1.	20.	33.	6.	100.
1972	32.	0.	1.	11.	52.	4.	100.
1973	56.	1.	3.	26.	6.	9.	100.
1974	53.	7.	3.	26.	3.	7.	100.
1975	55.	7.	4.	24.	4.	5.	100.
1976	57.	8.	5.	22.	4.	5.	100.

TRADE TRENDS-- % OF 1967 TRADE (VALUE BASIS)

YEAR	CANADA	NETHLNDS	W GERMNY	OTH EURO	ASIA	OTHERS	TOTAL
1967	100.	100.	100.	100.	100.	100.	100.
1968	72.	141.	118.	88.	88.	86.	78.
1969	81.	236.	120.	87.	139.	109.	90.
1970	65.	235.	145.	122.	343.	184.	108.
1971	74.	69.	149.	115.	481.	136.	120.
1972	133.	252.	162.	135.	1683.	199.	266.
1973	161.	977.	400.	228.	133.	290.	185.
1974	189.	5488.	541.	282.	95.	303.	226.
1975	185.	5562.	777.	248.	114.	194.	215.
1976	285.	8534.	1348.	336.	160.	272.	320.

HARDWOOD LUMBER (2433)

VALUE--MILLION CURRENT DOLLARS

YEAR	N ATLANTC	S ATLANTC	GULF	S PACIFIC	PACIFC NW	GRT LAKES	ALASKA	N CENTRAL	S CENTRAL	OTHERS	TOTAL
1967	5.067	0.792	4.648	0.634	0.648	14.474	0.000	0.828	0.092	0.045	27.227
1968	4.008	0.593	3.856	0.522	0.823	10.306	0.004	1.045	0.096	0.014	21.269
1969	4.207	0.517	3.465	1.834	0.662	12.412	0.000	1.213	0.085	0.031	24.425
1970	6.514	0.469	4.507	4.982	1.112	10.658	0.199	0.799	0.083	0.042	29.364
1971	9.478	0.839	3.120	3.444	2.238	12.517	0.105	0.956	0.065	0.013	32.775
1972	28.549	1.515	4.793	12.913	3.880	19.075	0.000	1.550	0.064	0.047	72.386
1973	10.370	1.746	4.390	3.619	2.072	25.328	0.003	2.316	0.110	0.123	50.280
1974	17.565	1.325	5.260	2.917	3.352	27.966	0.000	2.636	0.135	0.207	61.505
1975	14.639	2.396	5.361	0.965	3.119	29.143	0.000	2.455	0.145	0.259	58.481
1976	21.962	5.030	5.958	1.159	3.399	45.923	0.000	3.250	0.458	0.088	87.228

QUANTITY--MILLION BF

YEAR	N ATLANTC	S ATLANTC	GULF	S PACIFIC	PACIFC NW	GRT LAKES	ALASKA	N CENTRAL	S CENTRAL	OTHERS	TOTAL
1967	26.955	2.433	20.449	3.289	3.777	103.021	0.000	3.938	0.459	0.138	164.450
1968	18.574	2.051	14.905	1.787	4.761	62.920	0.013	5.658	0.521	0.037	111.129
1969	23.595	1.684	13.067	6.461	3.068	61.580	0.000	6.016	0.308	0.060	115.839
1970	21.264	1.393	15.195	16.167	5.478	50.338	2.824	3.709	0.385	0.116	116.870
1971	32.939	2.405	10.103	11.725	8.314	68.915	1.498	4.537	0.283	0.035	140.644
1972	62.274	4.163	12.308	19.292	9.931	98.536	0.000	6.098	0.235	0.156	212.893
1973	34.140	4.513	12.163	15.620	5.891	149.272	0.022	8.424	0.333	0.417	230.804
1974	40.005	2.581	10.266	10.709	9.445	98.616	0.000	6.362	0.592	0.428	179.004
1975	33.602	4.201	11.920	2.888	9.354	201.494	0.000	7.714	0.433	0.389	271.995
1976	56.249	6.180	11.385	3.070	10.204	133.744	0.000	8.575	1.502	0.268	233.176

EXPORTS
HARDWOOD LUMBER (2433)

TRADE STRUCTURE-- % OF TOTAL TRADE (VALUE BASIS)

YEAR	N ATLANTIC	S ATLANTC	GULF	S PACIFIC	PACIFC NW	GRT LAKES	ALASKA	N CENTRAL	S CENTRAL	OTHERS	TOTAL
1967	19.	3.	17.	2.	2.	52.	0.	3.	0.	0.	100.
1968	19.	3.	18.	2.	4.	48.	0.	5.	0.	0.	100.
1969	17.	2.	14.	8.	3.	51.	0.	5.	0.	0.	100.
1970	22.	2.	15.	17.	4.	36.	1.	3.	0.	0.	100.
1971	29.	3.	10.	11.	7.	38.	0.	3.	0.	0.	100.
1972	39.	2.	7.	19.	5.	26.	0.	2.	0.	0.	100.
1973	21.	3.	9.	8.	4.	50.	0.	5.	0.	0.	100.
1974	26.	2.	9.	5.	5.	45.	0.	5.	0.	0.	100.
1975	25.	4.	9.	2.	5.	50.	0.	4.	0.	0.	100.
1976	25.	6.	7.	1.	4.	53.	0.	4.	1.	0.	100.

TRADE TRENDS-- % OF 1967 TRADE (VALUE BASIS)

YEAR	N ATLANTIC	S ATLANTC	GULF	S PACIFIC	PACIFC NW	GRT LAKES	ALASKA	N CENTRAL	S CENTRAL	OTHERS	TOTAL
1967	100.	100.	100.	100.	100.	100.	******	100.	100.	100.	100.
1968	79.	75.	83.	82.	127.	71.	******	126.	104.	31.	78.
1969	83.	65.	75.	289.	102.	86.	******	147.	92.	69.	90.
1970	129.	59.	97.	786.	172.	74.	******	96.	90.	94.	108.
1971	187.	106.	67.	543.	346.	96.	******	115.	71.	29.	120.
1972	563.	191.	103.	2037.	599.	132.	******	187.	69.	105.	266.
1973	205.	221.	94.	603.	320.	175.	******	280.	119.	274.	185.
1974	345.	167.	113.	460.	517.	193.	******	343.	147.	458.	226.
1975	289.	302.	115.	152.	482.	201.	******	297.	158.	573.	215.
1976	433.	635.	128.	183.	525.	317.	******	393.	497.	195.	320.

EXPORTS
WOOD VENEER (6311)
UNITED STATES

VALUE--MILLION CURRENT DOLLARS

YEAR	CANADA	SWEDEN	W GERMNY	SWITZLND	OTH EURO	OTHERS	TOTAL
1967	5.762	0.151	1.212	0.218	0.217	0.125	7.685
1968	7.402	0.145	2.835	0.447	0.343	0.418	11.590
1969	8.756	0.330	4.412	0.212	0.465	0.317	14.492
1970	7.068	0.100	5.542	0.054	0.338	0.225	13.328
1971	7.605	0.133	6.101	0.086	0.714	0.223	14.861
1972	13.106	0.379	6.595	0.293	1.236	0.278	21.887
1973	19.105	0.976	9.581	1.765	3.064	0.724	35.216
1974	17.801	0.908	12.149	1.285	4.021	0.685	36.848
1975	19.104	1.110	15.501	2.983	3.671	0.585	42.953
1976	14.554	1.915	16.707	3.731	7.799	0.976	45.682

QUANTITY--MIL SQ FT

YEAR	CANADA	SWEDEN	W GERMNY	SWITZLND	OTH EURO	OTHERS	TOTAL
1967	163.214	0.109	1.364	0.311	0.274	0.646	165.919
1968	223.004	0.123	3.606	0.434	0.725	4.235	232.127
1969	261.733	0.324	4.323	0.208	0.471	1.516	268.575
1970	215.538	0.121	5.438	0.097	0.651	1.665	223.510
1971	451.751	0.166	5.337	0.054	0.915	1.038	459.262
1972	362.809	0.284	5.050	0.304	1.320	2.346	372.113
1973	419.337	0.787	8.503	1.498	3.107	6.255	439.487
1974	311.660	0.963	11.731	1.302	4.583	1.553	331.812
1975	403.377	1.035	12.751	3.005	4.353	2.139	426.661
1976	322.942	1.815	17.368	3.948	7.519	4.224	357.819

EXPORTS
WOOD VENEER (6311)
UNITED STATES

TRADE STRUCTURE-- % OF TOTAL TRADE (VALUE BASIS)

YEAR	CANADA	SWEDEN	W GERMNY	SWITZLND	OTH EURO	OTHERS	TOTAL
1967	75.	2.	16.	3.	3.	2.	100.
1968	64.	1.	24.	4.	3.	4.	100.
1969	60.	2.	30.	1.	3.	2.	100.
1970	53.	1.	42.	0.	3.	2.	100.
1971	51.	1.	41.	1.	5.	1.	100.
1972	60.	2.	30.	1.	6.	1.	100.
1973	54.	3.	27.	5.	9.	2.	100.
1974	48.	2.	33.	3.	11.	2.	100.
1975	44.	3.	36.	7.	9.	1.	100.
1976	32.	4.	37.	8.	17.	2.	100.

TRADE TRENDS-- % OF 1967 TRADE (VALUE BASIS)

YEAR	CANADA	SWEDEN	W GERMNY	SWITZLND	OTH EURO	OTHERS	TOTAL
1967	100.	100.	100.	100.	100.	100.	100.
1968	128.	96.	234.	205.	158.	336.	151.
1969	152.	219.	364.	97.	215.	255.	189.
1970	123.	67.	457.	25.	156.	181.	173.
1971	132.	88.	503.	39.	329.	179.	193.
1972	227.	251.	544.	134.	570.	224.	285.
1973	332.	648.	791.	809.	1412.	582.	458.
1974	309.	603.	1002.	589.	1853.	550.	480.
1975	332.	737.	1279.	1366.	1692.	470.	559.
1976	253.	1272.	1378.	1709.	3595.	784.	594.

EXPORTS
WOOD VENEER (6311)

VALUE--MILLION CURRENT DOLLARS

YEAR	N ATLANTC	S ATLANTC	GULF	S PACIFIC	PACIFC NW	GRT LAKES	ALASKA	N CENTRAL	S CENTRAL	OTHERS	TOTAL
1967	0.993	0.031	0.117	0.105	2.170	4.217	0.000	0.050	0.001	0.001	7.685
1968	1.821	1.152	0.201	0.038	3.178	5.122	0.000	0.077	0.001	0.000	11.590
1969	3.184	1.178	0.121	0.067	4.241	5.679	0.000	0.019	0.002	0.000	14.492
1970	4.574	0.402	0.176	0.007	3.603	4.481	0.000	0.071	0.002	0.012	13.328
1971	5.398	0.666	0.094	0.017	4.714	3.833	0.000	0.126	0.013	0.000	14.861
1972	5.747	2.510	0.165	0.018	8.946	4.437	0.000	0.043	0.018	0.002	21.887
1973	9.720	4.969	0.309	0.013	13.244	6.696	0.000	0.225	0.041	0.000	35.216
1974	15.358	3.435	0.052	0.039	9.861	7.789	0.000	0.295	0.018	0.000	36.848
1975	18.517	4.117	0.799	0.053	10.104	8.946	0.000	0.366	0.023	0.028	42.953
1976	24.817	3.591	2.541	0.307	8.080	5.810	0.000	0.528	0.008	0.000	45.682

QUANTITY--MIL SQ FT

YEAR	N ATLANTC	S ATLANTC	GULF	S PACIFIC	PACIFC NW	GRT LAKES	ALASKA	N CENTRAL	S CENTRAL	OTHERS	TOTAL
1967	2.844	0.364	0.274	0.175	95.103	66.263	0.000	1.193	0.002	0.002	165.919
1968	14.731	0.932	0.886	0.329	133.829	78.956	0.000	2.444	0.021	0.000	232.127
1969	7.000	1.052	0.809	0.092	165.288	94.002	0.000	0.280	0.053	0.000	268.575
1970	8.220	0.673	1.621	0.031	141.276	70.776	0.000	0.888	0.022	0.004	223.510
1971	5.900	0.502	0.671	0.033	388.787	58.828	0.000	4.320	0.221	0.000	459.262
1972	6.208	1.963	1.312	0.051	276.213	84.823	0.000	1.235	0.302	0.005	372.113
1973	13.689	4.013	5.094	0.194	312.787	99.512	0.000	3.655	0.543	0.000	439.487
1974	17.482	2.929	0.513	0.216	202.382	103.894	0.000	4.102	0.295	0.000	331.812
1975	18.915	3.217	1.089	0.429	250.935	148.058	0.000	3.254	0.745	0.020	426.661
1976	36.000	3.283	3.151	2.896	211.967	94.944	0.000	5.305	0.263	0.000	357.918

EXPORTS
WOOD VENEER (6311)

TRADE STRUCTURE-- % OF TOTAL TRADE (VALUE BASIS)

YEAR	N ATLANTC	S ATLANTC	GULF	S PACIFIC	PACIFC NW	GRT LAKES	ALASKA	N CENTRAL	S CENTRAL	OTHERS	TOTAL
1967	13.	0.	2.	1.	28.	55.	0.	1.	0.	0.	100.
1968	16.	10.	2.	0.	27.	44.	0.	1.	0.	0.	100.
1969	22.	6.	1.	0.	29.	39.	0.	0.	0.	0.	100.
1970	34.	3.	1.	0.	27.	34.	0.	1.	0.	0.	100.
1971	36.	4.	1.	0.	32.	26.	0.	1.	0.	0.	100.
1972	26.	11.	1.	0.	41.	20.	0.	0.	0.	0.	100.
1973	26.	14.	1.	0.	38.	19.	0.	1.	0.	0.	100.
1974	42.	9.	1.	0.	27.	21.	0.	1.	0.	0.	100.
1975	43.	10.	2.	0.	24.	21.	0.	1.	0.	0.	100.
1976	54.	8.	6.	1.	18.	13.	0.	1.	0.	0.	100.

TRADE TRENDS-- % OF 1967 TRADE (VALUE BASIS)

YEAR	N ATLANTC	S ATLANTC	GULF	S PACIFIC	PACIFC NW	GRT LAKES	ALASKA	N CENTRAL	S CENTRAL	OTHERS	TOTAL
1967	100.	100.	100.	100.	100.	100.	******	100.	100.	100.	100.
1968	183.	3752.	172.	36.	146.	121.	******	154.	126.	0.	151.
1969	321.	3838.	103.	64.	195.	135.	******	37.	197.	0.	189.
1970	461.	1311.	150.	6.	166.	106.	******	142.	209.	1155.	173.
1971	544.	2169.	90.	16.	217.	91.	******	252.	1236.	0.	193.
1972	579.	8174.	142.	17.	412.	105.	******	86.	1779.	157.	285.
1973	979.	16184.	264.	12.	610.	159.	******	448.	3953.	0.	458.
1974	1547.	11189.	44.	37.	454.	185.	******	589.	1759.	0.	480.
1975	1865.	13409.	681.	51.	466.	212.	******	729.	2259.	2849.	559.
1976	2500.	11696.	2168.	294.	372.	138.	******	1053.	791.	0.	594.

EXPORTS
PLYWOOD (6312)
UNITED STATES

VALUE--MILLION CURRENT DOLLARS

YEAR	CANADA	DENMARK	U KINGDM	NETHLNDS	OTH EURO	OTHERS	TOTAL
1967	1.456	1.569	1.832	0.302	1.718	4.202	11.079
1968	2.253	1.547	0.875	0.295	2.155	4.211	11.336
1969	7.279	7.775	2.081	1.300	4.419	4.749	27.603
1970	2.639	3.789	2.262	0.564	2.901	4.009	16.165
1971	3.170	3.813	1.910	0.412	2.446	3.636	15.387
1972	13.117	11.552	2.915	0.484	3.146	5.477	36.692
1973	17.031	21.589	8.240	3.150	9.976	10.795	70.780
1974	54.997	9.732	8.257	1.164	11.945	11.219	97.315
1975	75.953	19.410	11.493	5.589	13.619	10.243	136.307
1976	41.091	28.514	24.676	12.608	26.934	14.060	147.884

QUANTITY--MIL SQ FT

YEAR	CANADA	DENMARK	U KINGDM	NETHLNDS	OTH EURO	OTHERS	TOTAL
1967	6.381	13.239	17.243	2.786	12.184	39.495	91.328
1968	14.887	11.661	4.730	0.979	11.061	30.858	74.176
1969	58.642	64.902	17.362	10.777	29.832	29.552	211.067
1970	69.027	28.642	20.625	5.199	24.394	28.746	176.633
1971	23.002	24.025	15.108	2.950	14.373	25.973	105.431
1972	95.418	95.312	21.111	1.845	14.122	29.417	257.223
1973	139.402	112.207	51.155	14.379	46.092	99.888	463.123
1974	338.821	50.552	52.552	6.634	69.883	53.607	572.050
1975	455.017	108.357	69.310	32.191	79.286	50.741	794.902
1976	229.137	125.708	139.009	64.966	128.031	59.450	746.304

EXPORTS
PLYWOOD (6312)
UNITED STATES

TRADE STRUCTURE-- % OF TOTAL TRADE (VALUE BASIS)

YEAR	CANADA	DENMARK	U KINGDM	NETHLNDS	OTH EURO	OTHERS	TOTAL
1967	13.	14.	17.	3.	16.	38.	100.
1968	20.	14.	8.	3.	19.	37.	100.
1969	26.	28.	8.	5.	16.	17.	100.
1970	16.	23.	14.	3.	18.	25.	100.
1971	21.	25.	12.	3.	16.	24.	100.
1972	36.	31.	8.	1.	9.	15.	100.
1973	24.	31.	12.	4.	14.	15.	100.
1974	57.	10.	6.	1.	12.	12.	100.
1975	56.	14.	8.	4.	10.	8.	100.
1976	28.	19.	17.	9.	18.	10.	100.

TRADE TRENDS-- % OF 1967 TRADE (VALUE BASIS)

YEAR	CANADA	DENMARK	U KINGDM	NETHLNDS	OTH EURO	OTHERS	TOTAL
1967	100.	100.	100.	100.	100.	100.	100.
1968	155.	99.	48.	98.	125.	100.	102.
1969	500.	496.	114.	431.	257.	113.	249.
1970	181.	242.	123.	187.	169.	95.	146.
1971	218.	243.	104.	136.	142.	87.	139.
1972	901.	736.	159.	161.	183.	130.	331.
1973	1170.	1376.	450.	1044.	581.	257.	639.
1974	3778.	620.	451.	386.	695.	267.	678.
1975	5217.	1237.	627.	1852.	793.	244.	1230.
1976	2823.	1818.	1347.	4178.	1567.	335.	1335.

EXPORTS
PLYWOOD (6312)

VALUE--MILLION CURRENT DOLLARS

YEAR	N ATLANTC	S ATLANTC	GULF	S PACIFIC	PACIFC NW	GRT LAKES	ALASKA	N CENTRAL	S CENTRAL	OTHERS	TOTAL
1967	0.446	1.601	0.389	1.395	5.889	0.857	0.000	0.329	0.066	0.107	11.079
1968	0.448	2.299	0.206	1.510	5.044	1.005	0.000	0.674	0.073	0.077	11.336
1969	0.584	2.335	0.314	2.407	15.761	4.505	0.000	1.502	0.099	0.097	27.603
1970	0.467	1.746	0.339	1.317	10.331	1.037	0.000	0.629	0.205	0.095	16.165
1971	0.533	1.388	0.207	1.643	9.418	1.378	0.000	0.622	0.110	0.088	15.387
1972	0.528	1.443	0.727	3.204	19.901	8.661	0.000	1.922	0.127	0.180	36.692
1973	1.710	3.783	2.539	4.524	44.889	9.872	0.000	2.874	0.258	0.330	70.780
1974	1.453	4.605	3.885	3.692	39.990	32.856	0.000	9.841	0.574	0.420	97.315
1975	2.181	3.674	4.277	4.265	56.866	50.115	0.000	13.831	0.773	0.325	136.307
1976	2.031	6.288	6.473	4.824	93.407	24.537	0.000	8.884	0.869	0.570	147.884

QUANTITY--MIL SQ FT

YEAR	N ATLANTC	S ATLANTC	GULF	S PACIFIC	PACIFC NW	GRT LAKES	ALASKA	N CENTRAL	S CENTRAL	OTHERS	TOTAL
1967	2.597	12.485	3.148	10.175	57.259	3.907	0.000	0.615	0.573	0.568	91.328
1968	2.880	15.170	1.375	9.196	34.438	7.127	0.000	3.148	0.484	0.357	74.176
1969	2.058	14.753	2.178	11.995	130.850	39.404	0.000	8.666	0.716	0.447	211.067
1970	1.029	12.124	3.090	8.192	87.748	54.657	0.000	7.554	1.705	0.533	176.633
1971	1.055	9.635	1.832	10.974	66.405	8.657	0.000	5.509	0.985	0.380	105.431
1972	1.449	8.064	5.272	15.124	150.377	56.556	0.000	18.710	1.038	0.633	257.223
1973	6.403	18.678	15.955	17.035	296.481	81.874	0.000	23.836	1.670	1.191	463.123
1974	5.592	27.868	27.271	17.802	229.869	190.556	0.000	68.359	3.328	1.404	572.050
1975	11.198	18.644	24.965	19.835	339.568	283.458	0.000	92.061	4.174	0.999	794.902
1976	8.386	21.678	34.014	20.407	482.670	121.632	0.000	51.764	4.244	1.508	746.304

EXPORTS
PLYWOOD (6312)

TRADE STRUCTURE-- % OF TOTAL TRADE (VALUE BASIS)

YEAR	N ATLANTC	S ATLANTC	GULF	S PACIFIC	PACIFC NW	GRT LAKES	ALASKA	N CENTRAL	S CENTRAL	OTHERS	TOTAL
1967	4.	14.	4.	13.	53.	8.	0.	3.	1.	1.	100.
1968	4.	20.	2.	13.	44.	9.	0.	6.	1.	1.	100.
1969	2.	8.	1.	9.	57.	16.	0.	5.	0.	0.	100.
1970	3.	11.	2.	8.	64.	6.	0.	4.	1.	1.	100.
1971	3.	9.	1.	11.	61.	9.	0.	4.	1.	1.	100.
1972	1.	4.	2.	9.	54.	24.	0.	5.	0.	0.	100.
1973	2.	5.	4.	6.	63.	14.	0.	4.	0.	0.	100.
1974	1.	5.	4.	4.	41.	34.	0.	10.	1.	0.	100.
1975	2.	3.	3.	3.	42.	37.	0.	10.	1.	0.	100.
1976	1.	4.	4.	3.	63.	17.	0.	6.	1.	0.	100.

TRADE TRENDS-- % OF 1967 TRADE (VALUE BASIS)

YEAR	N ATLANTC	S ATLANTC	GULF	S PACIFIC	PACIFC NW	GRT LAKES	ALASKA	N CENTRAL	S CENTRAL	OTHERS	TOTAL
1967	100.	100.	100.	100.	100.	100.	******	100.	100.	100.	100.
1968	100.	144.	53.	108.	86.	117.	******	205.	111.	72.	102.
1969	131.	146.	81.	173.	268.	525.	******	456.	151.	91.	249.
1970	105.	109.	87.	94.	175.	121.	******	191.	313.	89.	146.
1971	120.	87.	53.	118.	160.	161.	******	189.	168.	82.	139.
1972	118.	90.	187.	230.	338.	1010.	******	584.	194.	169.	331.
1973	383.	236.	653.	324.	762.	1152.	******	873.	395.	310.	639.
1974	326.	288.	999.	265.	679.	3833.	******	2989.	876.	394.	878.
1975	489.	229.	1100.	306.	966.	5846.	******	4200.	1180.	305.	1230.
1976	455.	393.	1665.	346.	1586.	2862.	******	2698.	1327.	534.	1335.

EXPORTS
RECONSTITUTED WOOD (6314)
UNITED STATES

VALUE--MILLION CURRENT DOLLARS

YEAR	CANADA	OTH N AM	S AMERIC	EUROPE	ASIA	OTHERS	TOTAL
1967	0.735	0.135	0.000	0.047	0.020	0.015	0.951
1968	0.867	0.103	0.006	0.031	0.031	0.055	1.093
1969	2.035	0.119	0.008	0.083	0.000	0.023	2.269
1970	2.116	0.085	0.004	0.019	0.001	0.006	2.230
1971	4.634	0.127	0.000	0.111	0.081	0.012	4.965
1972	8.313	0.212	0.000	0.063	0.084	0.013	8.685
1973	14.612	0.644	0.005	0.042	0.273	0.082	15.656
1974	18.032	1.674	0.013	0.226	0.373	1.454	21.772
1975	15.705	0.810	0.029	0.165	0.370	0.127	17.206
1976	15.888	1.732	0.236	0.160	0.574	0.030	18.620

QUANTITY--MIL SQ FT

YEAR	CANADA	OTH N AM	S AMERIC	EUROPE	ASIA	OTHERS	TOTAL
1967	1.280	0.406	0.000	0.035	0.115	0.042	1.878
1968	6.630	0.542	0.017	0.016	0.050	0.066	7.320
1969	15.757	0.639	0.004	0.024	0.000	0.136	16.559
1970	13.099	0.478	0.011	0.014	0.008	0.019	13.629
1971	33.777	0.747	0.000	0.067	0.194	0.067	34.852
1972	69.792	1.061	0.000	0.048	0.068	0.117	71.086
1973	110.951	2.741	0.002	0.044	1.558	0.670	115.968
1974	131.490	13.024	0.030	0.187	1.646	9.606	155.982
1975	118.310	4.589	0.166	0.280	0.179	0.585	124.107
1976	107.893	9.795	1.537	0.107	2.488	0.119	121.940

EXPORTS
RECONSTITUTED WOOD (6314)
UNITED STATES

TRADE STRUCTURE-- % OF TOTAL TRADE (VALUE BASIS)

YEAR	CANADA	OTH N AM	S AMERIC	EUROPE	ASIA	OTHERS	TOTAL
1967	77.	14.	0.	5.	2.	2.	100.
1968	79.	9.	1.	3.	3.	5.	100.
1969	90.	5.	0.	4.	0.	1.	100.
1970	95.	4.	0.	1.	0.	0.	100.
1971	93.	3.	0.	2.	2.	0.	100.
1972	96.	2.	0.	1.	1.	0.	100.
1973	93.	4.	0.	0.	2.	1.	100.
1974	83.	8.	0.	1.	2.	7.	100.
1975	91.	5.	0.	1.	2.	1.	100.
1976	85.	9.	1.	1.	3.	0.	100.

TRADE TRENDS-- % OF 1967 TRADE (VALUE BASIS)

YEAR	CANADA	OTH N AM	S AMERIC	EUROPE	ASIA	OTHERS	TOTAL
1967	100.	100.	******	100.	100.	100.	100.
1968	118.	77.	******	66.	151.	373.	115.
1969	277.	89.	******	178.	0.	155.	239.
1970	288.	63.	******	40.	4.	39.	234.
1971	631.	94.	******	237.	398.	78.	522.
1972	1132.	157.	******	134.	413.	89.	913.
1973	1989.	478.	******	89.	1342.	551.	1646.
1974	2455.	1242.	******	483.	1837.	9807.	2288.
1975	2136.	601.	******	353.	1820.	857.	1809.
1976	2163.	1286.	******	341.	2826.	202.	1957.

EXPORTS
RECONSTITUTED WOOD (6314)

VALUE--MILLION CURRENT DOLLARS

YEAR	N ATLANTC	S ATLANTC	GULF	S PACIFIC	PACIFC NW	GRT LAKES	ALASKA	N CENTRAL	S CENTRAL	OTHERS	TOTAL
1967	0.065	0.023	0.008	0.083	0.402	0.290	0.000	0.079	0.001	0.000	0.951
1968	0.052	0.016	0.024	0.080	0.587	0.277	0.000	0.057	0.001	0.000	1.093
1969	0.091	0.018	0.008	0.081	0.485	1.423	0.000	0.153	0.010	0.000	2.269
1970	0.021	0.006	0.006	0.067	1.000	0.903	0.000	0.224	0.001	0.004	2.230
1971	0.151	0.010	0.024	0.114	1.986	1.733	0.000	0.945	0.003	0.000	4.965
1972	0.141	0.076	0.017	0.122	2.718	3.502	0.000	2.099	0.008	0.002	8.685
1973	0.332	0.094	0.054	0.265	5.765	5.624	0.000	3.219	0.297	0.007	15.656
1974	0.550	0.094	0.062	1.122	7.034	8.552	0.000	3.306	1.050	0.002	21.772
1975	0.510	0.074	0.500	0.454	5.211	6.134	0.000	4.093	0.187	0.042	17.206
1976	0.260	0.220	0.291	1.326	5.619	4.516	0.000	6.073	0.238	0.076	18.620

QUANTITY--MIL SQ FT

YEAR	N ATLANTC	S ATLANTC	GULF	S PACIFIC	PACIFC NW	GRT LAKES	ALASKA	N CENTRAL	S CENTRAL	OTHERS	TOTAL
1967	0.056	0.085	0.022	0.238	0.922	0.452	0.000	0.103	0.001	0.000	1.878
1968	0.110	0.109	0.054	0.385	4.520	1.567	0.000	0.571	0.003	0.000	7.320
1969	0.028	0.076	0.012	0.455	3.761	11.146	0.000	0.997	0.083	0.000	16.559
1970	0.022	0.027	0.016	0.384	5.039	6.714	0.000	1.401	0.003	0.024	13.629
1971	0.089	0.045	0.091	0.657	14.074	12.498	0.000	7.370	0.027	0.000	34.852
1972	0.111	0.305	0.069	0.684	23.686	29.539	0.000	16.632	0.060	0.001	71.086
1973	1.597	0.176	0.044	1.647	43.757	42.551	0.000	24.771	1.401	0.023	115.968
1974	1.628	0.256	0.222	7.761	52.640	61.056	0.000	22.896	9.522	0.002	155.982
1975	0.752	0.426	0.892	2.521	47.981	40.541	0.000	29.478	1.458	0.057	124.107
1976	0.757	0.971	1.677	7.098	44.481	21.010	0.000	44.084	1.651	0.211	121.940

EXPORTS
RECONSTITUTED WOOD (6314)

TRADE STRUCTURE-- % OF TOTAL TRADE (VALUE BASIS)

YEAR	N ATLANTIC	S ATLANTC	GULF	S PACIFIC	PACIFC NW	GRT LAKES	ALASKA	N CENTRAL	S CENTRAL	OTHERS	TOTAL
1967	7.	2.	1.	9.	42.	31.	0.	8.	0.	0.	100.
1968	5.	1.	2.	7.	54.	25.	0.	5.	0.	0.	100.
1969	4.	1.	0.	4.	21.	63.	0.	7.	0.	0.	100.
1970	1.	0.	0.	3.	45.	40.	0.	10.	0.	0.	100.
1971	3.	0.	0.	2.	40.	35.	0.	19.	0.	0.	100.
1972	2.	1.	0.	1.	31.	40.	0.	24.	0.	0.	100.
1973	2.	1.	0.	2.	37.	36.	0.	21.	2.	0.	100.
1974	3.	0.	0.	5.	32.	39.	0.	15.	5.	0.	100.
1975	3.	0.	3.	3.	30.	36.	0.	24.	1.	0.	100.
1976	1.	1.	2.	7.	30.	24.	0.	33.	1.	0.	100.

TRADE TRENDS-- % OF 1967 TRADE (VALUE BASIS)

YEAR	N ATLANTIC	S ATLANTC	GULF	S PACIFIC	PACIFC NW	GRT LAKES	ALASKA	N CENTRAL	S CENTRAL	OTHERS	TOTAL
1967	100.	100.	100.	100.	100.	100.	******	100.	100.	******	100.
1968	90.	70.	302.	96.	146.	95.	******	72.	51.	******	115.
1969	141.	82.	103.	97.	121.	490.	******	193.	1059.	******	239.
1970	32.	26.	70.	90.	249.	311.	******	283.	80.	******	234.
1971	232.	43.	303.	137.	494.	596.	******	1197.	339.	******	522.
1972	217.	335.	208.	146.	676.	1206.	******	2659.	820.	******	913.
1973	511.	418.	672.	317.	1434.	1936.	******	4078.	30353.	******	1646.
1974	846.	416.	780.	1344.	1750.	2944.	******	4188.	******	******	2288.
1975	785.	329.	6267.	544.	1296.	2112.	******	5185.	19123.	******	1809.
1976	400.	976.	3649.	1589.	1398.	1555.	******	7694.	24237.	******	1957.

EXPORTS
PULPWOOD CHIPS (631.8320)
UNITED STATES

VALUE--MILLION CURRENT DOLLARS

YEAR	CANADA	OTH N AM	EUROPE	JAPAN	OTH ASIA	OTHERS	TOTAL
1967	0.168	0.000	0.008	11.916	0.000	0.000	12.092
1968	0.247	0.001	0.001	26.208	0.168	0.000	26.625
1969	0.328	0.000	0.000	36.018	0.000	0.000	36.346
1970	0.016	0.003	0.000	36.611	0.005	0.000	36.634
1971	0.059	0.001	0.000	41.042	0.000	0.023	41.124
1972	0.254	0.002	0.000	56.859	0.015	0.000	57.130
1973	1.172	0.003	0.000	84.552	0.000	0.000	85.727
1974	0.901	0.003	0.000	113.025	0.000	0.000	113.930
1975	1.224	0.015	4.627	109.519	0.000	0.000	115.385
1976	2.254	0.003	2.210	147.734	0.800	0.001	153.002

QUANTITY--1000 STN

YEAR	CANADA	OTH N AM	EUROPE	JAPAN	OTH ASIA	OTHERS	TOTAL
1967	10.698	0.000	0.516	765.204	0.000	0.000	776.418
1968	16.663	0.041	0.043	1637.770	10.654	0.000	1665.171
1969	21.910	0.000	0.000	2390.696	0.000	0.000	2412.606
1970	0.649	0.105	0.000	2417.389	0.019	0.000	2418.161
1971	3.018	0.033	0.000	1975.262	0.000	0.418	1978.731
1972	14.219	0.065	0.000	2509.436	0.696	0.000	2524.415
1973	58.476	0.131	0.000	3422.233	0.000	0.000	3480.840
1974	56.684	0.131	0.000	3808.965	0.000	0.000	3865.780
1975	37.795	0.454	93.333	3045.622	0.000	0.000	3177.204
1976	57.555	0.052	62.916	3808.258	18.000	0.036	3946.817

EXPORTS
PULPWOOD CHIPS (631.8320)
UNITED STATES

TRADE STRUCTURE-- % OF TOTAL TRADE (VALUE BASIS)

YEAR	CANADA	OTH N AM	EUROPE	JAPAN	OTH ASIA	OTHERS	TOTAL
1967	1.	0.	0.	99.	0.	0.	100.
1968	1.	0.	0.	98.	1.	0.	100.
1969	1.	0.	0.	99.	0.	0.	100.
1970	0.	0.	0.	100.	0.	0.	100.
1971	0.	0.	0.	100.	0.	0.	100.
1972	0.	0.	0.	100.	0.	0.	100.
1973	1.	0.	0.	99.	0.	0.	100.
1974	1.	0.	0.	99.	0.	0.	100.
1975	1.	0.	4.	95.	0.	0.	100.
1976	1.	0.	1.	97.	1.	0.	100.

TRADE TRENDS-- % OF 1967 TRADE (VALUE BASIS)

YEAR	CANADA	OTH N AM	EUROPE	JAPAN	OTH ASIA	OTHERS	TOTAL
1967	100.	******	100.	100.	******	******	100.
1968	147.	******	9.	220.	******	******	220.
1969	195.	******	0.	302.	******	******	301.
1970	9.	******	0.	307.	******	******	303.
1971	35.	******	0.	344.	******	******	340.
1972	151.	******	0.	477.	******	******	472.
1973	698.	******	0.	710.	******	******	709.
1974	536.	******	0.	948.	******	******	942.
1975	729.	******	56513.	919.	******	******	954.
1976	1342.	******	26994.	1240.	******	******	1265.

EXPORTS
PULPWOOD CHIPS (631.8320)

VALUE--MILLION CURRENT DOLLARS

YEAR	N ATLANTC	S ATLANTC	GULF	S PACIFIC	PACIFC NW	GRT LAKES	ALASKA	N CENTRAL	S CENTRAL	OTHERS	TOTAL
1967	0.000	0.000	0.007	0.001	11.916	0.168	0.000	0.000	0.000	0.000	12.092
1968	0.005	0.000	0.000	1.231	25.163	0.226	0.000	0.000	0.000	0.000	26.625
1969	0.000	0.000	0.000	1.266	34.818	0.262	0.000	0.000	0.000	0.000	36.346
1970	0.000	0.000	0.180	0.516	35.650	0.016	0.270	0.000	0.002	0.000	36.634
1971	0.022	0.000	0.001	4.637	35.601	0.033	0.830	0.000	0.000	0.000	41.124
1972	0.000	0.000	0.002	6.275	50.249	0.024	0.580	0.000	0.000	0.000	57.130
1973	0.000	0.000	0.003	9.014	76.123	0.576	0.000	0.010	0.000	0.000	85.727
1974	0.048	0.663	0.002	7.428	104.706	0.058	1.007	0.018	0.000	0.000	113.930
1975	0.172	4.626	0.010	7.469	100.222	0.706	1.572	0.030	0.000	0.577	115.385
1976	0.715	1.600	0.000	12.748	130.510	1.468	4.078	0.041	0.512	1.330	153.002

QUANTITY--1000 STN

YEAR	N ATLANTC	S ATLANTC	GULF	S PACIFIC	PACIFC NW	GRT LAKES	ALASKA	N CENTRAL	S CENTRAL	OTHERS	TOTAL
1967	0.000	0.000	0.441	0.075	765.204	10.698	0.000	0.000	0.000	0.000	776.418
1968	0.411	0.000	0.000	98.186	1551.838	14.736	0.000	0.000	0.000	0.000	1665.171
1969	0.000	0.000	0.000	105.961	2289.885	16.760	0.000	0.000	0.000	0.000	2412.606
1970	0.000	0.000	9.900	48.041	2349.039	0.648	10.450	0.000	0.083	0.000	2418.151
1971	0.343	0.000	0.033	233.070	1716.108	1.377	27.800	0.000	0.000	0.000	1978.731
1972	0.000	0.000	0.065	253.401	2249.757	1.007	20.185	0.000	0.000	0.000	2524.415
1973	0.000	0.000	0.131	369.403	3091.537	19.338	0.000	0.431	0.000	0.000	3480.840
1974	1.972	16.212	0.090	242.058	3567.835	2.202	34.741	0.670	0.000	0.000	3865.780
1975	7.116	93.309	0.333	257.848	2762.913	13.222	32.399	0.725	0.000	9.339	3177.204
1976	24.303	30.883	0.000	366.730	3339.378	31.374	107.652	1.613	11.396	33.488	3946.817

EXPORTS
PULPWOOD CHIPS (631.8320)

TRADE STRUCTURE-- % OF TOTAL TRADE (VALUE BASIS)

YEAR	N ATLANTC	S ATLANTC	GULF	S PACIFIC	PACIFC NW	GRT LAKES	ALASKA	N CENTRAL	S CENTRAL	OTHERS	TOTAL
1967	0.	0.	0.	0.	99.	1.	0.	0.	0.	0.	100.
1968	0.	0.	0.	5.	95.	1.	0.	0.	0.	0.	100.
1969	0.	0.	0.	3.	96.	1.	0.	0.	0.	0.	100.
1970	0.	0.	0.	1.	97.	0.	1.	0.	0.	0.	100.
1971	0.	0.	0.	11.	87.	0.	2.	0.	0.	0.	100.
1972	0.	0.	0.	11.	88.	0.	1.	0.	0.	0.	100.
1973	0.	0.	0.	11.	89.	1.	0.	0.	0.	0.	100.
1974	0.	1.	0.	7.	92.	0.	1.	0.	0.	0.	100.
1975	0.	4.	0.	6.	87.	1.	1.	0.	0.	1.	100.
1976	0.	1.	0.	8.	85.	1.	3.	0.	0.	1.	100.

TRADE TRENDS-- % OF 1967 TRADE (VALUE BASIS)

YEAR	N ATLANTC	S ATLANTC	GULF	S PACIFIC	PACIFC NW	GRT LAKES	ALASKA	N CENTRAL	S CENTRAL	OTHERS	TOTAL
1967	******	******	100.	100.	100.	100.	******	******	******	******	100.
1968	******	******	0.	94692.	211.	134.	******	******	******	******	220.
1969	******	******	0.	97385.	292.	156.	******	******	******	******	301.
1970	******	******	2614.	39704.	299.	9.	******	******	******	******	303.
1971	******	******	12.	******	299.	20.	******	******	******	******	340.
1972	******	******	23.	******	422.	14.	******	******	******	******	472.
1973	******	******	46.	******	639.	343.	******	******	******	******	709.
1974	******	******	31.	******	879.	34.	******	******	******	******	942.
1975	******	******	146.	******	841.	420.	******	******	******	******	954.
1976	******	******	0.	******	1095.	874.	******	******	******	******	1265.

EXPORTS
PULPWOOD (EXC CHIPS)(2421)
UNITED STATES

VALUE--MILLION CURRENT DOLLARS

YEAR	CANADA	OTH N AM	EUROPE	JAPAN	OTH ASIA	OTHERS	TOTAL
1967	3.017	0.003	0.000	0.000	0.000	0.001	3.022
1968	2.366	0.000	0.000	0.005	0.000	0.019	2.389
1969	1.800	0.000	0.003	0.000	0.000	0.000	1.803
1970	1.602	0.000	0.002	0.315	0.000	0.000	1.919
1971	0.000	0.000	0.000	0.477	0.000	0.000	0.477
1972	2.908	0.000	0.000	0.383	0.005	0.000	3.295
1973	1.747	0.154	0.000	0.221	0.000	0.000	2.122
1974	5.686	0.021	0.001	1.392	0.000	0.000	7.099
1975	5.786	0.003	0.224	0.407	0.000	0.000	6.420
1976	6.745	0.055	0.026	0.399	0.006	0.000	7.230

QUANTITY--1000 CORDS

YEAR	CANADA	OTH N AM	EUROPE	JAPAN	OTH ASIA	OTHERS	TOTAL
1967	131.530	0.100	0.000	0.000	0.000	0.032	131.662
1968	95.689	0.000	0.000	0.314	0.000	0.588	96.592
1969	76.994	0.000	0.073	0.000	0.000	0.000	77.067
1970	64.165	0.000	0.034	39.224	0.000	0.000	103.423
1971	0.000	0.000	0.000	47.609	0.000	0.000	47.609
1972	103.997	0.000	0.000	33.535	0.137	0.000	137.669
1973	61.792	7.151	0.000	25.960	0.000	0.000	94.903
1974	169.144	0.976	0.029	67.718	0.000	0.000	237.867
1975	182.886	0.128	3.908	18.746	0.000	0.000	205.668
1976	161.618	2.530	0.737	13.400	0.041	0.000	178.326

EXPORTS
PULPWOOD (EXC CHIPS)(2421)
UNITED STATES

TRADE STRUCTURE-- % OF TOTAL TRADE (VALUE BASIS)

YEAR	CANADA	OTH N AM	EUROPE	JAPAN	OTH ASIA	OTHERS	TOTAL
1967	100.	0.	0.	0.	0.	0.	100.
1968	99.	0.	0.	0.	0.	1.	100.
1969	100.	0.	0.	0.	0.	0.	100.
1970	83.	0.	0.	16.	0.	0.	100.
1971	0.	0.	0.	100.	0.	0.	100.
1972	88.	0.	0.	12.	0.	0.	100.
1973	82.	7.	0.	10.	0.	0.	100.
1974	80.	0.	0.	20.	0.	0.	100.
1975	90.	0.	3.	6.	0.	0.	100.
1976	93.	1.	0.	6.	0.	0.	100.

TRADE TRENDS-- % OF 1967 TRADE (VALUE BASIS)

YEAR	CANADA	OTH N AM	EUROPE	JAPAN	OTH ASIA	OTHERS	TOTAL
1967	100.	100.	******	******	******	100.	100.
1968	78.	0.	******	******	******	1796.	79.
1969	60.	0.	******	******	******	0.	60.
1970	53.	0.	******	******	******	0.	64.
1971	0.	0.	******	******	******	0.	16.
1972	96.	0.	******	******	******	0.	109.
1973	58.	4716.	******	******	******	0.	70.
1974	188.	643.	******	******	******	0.	235.
1975	192.	84.	******	******	******	0.	212.
1976	224.	1668.	******	******	******	0.	239.

EXPORTS
PULPWOOD (EXC CHIPS)(2421)

VALUE--MILLION CURRENT DOLLARS

YEAR	N ATLANTC	S ATLANTC	GULF	S PACIFIC	PACIFC NW	GRT LAKES	ALASKA	N CENTRAL	S CENTRAL	OTHERS	TOTAL
1967	1.162	0.001	0.000	0.003	0.004	1.346	0.000	0.505	0.000	0.000	3.022
1968	0.537	0.000	0.019	0.000	0.005	1.447	0.000	0.381	0.000	0.000	2.389
1969	0.748	0.000	0.000	0.000	0.000	0.803	0.000	0.252	0.000	0.000	1.903
1970	0.350	0.000	0.000	0.000	0.320	0.774	0.000	0.475	0.000	0.000	1.919
1971	0.000	0.000	0.000	0.000	0.477	0.000	0.000	0.000	0.000	0.000	0.477
1972	0.220	0.000	0.000	0.000	0.399	2.412	0.000	0.264	0.000	0.000	3.295
1973	0.197	0.000	0.000	0.000	0.230	1.297	0.000	0.245	0.154	0.000	2.122
1974	1.523	0.000	0.022	0.791	0.658	3.785	0.000	0.321	0.000	0.000	7.099
1975	0.832	0.000	0.225	0.287	0.139	4.179	0.000	0.758	0.000	0.000	6.420
1976	0.092	0.026	0.054	0.287	0.113	6.310	0.000	0.348	0.000	0.000	7.230

QUANTITY--1000 CORDS

YEAR	N ATLANTC	S ATLANTC	GULF	S PACIFIC	PACIFC NW	GRT LAKES	ALASKA	N CENTRAL	S CENTRAL	OTHERS	TOTAL
1967	60.356	0.032	0.000	0.100	0.420	48.634	0.000	22.120	0.000	0.000	131.662
1968	26.346	0.000	0.588	0.000	0.314	53.574	0.000	15.769	0.000	0.000	96.592
1969	37.960	0.000	0.000	0.000	0.000	28.323	0.000	10.784	0.000	0.000	77.067
1970	17.819	0.000	0.000	0.000	39.444	26.007	0.000	20.153	0.000	0.000	103.423
1971	0.000	0.000	0.000	0.000	47.609	0.000	0.000	0.000	0.000	0.000	47.609
1972	11.260	0.000	0.000	0.000	34.166	80.870	0.000	11.373	0.000	0.000	137.669
1973	7.205	0.000	0.000	0.000	26.120	43.843	0.000	10.584	7.151	0.000	94.913
1974	44.927	0.000	1.005	20.383	48.533	111.500	0.000	11.519	0.000	0.000	237.867
1975	24.181	0.000	3.952	7.901	11.742	125.245	0.000	32.647	0.000	0.000	205.668
1976	2.486	0.737	2.502	7.868	5.583	152.630	0.000	6.520	0.000	0.000	179.326

EXPORTS
PULPWOOD (EXC CHIPS)(2421)

TRADE STRUCTURE-- % OF TOTAL TRADE (VALUE BASIS)

YEAR	N ATLANTC	S ATLANTC	GULF	S PACIFIC	PACIFC NW	GRT LAKES	ALASKA	N CENTRAL	S CENTRAL	OTHERS	TOTAL
1967	38.	0.	0.	0.	0.	45.	0.	17.	0.	0.	100.
1968	22.	0.	1.	0.	0.	61.	0.	16.	0.	0.	100.
1969	41.	0.	0.	0.	0.	45.	0.	14.	0.	0.	100.
1970	18.	0.	0.	0.	17.	40.	0.	25.	0.	0.	100.
1971	0.	0.	0.	0.	100.	0.	0.	0.	0.	0.	100.
1972	7.	0.	0.	0.	12.	73.	0.	8.	0.	0.	100.
1973	9.	0.	0.	0.	11.	61.	0.	12.	7.	0.	100.
1974	21.	0.	0.	11.	9.	53.	0.	5.	0.	0.	100.
1975	13.	0.	4.	4.	2.	65.	0.	12.	0.	0.	100.
1976	1.	0.	1.	4.	2.	87.	0.	5.	0.	0.	100.

TRADE TRENDS-- % OF 1967 TRADE (VALUE BASIS)

YEAR	N ATLANTC	S ATLANTC	GULF	S PACIFIC	PACIFC NW	GRT LAKES	ALASKA	N CENTRAL	S CENTRAL	OTHERS	TOTAL
1967	100.	100.	******	100.	100.	100.	******	100.	******	******	100.
1968	46.	0.	******	0.	121.	108.	******	75.	******	******	79.
1969	64.	0.	******	0.	0.	60.	******	50.	******	******	67.
1970	30.	0.	******	0.	8065.	58.	******	94.	******	******	64.
1971	0.	0.	******	0.	12015.	0.	******	0.	******	******	16.
1972	19.	0.	******	0.	10056.	179.	******	52.	******	******	109.
1973	17.	0.	******	0.	5793.	96.	******	48.	******	******	79.
1974	131.	0.	******	24183.	16581.	281.	******	64.	******	******	235.
1975	72.	0.	******	8793.	3499.	310.	******	150.	******	******	212.
1976	8.	2532.	******	8781.	2846.	469.	******	69.	******	******	239.

EXPORTS
WOOD PULP (2510)
UNITED STATES

VALUE--MILLION CURRENT DOLLARS

YEAR	N AMERIC	S AMERIC	EUROPE	JAPAN	OTH ASIA	OTHERS	TOTAL
1967	13.193	16.459	128.738	49.639	30.716	5.672	244.417
1968	14.080	23.565	148.588	43.942	29.483	5.528	265.186
1969	15.768	20.925	154.862	52.349	35.996	8.965	288.865
1970	32.762	37.739	267.191	60.023	48.325	22.336	468.376
1971	27.167	27.059	199.032	51.947	35.551	15.252	355.949
1972	30.772	32.542	195.412	61.944	27.934	20.276	368.880
1973	43.021	34.955	219.063	80.176	37.613	20.703	434.531
1974	78.808	68.365	403.684	186.440	55.443	40.833	833.573
1975	69.185	66.718	512.491	155.664	69.426	34.582	908.066
1976	70.954	52.646	488.414	145.168	67.282	41.932	866.396

QUANTITY--1000 STN

YEAR	N AMERIC	S AMERIC	EUROPE	JAPAN	OTH ASIA	OTHERS	TOTAL
1967	92.364	132.225	867.775	385.844	251.899	42.968	1773.075
1968	98.007	201.962	1021.076	346.965	259.428	43.119	1970.578
1969	107.569	173.136	1102.227	398.532	297.592	74.015	2153.071
1970	211.897	279.015	1751.773	394.187	341.719	154.919	3133.509
1971	163.616	188.706	1178.290	304.788	247.134	107.419	2189.951
1972	200.438	236.156	1179.820	369.883	199.426	138.624	2324.349
1973	283.639	198.583	1213.949	404.780	207.370	120.004	2428.325
1974	290.941	214.337	1391.112	593.000	180.210	129.188	2798.788
1975	205.191	188.705	1489.503	438.262	212.147	130.287	2664.094
1976	203.568	162.997	1428.998	423.223	222.894	157.094	2598.775

EXPORTS
WOOD PULP (2510)
UNITED STATES

TRADE STRUCTURE-- % OF TOTAL TRADE (VALUE BASIS)

YEAR	N AMERIC	S AMERIC	EUROPE	JAPAN	OTH ASIA	OTHERS	TOTAL
1967	5.	7.	53.	20.	13.	2.	100.
1968	5.	9.	56.	17.	11.	2.	100.
1969	5.	7.	54.	18.	12.	3.	100.
1970	7.	8.	57.	13.	10.	5.	100.
1971	8.	8.	56.	15.	10.	4.	100.
1972	8.	9.	53.	17.	8.	5.	100.
1973	10.	8.	50.	18.	9.	5.	100.
1974	9.	8.	48.	22.	7.	5.	100.
1975	8.	7.	56.	17.	8.	4.	100.
1976	8.	6.	56.	17.	8.	5.	100.

TRADE TRENDS-- % OF 1967 TRADE (VALUE BASIS)

YEAR	N AMERIC	S AMERIC	EUROPE	JAPAN	OTH ASIA	OTHERS	TOTAL
1967	100.	100.	100.	100.	100.	100.	100.
1968	107.	143.	115.	89.	96.	97.	108.
1969	120.	127.	120.	105.	117.	158.	118.
1970	248.	229.	208.	121.	157.	394.	192.
1971	205.	164.	155.	105.	116.	269.	146.
1972	233.	198.	152.	125.	91.	357.	151.
1973	326.	212.	169.	162.	122.	365.	178.
1974	597.	415.	314.	376.	181.	720.	341.
1975	524.	405.	398.	314.	226.	610.	372.
1976	538.	320.	379.	292.	219.	739.	354.

EXPORTS
WOOD PULP (2510)

VALUE--MILLION CURRENT DOLLARS

YEAR	N ATLANTC	S ATLANTC	GULF	S PACIFIC	PACIFC NW	GRT LAKES	ALASKA	N CENTRAL	S CENTRAL	OTHERS	TOTAL
1967	13.755	64.090	61.418	10.477	65.532	2.294	26.087	0.040	0.724	0.000	244.417
1968	14.937	70.424	73.214	13.098	65.194	3.608	23.991	0.005	0.813	0.002	265.186
1969	13.970	64.516	79.068	18.520	81.242	2.371	28.112	0.011	1.053	0.003	288.865
1970	22.835	113.148	146.179	37.629	108.971	6.586	31.939	0.067	0.990	0.031	468.376
1971	21.814	88.819	99.950	28.176	77.436	7.175	31.416	0.109	1.043	0.011	355.949
1972	22.662	82.255	93.348	37.563	89.446	7.341	35.147	0.005	1.097	0.017	368.880
1973	23.761	99.781	104.918	58.640	107.501	7.018	30.880	0.195	1.710	0.127	434.531
1974	48.676	195.244	209.459	120.228	180.275	10.316	65.290	1.933	1.888	0.264	833.573
1975	46.119	216.355	260.862	105.239	208.255	9.662	55.091	3.785	2.569	0.127	908.066
1976	47.044	242.225	229.522	85.315	173.542	17.058	66.919	3.201	1.553	0.016	866.396

QUANTITY--1000 STN

YEAR	N ATLANTC	S ATLANTC	GULF	S PACIFIC	PACIFC NW	GRT LAKES	ALASKA	N CENTRAL	S CENTRAL	OTHERS	TOTAL
1967	72.704	450.476	404.366	99.306	541.570	14.784	185.274	0.547	4.048	0.000	1773.075
1968	90.240	499.776	505.755	120.124	559.167	23.409	169.650	0.017	4.423	0.016	1970.578
1969	77.123	425.825	621.821	159.887	659.474	15.040	189.750	0.066	5.974	0.111	2153.071
1970	130.228	726.681	981.021	260.198	786.920	45.581	196.729	0.405	5.476	0.270	3133.509
1971	122.020	531.266	612.659	178.044	509.082	47.129	164.947	0.653	5.174	0.078	2189.951
1972	136.967	477.479	593.750	249.260	606.160	52.314	202.973	0.063	5.270	0.114	2324.349
1973	102.354	508.965	620.003	315.635	658.759	55.445	152.930	3.140	10.370	0.623	2428.325
1974	127.413	652.630	786.026	353.533	629.995	43.045	188.343	9.829	6.337	1.637	2798.788
1975	111.090	667.481	770.214	298.447	622.040	29.947	143.796	14.015	6.856	0.208	2664.094
1976	122.349	708.225	706.622	267.594	529.999	48.132	197.497	13.796	4.525	0.036	2598.775

EXPORTS
WOOD PULP (2510)

TRADE STRUCTURE-- % OF TOTAL TRADE (VALUE BASIS)

YEAR	N ATLANTIC	S ATLANTC	GULF	S PACIFIC	PACIFC NW	GRT LAKES	ALASKA	N CENTRAL	S CENTRAL	OTHERS	TOTAL
1967	6.	26.	25.	4.	27.	1.	11.	0.	0.	0.	100.
1968	6.	27.	23.	5.	25.	1.	9.	0.	0.	0.	100.
1969	5.	22.	27.	6.	28.	1.	10.	0.	0.	0.	100.
1970	5.	24.	31.	8.	23.	1.	7.	0.	0.	0.	100.
1971	6.	25.	28.	9.	22.	2.	9.	0.	0.	0.	100.
1972	6.	22.	25.	10.	24.	2.	10.	0.	0.	0.	100.
1973	5.	23.	24.	13.	25.	2.	7.	0.	0.	0.	100.
1974	6.	23.	25.	14.	22.	1.	8.	0.	0.	0.	100.
1975	5.	24.	29.	12.	23.	1.	6.	0.	0.	0.	100.
1976	5.	28.	26.	10.	20.	2.	8.	0.	0.	0.	100.

TRADE TRENDS-- % OF 1967 TRADE (VALUE BASIS)

YEAR	N ATLANTC	S ATLANTC	GULF	S PACIFIC	PACIFC NW	GRT LAKES	ALASKA	N CENTRAL	S CENTRAL	OTHERS	TOTAL
1967	100.	100.	100.	100.	100.	100.	100.	100.	100.	******	100.
1968	109.	110.	119.	125.	99.	157.	92.	13.	112.	******	108.
1969	102.	101.	129.	177.	124.	163.	108.	26.	145.	******	118.
1970	166.	177.	238.	359.	166.	267.	132.	167.	137.	******	192.
1971	159.	139.	163.	269.	118.	313.	120.	271.	144.	******	146.
1972	165.	128.	152.	359.	136.	320.	135.	12.	152.	******	151.
1973	173.	156.	171.	560.	164.	306.	118.	483.	236.	******	178.
1974	354.	305.	341.	1147.	275.	450.	250.	4789.	261.	******	341.
1975	335.	338.	425.	1004.	318.	421.	211.	9379.	355.	******	372.
1976	342.	378.	374.	814.	265.	743.	257.	7933.	214.	******	354.

EXPORTS
NEWSPRINT (6411)
UNITED STATES

VALUE--MILLION CURRENT DOLLARS

YEAR	MEXICO	VENEZUEL	EUROPE	ASIA	AFRICA	OTHERS	TOTAL
1967	4.669	1.183	0.819	2.751	0.005	2.017	11.444
1968	7.130	1.760	0.396	4.683	0.015	3.100	17.084
1969	3.705	1.857	0.790	5.650	0.024	3.825	15.851
1970	7.861	1.837	2.309	4.506	0.033	2.128	18.674
1971	5.592	1.027	0.968	8.907	0.052	4.125	20.671
1972	6.167	2.964	2.267	4.126	0.036	4.735	20.336
1973	6.052	3.623	1.842	0.984	0.008	4.394	16.903
1974	19.303	4.694	6.859	8.673	5.129	9.488	54.146
1975	15.792	7.140	5.337	6.201	5.940	12.361	52.771
1976	16.818	5.998	5.505	12.140	1.381	7.086	48.928

QUANTITY--1000 STN

YEAR	MEXICO	VENEZUEL	EUROPE	ASIA	AFRICA	OTHERS	TOTAL
1967	34.153	9.541	6.858	24.133	0.023	16.025	90.732
1968	45.938	14.177	3.234	41.495	0.121	24.779	129.743
1969	26.586	14.649	6.213	50.792	0.188	29.588	128.014
1970	55.172	13.388	20.482	39.051	0.284	15.550	143.927
1971	36.179	7.045	7.241	82.066	0.382	33.743	166.655
1972	39.420	17.623	17.204	34.764	0.096	35.688	144.794
1973	34.936	19.337	9.332	5.264	0.041	28.908	97.818
1974	73.353	19.592	22.796	25.143	15.433	33.463	189.781
1975	59.214	24.885	15.099	20.915	14.854	34.445	169.413
1976	56.116	18.617	16.138	38.858	4.314	18.876	152.919

EXPORTS
NEWSPRINT (6411)
UNITED STATES

TRADE STRUCTURE-- % OF TOTAL TRADE (VALUE BASIS)

YEAR	MEXICO	VENEZUEL	EUROPE	ASIA	AFRICA	OTHERS	TOTAL
1967	41.	10.	7.	24.	0.	18.	100.
1968	42.	10.	2.	27.	0.	18.	100.
1969	23.	12.	5.	36.	0.	24.	100.
1970	42.	10.	12.	24.	0.	11.	100.
1971	27.	5.	5.	43.	0.	20.	100.
1972	30.	15.	11.	20.	0.	23.	100.
1973	36.	21.	11.	6.	0.	26.	100.
1974	36.	9.	13.	16.	9.	18.	100.
1975	30.	14.	10.	12.	11.	23.	100.
1976	34.	12.	11.	25.	3.	14.	100.

TRADE TRENDS-- % OF 1967 TRADE (VALUE BASIS)

YEAR	MEXICO	VENEZUEL	EUROPE	ASIA	AFRICA	OTHERS	TOTAL
1967	100.	100.	100.	100.	100.	100.	100.
1968	153.	149.	48.	170.	270.	154.	149.
1969	79.	157.	96.	205.	437.	190.	139.
1970	168.	155.	282.	164.	599.	105.	163.
1971	120.	87.	118.	324.	959.	205.	181.
1972	133.	250.	279.	150.	667.	235.	178.
1973	130.	306.	225.	36.	149.	218.	148.
1974	413.	397.	838.	315.	93930.	470.	473.
1975	338.	693.	652.	225.	******	613.	461.
1976	360.	507.	673.	441.	25296.	351.	428.

EXPORTS
NEWSPRINT (6411)

VALUE--MILLION CURRENT DOLLARS

YEAR	N ATLANTC	S ATLANTC	GULF	S PACIFIC	PACIFC NW	GRT LAKES	ALASKA	N CENTRAL	S CENTRAL	OTHERS	TOTAL
1967	3.613	0.290	4.040	0.216	1.433	0.134	0.000	0.006	1.711	0.000	11.444
1968	3.127	1.557	8.988	0.344	1.173	0.205	0.000	0.008	1.681	0.000	17.084
1969	2.310	1.086	9.108	0.535	2.129	0.242	0.000	0.003	0.433	0.004	15.851
1970	2.535	0.288	10.361	0.375	3.084	0.196	0.000	0.030	1.802	0.004	18.674
1971	3.734	0.105	10.170	0.419	5.418	0.179	0.000	0.017	0.623	0.007	20.671
1972	8.424	0.109	9.650	0.332	0.677	0.671	0.000	0.006	0.444	0.023	20.336
1973	7.718	0.215	7.950	0.356	0.068	0.263	0.000	0.014	0.285	0.035	16.903
1974	29.160	0.521	20.458	0.532	0.728	1.304	0.000	0.069	1.138	0.235	54.146
1975	24.245	3.714	19.469	0.491	1.822	1.768	0.000	0.089	1.126	0.048	52.771
1976	14.995	0.342	17.256	1.394	11.721	1.717	0.000	0.114	1.369	0.021	48.928

QUANTITY--1000 STN

YEAR	N ATLANTC	S ATLANTC	GULF	S PACIFIC	PACIFC NW	GRT LAKES	ALASKA	N CENTRAL	S CENTRAL	OTHERS	TOTAL
1967	30.352	2.300	31.495	1.452	12.232	0.574	0.000	0.016	12.311	0.000	90.732
1968	24.215	13.854	65.205	2.396	10.807	1.041	0.000	0.034	12.191	0.000	129.743
1969	16.927	10.412	72.768	4.323	19.104	1.392	0.000	0.007	3.057	0.024	128.014
1970	19.692	2.266	79.007	2.691	26.333	0.820	0.000	0.141	12.951	0.025	143.927
1971	30.068	0.583	78.026	2.922	49.826	1.049	0.000	0.126	4.023	0.031	166.655
1972	62.088	0.901	67.770	2.260	5.032	3.769	0.000	0.033	2.817	0.124	144.794
1973	40.098	1.053	50.489	2.150	0.527	1.524	0.000	0.081	1.709	0.189	97.818
1974	92.877	1.866	78.909	2.244	2.822	4.585	0.000	0.206	5.607	0.665	189.781
1975	73.069	9.042	72.445	1.144	5.835	3.568	0.000	0.140	4.049	0.121	169.413
1976	44.941	0.996	59.180	3.712	36.657	3.291	0.000	0.183	3.906	0.052	152.919

EXPORTS
NEWSPRINT (6411)

TRADE STRUCTURE-- % OF TOTAL TRADE (VALUE BASIS)

YEAR	N ATLANTC	S ATLANTC	GULF	S PACIFIC	PACIFC NW	GRT LAKES	ALASKA	N CENTRAL	S CENTRAL	OTHERS	TOTAL
1967	32.	3.	35.	2.	13.	1.	0.	0.	15.	0.	100.
1968	18.	9.	53.	2.	7.	1.	0.	0.	10.	0.	100.
1969	15.	7.	57.	3.	13.	2.	0.	0.	3.	0.	100.
1970	14.	2.	55.	2.	17.	1.	0.	0.	10.	0.	100.
1971	18.	1.	49.	2.	26.	1.	0.	0.	3.	0.	100.
1972	41.	1.	47.	2.	3.	3.	0.	0.	2.	0.	100.
1973	46.	1.	47.	2.	0.	2.	0.	0.	2.	0.	100.
1974	54.	1.	38.	1.	1.	2.	0.	0.	2.	0.	100.
1975	46.	7.	37.	1.	3.	3.	0.	0.	2.	0.	100.
1976	31.	1.	35.	3.	24.	4.	0.	0.	3.	0.	100.

TRADE TRENDS-- % OF 1967 TRADE (VALUE BASIS)

YEAR	N ATLANTC	S ATLANTC	GULF	S PACIFIC	PACIFC NW	GRT LAKES	ALASKA	N CENTRAL	S CENTRAL	OTHERS	TOTAL
1967	100.	100.	100.	100.	100.	100.	******	100.	100.	******	100.
1968	87.	537.	222.	159.	82.	153.	******	122.	98.	******	149.
1969	64.	375.	225.	248.	149.	181.	******	43.	25.	******	139.
1970	70.	99.	256.	173.	215.	146.	******	475.	105.	******	163.
1971	103.	36.	252.	194.	378.	134.	******	265.	36.	******	181.
1972	233.	38.	239.	153.	47.	501.	******	93.	26.	******	178.
1973	214.	74.	197.	165.	5.	196.	******	226.	17.	******	148.
1974	807.	180.	506.	246.	51.	973.	******	1086.	66.	******	473.
1975	671.	1281.	482.	227.	127.	1320.	******	1394.	66.	******	461.
1976	415.	118.	427.	645.	818.	1281.	******	1783.	80.	******	428.

EXPORTS
BUILDING BOARD (6416)
UNITED STATES

VALUE--MILLION CURRENT DOLLARS

YEAR	CANADA	OTH N AM	EUROPE	ASIA	AFRICA	OTHERS	TOTAL
1967	3.431	0.857	2.930	0.663	0.053	0.321	8.255
1968	3.609	1.159	2.583	0.793	0.205	0.309	8.659
1969	5.691	0.865	3.041	0.691	0.166	0.624	11.078
1970	5.708	0.915	2.808	0.451	0.131	0.211	10.225
1971	6.488	0.883	2.820	0.638	0.869	0.226	11.923
1972	8.489	0.919	2.322	0.778	0.101	0.490	13.098
1973	18.768	1.558	3.098	1.002	0.313	0.281	25.020
1974	17.946	2.354	4.345	2.190	0.177	0.839	27.850
1975	17.269	2.360	2.862	2.145	0.185	0.453	25.293
1976	21.489	2.370	3.501	3.242	0.145	0.943	31.689

QUANTITY--1000 STN

YEAR	CANADA	OTH N AM	EUROPE	ASIA	AFRICA	OTHERS	TOTAL
1967	13.344	3.475	12.271	2.283	0.176	1.210	32.759
1968	17.284	4.123	10.363	2.531	0.903	1.341	36.544
1969	26.552	3.403	11.549	2.330	0.474	1.782	46.089
1970	30.055	3.834	9.614	1.661	0.353	0.681	46.218
1971	33.434	3.370	11.184	1.943	2.403	0.772	53.106
1972	44.230	3.615	8.549	2.277	0.344	1.817	60.832
1973	90.555	6.634	10.221	2.622	0.787	0.671	111.489
1974	72.187	8.603	12.521	5.135	0.376	1.801	100.622
1975	64.103	6.750	7.445	8.760	0.320	0.928	88.306
1976	74.403	7.061	9.183	6.586	0.240	1.709	99.183

EXPORTS
BUILDING BOARD (6416)
UNITED STATES

TRADE STRUCTURE-- % OF TOTAL TRADE (VALUE BASIS)

YEAR	CANADA	OTH N AM	EUROPE	ASIA	AFRICA	OTHERS	TOTAL
1967	42.	10.	35.	8.	1.	4.	100.
1968	42.	13.	30.	9.	2.	4.	100.
1969	51.	8.	27.	6.	1.	6.	100.
1970	56.	9.	27.	4.	1.	2.	100.
1971	54.	7.	24.	5.	7.	2.	100.
1972	65.	7.	18.	6.	1.	4.	100.
1973	75.	6.	12.	4.	1.	1.	100.
1974	64.	8.	16.	8.	1.	3.	100.
1975	68.	9.	11.	8.	1.	2.	100.
1976	68.	7.	11.	10.	0.	3.	100.

TRADE TRENDS-- % OF 1967 TRADE (VALUE BASIS)

YEAR	CANADA	OTH N AM	EUROPE	ASIA	AFRICA	OTHERS	TOTAL
1967	100.	100.	100.	100.	100.	100.	100.
1968	105.	135.	88.	120.	390.	96.	105.
1969	166.	101.	104.	104.	315.	195.	134.
1970	166.	107.	96.	68.	249.	66.	124.
1971	189.	103.	96.	96.	1650.	70.	144.
1972	247.	107.	79.	117.	191.	153.	159.
1973	547.	182.	106.	151.	595.	88.	303.
1974	523.	275.	148.	330.	337.	261.	337.
1975	503.	275.	98.	323.	351.	141.	306.
1976	626.	276.	120.	489.	275.	294.	384.

EXPORTS
BUILDING BOARD (6416)

VALUE--MILLION CURRENT DOLLARS

YEAR	N ATLANTC	S ATLANTC	GULF	S PACIFIC	PACIFC NW	GRT LAKES	ALASKA	N CENTRAL	S CENTRAL	OTHERS	TOTAL
1967	1.263	0.577	2.394	0.408	0.556	2.694	0.000	0.337	0.011	0.017	8.255
1968	1.177	0.545	2.554	0.221	0.786	3.136	0.000	0.204	0.009	0.028	8.659
1969	1.960	0.891	1.651	0.344	1.104	4.581	0.000	0.506	0.014	0.027	11.078
1970	2.169	0.482	0.896	0.316	1.392	4.550	0.000	0.370	0.016	0.034	10.225
1971	1.730	0.779	1.407	0.205	1.360	5.604	0.000	0.805	0.012	0.021	11.923
1972	1.455	0.856	1.422	0.397	1.697	6.308	0.000	0.890	0.020	0.054	13.098
1973	2.535	1.057	1.789	0.456	3.654	13.303	0.000	2.031	0.099	0.096	25.020
1974	3.503	2.104	2.464	1.804	4.370	11.409	0.000	2.028	0.034	0.134	27.850
1975	3.316	1.555	2.327	0.927	2.453	11.851	0.000	2.724	0.061	0.079	25.293
1976	4.354	1.947	2.540	1.598	3.487	14.505	0.033	3.145	0.042	0.039	31.689

QUANTITY--1000 STN

YEAR	N ATLANTC	S ATLANTC	GULF	S PACIFIC	PACIFC NW	GRT LAKES	ALASKA	N CENTRAL	S CENTRAL	OTHERS	TOTAL
1967	3.998	3.105	10.532	1.063	2.236	10.814	0.000	0.919	0.027	0.065	32.759
1968	3.820	2.162	10.359	0.853	3.456	14.960	0.000	0.828	0.041	0.066	36.544
1969	6.221	2.887	7.314	0.870	5.006	21.742	0.000	1.902	0.065	0.083	46.089
1970	6.691	2.232	3.719	0.961	5.751	24.980	0.000	1.739	0.045	0.100	46.218
1971	6.513	3.223	5.046	0.934	5.764	28.153	0.000	3.384	0.031	0.058	53.106
1972	4.737	3.391	5.255	1.502	7.804	33.553	0.000	4.243	0.130	0.218	60.832
1973	7.077	4.308	6.684	1.423	16.402	66.771	0.000	7.919	0.484	0.422	111.489
1974	9.646	6.467	7.189	5.730	17.544	46.346	0.000	7.272	0.123	0.305	100.622
1975	7.099	4.041	10.395	3.379	8.695	44.873	0.000	9.502	0.225	0.096	88.306
1976	9.387	4.306	7.054	5.361	10.675	51.021	0.030	11.134	0.131	0.085	99.183

EXPORTS
BUILDING BOARD (6416)

TRADE STRUCTURE-- % OF TOTAL TRADE (VALUE BASIS)

YEAR	N ATLANTC	S ATLANTC	GULF	S PACIFIC	PACIFC NW	GRT LAKES	ALASKA	N CENTRAL	S CENTRAL	OTHERS	TOTAL
1967	15.	7.	29.	5.	7.	33.	0.	4.	0.	0.	100.
1968	14.	6.	29.	3.	9.	36.	0.	2.	0.	0.	100.
1969	18.	8.	15.	3.	10.	41.	0.	5.	0.	0.	100.
1970	21.	5.	9.	3.	14.	45.	0.	4.	0.	0.	100.
1971	15.	7.	12.	2.	11.	47.	0.	7.	0.	0.	100.
1972	11.	7.	11.	3.	13.	48.	0.	7.	0.	0.	100.
1973	10.	4.	7.	2.	15.	53.	0.	8.	0.	0.	100.
1974	13.	8.	9.	6.	16.	41.	0.	7.	0.	0.	100.
1975	13.	6.	9.	4.	10.	47.	0.	11.	0.	0.	100.
1976	14.	6.	8.	5.	11.	46.	0.	10.	0.	0.	100.

TRADE TRENDS-- % OF 1967 TRADE (VALUE BASIS)

YEAR	N ATLANTC	S ATLANTC	GULF	S PACIFIC	PACIFC NW	GRT LAKES	ALASKA	N CENTRAL	S CENTRAL	OTHERS	TOTAL
1967	100.	100.	100.	100.	100.	100.	******	100.	100.	100.	100.
1968	93.	94.	107.	54.	141.	116.	******	61.	82.	165.	105.
1969	155.	155.	69.	84.	198.	170.	******	150.	120.	162.	134.
1970	172.	84.	37.	78.	250.	169.	******	110.	138.	202.	124.
1971	137.	135.	59.	50.	245.	208.	******	239.	108.	124.	144.
1972	115.	148.	59.	98.	305.	234.	******	264.	174.	322.	159.
1973	201.	183.	75.	112.	657.	494.	******	603.	876.	573.	303.
1974	277.	365.	103.	443.	786.	424.	******	603.	301.	803.	337.
1975	263.	270.	97.	227.	441.	440.	******	809.	543.	476.	306.
1976	345.	338.	106.	392.	627.	538.	******	934.	374.	235.	384.

EXPORTS
PAPER & PAPERBOARD (6410)
UNITED STATES

VALUE--MILLION CURRENT DOLLARS

YEAR	CANADA	S AMERIC	EUROPE	ASIA	AFRICA	OTHERS	TOTAL
1967	60.848	35.000	135.031	37.131	22.193	67.482	357.685
1968	65.880	39.603	173.235	43.717	20.978	79.198	422.612
1969	75.621	36.318	194.979	43.318	22.877	78.175	451.289
1970	73.534	42.267	208.911	47.967	28.622	86.778	488.080
1971	89.643	42.292	225.943	52.092	27.671	89.357	526.998
1972	103.147	44.070	220.617	57.357	29.104	100.569	554.864
1973	129.578	51.055	245.244	85.665	45.546	120.170	677.257
1974	188.877	106.445	367.124	166.203	87.864	203.404	1119.917
1975	254.677	89.590	305.517	127.851	71.269	163.232	1012.137
1976	316.738	101.933	322.446	141.057	79.147	221.498	1182.820

QUANTITY--1000 STN

YEAR	CANADA	S AMERIC	EUROPE	ASIA	AFRICA	OTHERS	TOTAL
1967	177.051	205.361	800.194	219.113	117.785	374.868	1894.372
1968	212.556	213.937	1049.015	286.556	109.921	457.365	2329.350
1969	235.242	187.578	1202.453	277.148	122.214	430.721	2455.356
1970	225.626	203.957	1193.991	280.056	160.083	477.138	2540.852
1971	285.782	212.602	1336.854	337.133	152.541	478.435	2803.348
1972	324.816	205.093	1242.473	324.700	156.241	520.262	2773.585
1973	356.240	214.984	1062.324	367.557	188.539	512.493	2702.136
1974	476.889	293.558	1201.361	474.277	267.059	583.135	3296.279
1975	621.526	232.204	780.010	372.214	213.451	439.928	2659.333
1976	693.864	245.580	866.405	394.518	229.471	596.436	3026.275

EXPORTS
PAPER & PAPERBOARD (6410)
UNITED STATES

TRADE STRUCTURE-- % OF TOTAL TRADE (VALUE BASIS)

YEAR	CANADA	S AMERIC	EUROPE	ASIA	AFRICA	OTHERS	TOTAL
1967	17.	10.	38.	10.	6.	19.	100.
1968	16.	9.	41.	10.	5.	19.	100.
1969	17.	8.	43.	10.	5.	17.	100.
1970	15.	9.	43.	10.	6.	18.	100.
1971	17.	8.	43.	10.	5.	17.	100.
1972	19.	8.	40.	10.	5.	18.	100.
1973	19.	8.	36.	13.	7.	18.	100.
1974	17.	10.	33.	15.	8.	18.	100.
1975	25.	9.	30.	13.	7.	16.	100.
1976	27.	9.	27.	12.	7.	19.	100.

TRADE TRENDS-- % OF 1967 TRADE (VALUE BASIS)

YEAR	CANADA	S AMERIC	EUROPE	ASIA	AFRICA	OTHERS	TOTAL
1967	100.	100.	100.	100.	100.	100.	100.
1968	108.	113.	128.	118.	95.	117.	116.
1969	124.	104.	144.	117.	103.	116.	126.
1970	121.	121.	155.	129.	129.	129.	136.
1971	147.	121.	167.	140.	125.	132.	147.
1972	170.	126.	163.	154.	131.	149.	155.
1973	213.	146.	182.	231.	205.	178.	189.
1974	310.	304.	272.	448.	396.	301.	313.
1975	419.	256.	226.	344.	321.	242.	283.
1976	521.	291.	239.	380.	357.	328.	331.

EXPORTS
PAPER & PAPERBOARD (6410)

VALUE--MILLION CURRENT DOLLARS

YEAR	N ATLANTC	S ATLANTC	GULF	S PACIFIC	PACIFC NW	GRT LAKES	ALASKA	N CENTRAL	S CENTRAL	OTHERS	TOTAL
1967	105.255	68.760	81.446	9.874	43.851	43.438	0.000	2.006	0.123	2.932	357.685
1968	125.100	75.794	111.490	10.343	46.091	47.565	0.011	1.917	0.283	4.018	422.612
1969	126.426	73.183	122.156	12.679	54.872	55.014	0.003	2.578	0.215	4.163	451.289
1970	141.694	93.540	132.841	10.781	49.440	52.088	0.000	2.801	1.160	3.735	488.080
1971	141.014	96.753	159.923	9.155	41.843	66.657	0.000	3.470	4.175	4.008	526.998
1972	145.045	91.777	159.861	12.085	59.215	74.233	0.000	3.873	5.499	3.295	554.884
1973	200.152	116.682	160.259	16.404	80.621	88.927	0.000	4.818	4.023	5.370	677.257
1974	319.490	194.967	293.887	37.388	124.138	127.022	0.000	7.319	3.731	11.974	1119.917
1975	269.305	188.177	250.325	22.620	89.023	170.710	0.000	12.102	5.971	3.905	1012.137
1976	306.955	206.360	255.858	31.368	135.345	220.508	0.000	15.947	7.126	3.354	1182.820

QUANTITY--1000 STN

YEAR	N ATLANTC	S ATLANTC	GULF	S PACIFIC	PACIFC NW	GRT LAKES	ALASKA	N CENTRAL	S CENTRAL	OTHERS	TOTAL
1967	290.906	582.742	536.467	43.247	300.125	108.112	0.000	5.834	0.264	26.675	1894.372
1968	314.884	674.405	792.998	39.945	319.887	144.805	0.036	5.378	0.678	36.333	2329.350
1969	304.281	624.461	902.906	37.455	381.617	155.530	0.007	7.789	0.532	40.780	2455.356
1970	322.922	722.160	958.504	38.295	319.928	134.034	0.000	10.132	1.383	33.493	2540.852
1971	313.078	780.638	1155.518	32.127	289.461	177.795	0.000	12.476	6.814	35.440	2803.348
1972	313.444	672.649	1104.398	38.772	390.468	199.605	0.000	12.905	14.613	26.731	2773.585
1973	351.115	672.275	929.863	45.894	417.829	226.973	0.000	17.624	16.311	24.252	2702.136
1974	504.230	767.329	1146.381	88.501	410.336	308.043	0.000	18.954	13.652	38.854	3296.279
1975	385.856	662.448	824.803	42.193	277.503	405.127	0.000	30.769	19.102	11.531	2659.333
1976	412.834	738.118	872.084	54.553	410.540	458.557	0.000	32.426	21.379	25.785	3026.275

EXPORTS
PAPER & PAPERBOARD (6410)

TRADE STRUCTURE-- % OF TOTAL TRADE (VALUE BASIS)

YEAR	N ATLANTC	S ATLANTC	GULF	S PACIFIC	PACIFC NW	GRT LAKES	ALASKA	N CENTRAL	S CENTRAL	OTHERS	TOTAL
1967	29.	19.	23.	3.	12.	12.	0.	1.	0.	1.	100.
1968	30.	18.	26.	2.	11.	11.	0.	0.	0.	1.	100.
1969	28.	16.	27.	3.	12.	12.	0.	1.	0.	1.	100.
1970	29.	19.	27.	2.	10.	11.	0.	1.	0.	1.	100.
1971	27.	18.	30.	2.	8.	13.	0.	1.	1.	1.	100.
1972	26.	17.	29.	2.	11.	13.	0.	1.	1.	1.	100.
1973	30.	17.	24.	2.	12.	13.	0.	1.	1.	1.	100.
1974	29.	17.	26.	3.	11.	11.	0.	1.	0.	1.	100.
1975	27.	19.	25.	2.	9.	17.	0.	1.	1.	0.	100.
1976	26.	17.	22.	3.	11.	19.	0.	1.	1.	0.	100.

TRADE TRENDS-- % OF 1967 TRADE (VALUE BASIS)

YEAR	N ATLANTC	S ATLANTC	GULF	S PACIFIC	PACIFC NW	GRT LAKES	ALASKA	N CENTRAL	S CENTRAL	OTHERS	TOTAL
1967	100.	100.	100.	100.	100.	100.	******	100.	100.	100.	100.
1968	119.	110.	137.	105.	105.	110.	******	96.	231.	137.	118.
1969	120.	106.	150.	128.	125.	127.	******	129.	176.	142.	126.
1970	135.	136.	163.	109.	113.	120.	******	140.	946.	127.	136.
1971	134.	141.	196.	93.	95.	153.	******	173.	3406.	137.	147.
1972	138.	133.	196.	122.	135.	171.	******	193.	4486.	112.	155.
1973	190.	170.	197.	166.	184.	205.	******	240.	3282.	183.	189.
1974	304.	284.	361.	379.	283.	292.	******	365.	3044.	408.	313.
1975	256.	274.	307.	229.	203.	393.	******	603.	4871.	133.	283.
1976	292.	300.	314.	318.	309.	508.	******	795.	5814.	114.	331.

IMPORTS
SOFTWOOD LOGS (2422)
UNITED STATES

VALUE--MILLION CURRENT DOLLARS

YEAR	CANADA	OTH N AM	EUROPE	ASIA	AFRICA	OTHERS	TOTAL
1967	1.690	0.000	0.000	0.000	0.000	0.001	1.691
1968	1.383	0.001	0.000	0.000	0.000	0.000	1.384
1969	3.051	0.005	0.000	0.000	0.000	0.000	3.056
1970	8.559	0.010	0.000	0.001	0.000	0.029	8.599
1971	3.939	0.004	0.000	0.000	0.000	0.000	3.943
1972	0.697	0.039	0.001	0.000	0.000	0.000	0.737
1973	0.495	0.054	0.000	0.000	0.000	0.007	0.557
1974	8.494	0.055	0.000	0.000	0.020	0.001	8.571
1975	12.425	0.024	0.000	0.000	0.000	0.000	12.449
1976	7.424	0.000	0.001	0.002	0.000	0.000	7.427

QUANTITY--MILLION BF

YEAR	CANADA	OTH N AM	EUROPE	ASIA	AFRICA	OTHERS	TOTAL
1967	33.856	0.000	0.000	0.000	0.000	0.004	33.860
1968	39.397	0.006	0.000	0.000	0.000	0.000	39.403
1969	41.666	0.022	0.000	0.000	0.000	0.000	41.688
1970	106.391	0.028	0.000	0.001	0.000	0.048	106.468
1971	55.399	0.233	0.000	0.000	0.000	0.000	55.632
1972	8.902	1.577	0.003	0.000	0.000	0.000	10.481
1973	6.074	1.793	0.000	0.000	0.000	0.078	7.945
1974	42.953	0.967	0.000	0.000	0.617	0.003	44.541
1975	67.900	0.332	0.000	0.000	0.000	0.000	68.232
1976	67.355	0.000	0.008	0.007	0.000	0.000	67.370

IMPORTS
SOFTWOOD LOGS (2422)
UNITED STATES

TRADE STRUCTURE-- % OF TOTAL TRADE (VALUE BASIS)

YEAR	CANADA	OTH N AM	EUROPE	ASIA	AFRICA	OTHERS	TOTAL
1967	100.	0.	0.	0.	0.	0.	100.
1968	100.	0.	0.	0.	0.	0.	100.
1969	100.	0.	0.	0.	0.	0.	100.
1970	100.	0.	0.	0.	0.	0.	100.
1971	100.	0.	0.	0.	0.	0.	100.
1972	95.	5.	0.	0.	0.	0.	100.
1973	89.	10.	0.	0.	0.	1.	100.
1974	99.	1.	0.	0.	0.	0.	100.
1975	100.	0.	0.	0.	0.	0.	100.
1976	100.	0.	0.	0.	0.	0.	100.

TRADE TRENDS-- % OF 1967 TRADE (VALUE BASIS)

YEAR	CANADA	OTH N AM	EUROPE	ASIA	AFRICA	OTHERS	TOTAL
1967	100.	******	******	******	******	100.	100.
1968	82.	******	******	******	******	0.	82.
1969	181.	******	******	******	******	0.	181.
1970	507.	******	******	******	******	2056.	509.
1971	233.	******	******	******	******	0.	233.
1972	41.	******	******	******	******	0.	44.
1973	29.	******	******	******	******	511.	33.
1974	503.	******	******	******	******	92.	507.
1975	735.	******	******	******	******	0.	736.
1976	439.	******	******	******	******	0.	439.

IMPORTS
SOFTWOOD LOGS (2422)

VALUE--MILLION CURRENT DOLLARS

YEAR	N ATLANTC	S ATLANTC	GULF	S PACIFIC	PACIFC NW	GRT LAKES	ALASKA	N CENTRAL	S CENTRAL	OTHERS	TOTAL
1967	0.025	0.001	0.000	0.000	1.635	0.001	0.000	0.028	0.000	0.000	1.691
1968	0.003	0.000	0.001	0.000	1.325	0.000	0.000	0.054	0.000	0.000	1.384
1969	0.008	0.000	0.000	0.000	2.981	0.001	0.008	0.058	0.000	0.000	3.056
1970	0.013	0.000	0.030	0.000	8.212	0.013	0.139	0.191	0.000	0.000	8.599
1971	0.019	0.004	0.003	0.000	3.904	0.005	0.003	0.006	0.000	0.000	3.943
1972	0.005	0.037	0.000	0.002	0.630	0.005	0.000	0.009	0.000	0.000	0.737
1973	0.203	0.050	0.003	0.008	0.262	0.006	0.021	0.003	0.000	0.000	0.557
1974	0.626	0.077	0.000	0.000	7.865	0.002	0.002	0.000	0.000	0.000	8.571
1975	0.915	0.021	0.002	0.001	11.494	0.004	0.009	0.003	0.001	0.000	12.449
1976	1.896	0.003	0.004	0.000	5.447	0.000	0.075	0.000	0.000	0.002	7.427

QUANTITY--MILLION BF

YEAR	N ATLANTC	S ATLANTC	GULF	S PACIFIC	PACIFC NW	GRT LAKES	ALASKA	N CENTRAL	S CENTRAL	OTHERS	TOTAL
1967	0.697	0.004	0.000	0.000	32.557	0.186	0.000	0.406	0.000	0.000	33.850
1968	0.032	0.000	0.006	0.000	38.660	0.000	0.000	0.705	0.000	0.000	39.403
1969	0.058	0.000	0.005	0.000	40.445	0.004	0.029	1.147	0.000	0.000	41.688
1970	0.063	0.000	0.053	0.000	99.452	0.135	1.987	4.758	0.000	0.000	106.458
1971	0.367	0.233	0.100	0.000	54.689	0.044	0.029	0.170	0.000	0.000	55.632
1972	0.079	1.558	0.000	0.019	8.451	0.169	0.000	0.206	0.000	0.000	10.481
1973	3.658	1.779	0.010	0.082	2.102	0.044	0.157	0.113	0.000	0.000	7.945
1974	11.306	1.588	0.000	0.000	31.625	0.012	0.010	0.000	0.000	0.000	44.541
1975	12.322	0.315	0.002	0.012	55.494	0.019	0.053	0.012	0.003	0.000	68.232
1976	22.603	0.007	0.017	0.000	44.437	0.000	0.302	0.000	0.000	0.005	67.370

IMPORTS
SOFTWOOD LOGS (2422)

TRADE STRUCTURE-- % OF TOTAL TRADE (VALUE BASIS)

YEAR	N ATLANTC	S ATLANTC	GULF	S PACIFIC	PACIFC NW	GRT LAKES	ALASKA	N CENTRAL	S CENTRAL	OTHERS	TOTAL
1967	1.	0.	0.	0.	97.	0.	0.	2.	0.	0.	100.
1968	0.	0.	0.	0.	96.	0.	0.	4.	0.	0.	100.
1969	0.	0.	0.	0.	98.	0.	0.	2.	0.	0.	100.
1970	0.	0.	0.	0.	96.	0.	2.	2.	0.	0.	100.
1971	0.	0.	0.	0.	99.	0.	0.	0.	0.	0.	100.
1972	1.	5.	0.	0.	92.	1.	0.	1.	0.	0.	100.
1973	36.	9.	1.	1.	47.	1.	4.	1.	0.	0.	100.
1974	7.	1.	0.	0.	92.	0.	0.	0.	0.	0.	100.
1975	7.	0.	0.	0.	92.	0.	0.	0.	0.	0.	100.
1976	26.	0.	0.	0.	73.	0.	1.	0.	0.	0.	100.

TRADE TRENDS-- % OF 1967 TRADE (VALUE BASIS)

YEAR	N ATLANTC	S ATLANTC	GULF	S PACIFIC	PACIFC NW	GRT LAKES	ALASKA	N CENTRAL	S CENTRAL	OTHERS	TOTAL
1967	100.	100.	******	******	100.	100.	******	100.	******	******	100.
1968	13.	0.	******	******	81.	0.	******	192.	******	******	82.
1969	32.	0.	******	******	192.	134.	******	206.	******	******	191.
1970	51.	0.	******	******	502.	1963.	******	675.	******	******	509.
1971	77.	246.	******	******	239.	758.	******	20.	******	******	233.
1972	18.	2589.	******	******	42.	747.	******	31.	******	******	44.
1973	804.	3534.	******	******	16.	953.	******	11.	******	******	33.
1974	2479.	5406.	******	******	481.	258.	******	0.	******	******	507.
1975	3626.	1475.	******	******	703.	549.	******	10.	******	******	735.
1976	7513.	205.	******	******	333.	0.	******	0.	******	******	439.

IMPORTS
HARDWOOD LOGS (2423)
UNITED STATES

VALUE--MILLION CURRENT DOLLARS

YEAR	CANADA	PHIL REP	OTH ASIA	IVRY CST	OTH AFRI	OTHERS	TOTAL
1967	1.214	0.309	0.099	0.537	0.588	2.855	5.602
1968	1.259	0.188	0.088	0.400	0.414	3.687	6.036
1969	1.465	0.278	0.125	0.536	0.606	1.829	4.839
1970	1.357	0.065	0.236	0.627	0.459	1.778	4.542
1971	1.536	0.058	0.094	0.200	0.229	1.370	3.468
1972	1.516	0.079	0.590	0.199	0.336	1.178	3.898
1973	1.452	0.244	0.492	0.423	0.446	0.585	3.643
1974	1.757	0.123	1.166	0.592	0.571	0.350	4.558
1975	2.326	0.047	0.272	0.307	0.210	0.287	3.449
1976	1.552	0.105	0.134	0.336	0.303	0.361	2.790

QUANTITY--MILLION BF

YEAR	CANADA	PHIL REP	OTH ASIA	IVRY CST	OTH AFRI	OTHERS	TOTAL
1967	6.788	2.905	0.257	2.929	2.751	17.427	33.057
1968	6.786	1.254	0.061	2.041	1.621	25.875	37.638
1969	7.809	2.012	0.074	2.613	2.203	14.636	29.346
1970	8.525	0.483	0.160	3.347	1.832	15.486	29.953
1971	9.225	0.117	0.238	1.056	0.906	12.283	23.824
1972	8.378	0.435	0.427	0.798	1.215	9.993	21.246
1973	11.719	1.795	1.335	1.045	1.666	2.741	20.301
1974	14.065	0.534	4.383	1.752	0.881	2.671	24.288
1975	12.379	0.185	0.095	0.558	0.459	1.060	14.736
1976	10.687	0.388	0.078	0.507	0.570	0.296	12.527

IMPORTS
HARDWOOD LOGS (2423)
UNITED STATES

TRADE STRUCTURE-- % OF TOTAL TRADE (VALUE BASIS)

YEAR	CANADA	PHIL REP	OTH ASIA	IVRY CST	OTH AFRI	OTHERS	TOTAL
1967	22.	6.	2.	10.	11.	51.	100.
1968	21.	3.	1.	7.	7.	61.	100.
1969	30.	6.	3.	11.	13.	38.	100.
1970	30.	2.	5.	14.	10.	39.	100.
1971	44.	2.	3.	6.	7.	39.	100.
1972	39.	2.	15.	5.	9.	30.	100.
1973	40.	7.	14.	12.	12.	16.	100.
1974	39.	3.	26.	13.	13.	8.	100.
1975	67.	1.	8.	9.	6.	8.	100.
1976	56.	4.	5.	12.	11.	13.	100.

TRADE TRENDS-- % OF 1967 TRADE (VALUE BASIS)

YEAR	CANADA	PHIL REP	OTH ASIA	IVRY CST	OTH AFRI	OTHERS	TOTAL
1967	100.	100.	100.	100.	100.	100.	100.
1968	104.	61.	89.	75.	70.	129.	106.
1969	121.	90.	126.	100.	103.	64.	86.
1970	112.	27.	240.	117.	78.	62.	81.
1971	127.	19.	96.	37.	39.	48.	62.
1972	125.	26.	599.	37.	57.	41.	70.
1973	120.	79.	499.	79.	76.	20.	65.
1974	145.	40.	1182.	110.	97.	12.	81.
1975	192.	15.	276.	57.	36.	10.	62.
1976	128.	34.	135.	63.	51.	13.	50.

IMPORTS
HARDWOOD LOGS (2423)

VALUE--MILLION CURRENT DOLLARS

YEAR	N ATLANTC	S ATLANTC	GULF	S PACIFIC	PACIFC NW	GRT LAKES	ALASKA	N CENTRAL	S CENTRAL	OTHERS	TOTAL
1967	1.933	0.234	2.037	0.008	0.306	1.060	0.000	0.001	0.008	0.016	5.602
1968	1.684	0.226	2.804	0.020	0.170	1.099	0.000	0.012	0.000	0.021	6.036
1969	1.523	0.020	1.709	0.006	0.233	1.316	0.000	0.000	0.000	0.031	4.839
1970	1.831	0.014	1.590	0.008	0.107	0.974	0.000	0.000	0.000	0.018	4.542
1971	1.301	0.043	1.155	0.026	0.036	0.903	0.000	0.000	0.001	0.022	3.488
1972	1.334	0.066	1.404	0.013	0.054	0.972	0.000	0.000	0.052	0.004	3.898
1973	1.705	0.252	0.320	0.178	0.470	0.715	0.000	0.000	0.002	0.000	3.643
1974	2.455	0.128	0.408	0.379	0.550	0.551	0.000	0.000	0.032	0.056	4.558
1975	1.636	0.223	0.049	0.028	1.239	0.261	0.000	0.000	0.000	0.013	3.449
1976	2.184	0.046	0.050	0.016	0.276	0.218	0.000	0.000	0.000	0.000	2.790

QUANTITY--MILLION BF

YEAR	N ATLANTC	S ATLANTC	GULF	S PACIFIC	PACIFC NW	GRT LAKES	ALASKA	N CENTRAL	S CENTRAL	OTHERS	TOTAL
1967	8.977	1.248	14.386	0.106	2.832	5.447	0.000	0.004	0.007	0.049	33.057
1968	6.361	0.870	23.548	0.059	1.294	5.333	0.000	0.063	0.000	0.110	37.638
1969	6.589	0.121	14.205	0.036	1.853	6.298	0.000	0.000	0.000	0.244	29.346
1970	9.192	0.067	14.009	0.031	1.821	4.711	0.000	0.000	0.000	0.021	29.853
1971	7.529	0.068	11.806	0.046	0.359	3.969	0.000	0.003	0.010	0.034	23.824
1972	6.311	0.251	9.868	0.015	0.399	4.342	0.000	0.000	0.051	0.008	21.246
1973	8.468	0.796	2.270	0.958	4.131	3.675	0.000	0.000	0.002	0.000	20.301
1974	12.520	0.293	2.534	1.820	3.911	2.969	0.000	0.000	0.122	0.118	24.288
1975	8.441	0.314	0.080	0.100	4.663	1.121	0.000	0.008	0.000	0.009	14.736
1976	9.535	0.049	0.088	0.017	1.778	1.059	0.000	0.000	0.000	0.000	12.527

IMPORTS
HARDWOOD LOGS (2423)

TRADE STRUCTURE-- % OF TOTAL TRADE (VALUE BASIS)

YEAR	N ATLANTC	S ATLANTC	GULF	S PACIFIC	PACIFC NW	GRT LAKES	ALASKA	N CENTRAL	S CENTRAL	OTHERS	TOTAL
1967	34.	4.	36.	0.	5.	19.	0.	0.	0.	0.	100.
1968	28.	4.	46.	0.	3.	18.	0.	0.	0.	0.	100.
1969	31.	0.	35.	0.	5.	27.	0.	0.	0.	1.	100.
1970	40.	0.	35.	0.	2.	21.	0.	0.	0.	0.	100.
1971	37.	1.	33.	1.	1.	26.	0.	0.	0.	1.	100.
1972	34.	2.	36.	0.	1.	25.	0.	0.	1.	0.	100.
1973	47.	7.	9.	5.	13.	20.	0.	0.	0.	0.	100.
1974	54.	3.	9.	8.	12.	12.	0.	0.	1.	1.	100.
1975	47.	6.	1.	1.	36.	8.	0.	0.	0.	0.	100.
1976	78.	2.	2.	1.	10.	8.	0.	0.	0.	0.	100.

TRADE TRENDS-- % OF 1967 TRADE (VALUE BASIS)

YEAR	N ATLANTC	S ATLANTC	GULF	S PACIFIC	PACIFC NW	GRT LAKES	ALASKA	N CENTRAL	S CENTRAL	OTHERS	TOTAL
1967	100.	100.	100.	100.	100.	100.	******	100.	100.	100.	100.
1968	87.	96.	138.	265.	56.	104.	******	1753.	0.	136.	108.
1969	79.	9.	84.	75.	76.	124.	******	0.	0.	200.	86.
1970	95.	6.	78.	101.	35.	92.	******	0.	0.	112.	81.
1971	67.	19.	57.	339.	12.	85.	******	41.	14.	137.	62.
1972	69.	28.	69.	166.	18.	92.	******	0.	658.	28.	70.
1973	88.	108.	16.	2352.	154.	67.	******	0.	25.	0.	65.
1974	127.	55.	20.	5006.	180.	52.	******	0.	413.	356.	81.
1975	85.	95.	2.	369.	405.	25.	******	43.	0.	86.	62.
1976	113.	20.	2.	212.	90.	21.	******	0.	0.	0.	50.

IMPORTS
SOFTWOOD LUMBER (2432)
UNITED STATES

VALUE--MILLION CURRENT DOLLARS

YEAR	CANADA	HONDURAS	BRAZIL	OTH S AM	EUROPE	OTHERS	TOTAL
1967	322.200	1.979	2.320	0.050	0.020	0.798	327.366
1968	488.048	1.918	3.110	0.383	0.026	0.842	494.327
1969	537.095	2.442	4.270	0.753	0.043	1.323	545.927
1970	431.158	2.242	3.069	0.266	0.069	1.732	438.538
1971	673.944	2.511	2.731	0.286	0.083	2.396	681.951
1972	1048.258	4.843	4.256	0.641	0.012	4.603	1062.613
1973	1405.457	7.279	6.988	0.296	1.582	8.667	1430.269
1974	906.034	3.343	2.439	0.821	2.306	5.906	920.850
1975	712.275	4.410	1.486	0.309	0.053	2.196	720.730
1976	1346.043	3.473	1.870	0.703	0.089	2.470	1354.648

QUANTITY--MILLION BF

YEAR	CANADA	HONDURAS	BRAZIL	OTH S AM	EUROPE	OTHERS	TOTAL
1967	8845.325	26.154	25.160	0.162	0.067	5.397	8902.264
1968	5749.188	23.105	32.615	4.143	0.054	5.306	5814.411
1969	5790.228	27.605	33.637	6.451	0.121	8.363	5866.404
1970	5709.271	23.525	20.657	1.815	1.017	12.188	5768.473
1971	7210.888	27.271	17.765	2.081	1.077	17.212	7276.293
1972	8830.248	44.760	26.670	5.564	0.050	37.574	8944.865
1973	8629.354	61.698	29.678	1.136	9.558	52.094	8983.518
1974	6724.791	39.744	6.277	2.678	3.614	26.123	6802.228
1975	5666.313	35.397	4.468	0.191	0.639	10.953	5717.961
1976	7904.482	30.292	5.084	1.516	0.314	12.596	7954.285

IMPORTS
SOFTWOOD LUMBER (2432)
UNITED STATES

TRADE STRUCTURE-- % OF TOTAL TRADE (VALUE BASIS)

YEAR	CANADA	HONDURAS	BRAZIL	OTH S AM	EUROPE	OTHERS	TOTAL
1967	98.	1.	1.	0.	0.	0.	100.
1968	99.	0.	1.	0.	0.	0.	100.
1969	98.	0.	1.	0.	0.	0.	100.
1970	98.	1.	1.	0.	0.	0.	100.
1971	99.	0.	0.	0.	0.	0.	100.
1972	99.	0.	0.	0.	0.	0.	100.
1973	98.	1.	0.	0.	0.	1.	100.
1974	98.	0.	0.	0.	0.	1.	100.
1975	99.	1.	0.	0.	0.	0.	100.
1976	99.	0.	0.	0.	0.	0.	100.

TRADE TRENDS-- % OF 1967 TRADE (VALUE BASIS)

YEAR	CANADA	HONDURAS	BRAZIL	OTH S AM	EUROPE	OTHERS	TOTAL
1967	100.	100.	100.	100.	100.	100.	100.
1968	151.	97.	134.	767.	129.	106.	151.
1969	167.	123.	184.	1509.	212.	166.	167.
1970	134.	113.	132.	534.	342.	217.	134.
1971	209.	127.	118.	572.	408.	300.	208.
1972	325.	245.	183.	1284.	58.	577.	325.
1973	436.	368.	301.	593.	7794.	1086.	437.
1974	281.	169.	105.	1645.	11362.	740.	281.
1975	221.	223.	64.	619.	261.	275.	220.
1976	418.	175.	81.	1408.	439.	310.	414.

IMPORTS
SOFTWOOD LUMBER (2432)

VALUE--MILLION CURRENT DOLLARS

YEAR	N ATLANTC	S ATLANTC	GULF	S PACIFIC	PACIFC NW	GRT LAKES	ALASKA	N CENTRAL	S CENTRAL	OTHERS	TOTAL
1967	90.501	0.544	3.858	1.993	46.165	112.936	0.159	62.311	0.563	8.336	327.366
1968	135.648	1.399	7.025	3.091	70.738	175.801	0.218	89.986	0.540	9.880	494.327
1969	142.604	1.140	9.356	4.373	76.971	195.678	0.400	100.625	0.803	13.977	545.927
1970	104.285	1.402	6.344	3.502	58.833	167.633	0.365	83.735	0.780	11.660	438.538
1971	153.071	4.036	14.122	3.361	85.121	265.286	0.448	139.010	0.761	16.735	681.951
1972	207.275	5.667	26.697	5.599	163.419	410.993	0.361	217.665	1.993	22.944	1062.613
1973	268.925	9.538	36.160	10.626	216.426	545.049	0.205	253.183	2.752	27.406	1370.269
1974	161.620	2.606	18.765	8.180	208.788	347.153	0.321	153.510	0.645	19.262	920.850
1975	114.096	1.300	9.253	3.593	185.598	284.382	0.699	108.531	0.290	12.987	720.730
1976	195.588	1.159	12.983	15.081	338.810	549.729	0.837	227.917	0.638	11.905	1354.648

QUANTITY--MILLION BF

YEAR	N ATLANTC	S ATLANTC	GULF	S PACIFIC	PACIFC NW	GRT LAKES	ALASKA	N CENTRAL	S CENTRAL	OTHERS	TOTAL
1967	5553.099	12.438	54.357	28.044	612.479	1618.990	1.929	900.643	3.677	116.608	8902.264
1968	1745.295	19.636	85.376	32.272	779.545	2002.985	2.166	1017.941	3.967	125.228	5814.411
1969	1616.880	12.233	101.501	45.967	789.963	2075.605	3.006	1076.151	4.690	140.407	5866.404
1970	1495.971	28.484	94.939	42.008	709.752	2168.596	3.637	1091.773	4.073	129.240	5768.473
1971	1826.406	64.094	184.640	37.639	834.781	2721.653	3.833	1416.928	5.453	180.866	7276.293
1972	1973.475	64.943	239.496	55.114	1295.775	3312.279	3.015	1768.518	18.481	213.770	8944.865
1973	1830.867	73.163	227.994	80.525	1385.149	3569.180	1.207	1619.603	17.480	178.349	8983.518
1974	1165.316	26.443	120.694	53.219	1538.363	2630.541	2.060	1119.286	3.928	142.378	6802.228
1975	893.073	9.459	69.647	22.492	1498.637	2295.542	4.737	815.528	1.495	107.350	5717.961
1976	1261.118	4.665	79.155	82.415	1963.271	3234.660	4.552	1221.315	2.455	100.679	7954.285

IMPORTS
SOFTWOOD LUMBER (2432)

TRADE STRUCTURE-- % OF TOTAL TRADE (VALUE BASIS)

YEAR	N ATLANTC	S ATLANTC	GULF	S PACIFIC	PACIFC NW	GRT LAKES	ALASKA	N CENTRAL	S CENTRAL	OTHERS	TOTAL
1967	28.	0.	1.	1.	14.	34.	0.	19.	0.	3.	100.
1968	27.	0.	1.	1.	14.	36.	0.	18.	0.	2.	100.
1969	26.	0.	2.	1.	14.	36.	0.	18.	0.	3.	100.
1970	24.	0.	1.	1.	13.	38.	0.	19.	0.	3.	100.
1971	22.	1.	2.	0.	12.	39.	0.	20.	0.	2.	100.
1972	20.	1.	3.	1.	15.	39.	0.	20.	0.	2.	100.
1973	20.	1.	3.	1.	16.	40.	0.	18.	0.	2.	100.
1974	18.	0.	2.	1.	23.	38.	0.	17.	0.	2.	100.
1975	16.	0.	1.	0.	26.	39.	0.	15.	0.	2.	100.
1976	14.	0.	1.	1.	25.	41.	0.	17.	0.	1.	100.

TRADE TRENDS-- % OF 1967 TRADE (VALUE BASIS)

YEAR	N ATLANTC	S ATLANTC	GULF	S PACIFIC	PACIFC NW	GRT LAKES	ALASKA	N CENTRAL	S CENTRAL	OTHERS	TOTAL
1967	100.	100.	100.	100.	100.	100.	100.	100.	100.	100.	100.
1968	150.	257.	182.	155.	153.	156.	137.	144.	96.	119.	151.
1969	158.	209.	243.	219.	167.	173.	252.	161.	143.	168.	167.
1970	115.	257.	164.	176.	127.	148.	230.	134.	138.	140.	134.
1971	169.	741.	366.	169.	184.	235.	282.	223.	135.	201.	208.
1972	229.	1041.	692.	281.	354.	364.	227.	349.	354.	275.	325.
1973	297.	1752.	937.	533.	469.	483.	129.	406.	489.	329.	419.
1974	179.	479.	486.	411.	452.	307.	202.	246.	115.	231.	281.
1975	126.	239.	240.	180.	402.	252.	440.	174.	51.	156.	220.
1976	216.	213.	337.	757.	734.	487.	527.	366.	113.	143.	414.

IMPORTS
HARDWOOD LUMBER (2433)
UNITED STATES

VALUE--MILLION CURRENT DOLLARS

YEAR	CANADA	BRAZIL	OTH S AM	MALAYSIA	PHIL REP	OTHERS	TOTAL
1967	31.635	2.891	5.454	3.525	2.827	15.062	61.394
1968	31.355	3.645	4.602	5.567	3.470	14.393	63.031
1969	35.695	5.386	6.552	9.546	4.734	15.287	77.200
1970	29.118	6.558	4.193	5.911	2.673	12.774	61.227
1971	30.928	6.310	4.487	7.801	2.526	12.611	64.662
1972	38.208	9.676	5.848	10.318	4.655	19.300	88.006
1973	43.944	18.913	11.420	10.764	8.386	27.081	120.508
1974	36.375	17.703	12.038	19.879	10.120	34.906	131.020
1975	20.118	10.811	10.192	5.922	3.744	16.881	67.669
1976	28.002	14.786	10.035	11.644	8.418	23.599	96.484

QUANTITY--MILLION BF

YEAR	CANADA	BRAZIL	OTH S AM	MALAYSIA	PHIL REP	OTHERS	TOTAL
1967	153.256	19.031	28.267	28.010	17.866	68.906	315.335
1968	149.161	23.216	22.694	36.667	21.120	70.524	323.382
1969	178.977	30.475	39.849	66.323	31.721	68.305	415.650
1970	145.026	44.837	22.769	35.433	14.821	51.556	314.443
1971	142.409	46.886	23.447	45.001	13.951	53.690	325.385
1972	151.343	68.298	30.271	57.916	25.925	77.016	410.770
1973	155.200	123.666	54.060	46.970	41.774	70.740	492.450
1974	115.627	77.744	42.705	73.064	35.653	69.885	414.677
1975	62.232	43.881	20.304	20.810	11.357	33.788	192.373
1976	82.986	62.316	18.325	31.624	24.271	40.080	259.602

IMPORTS
HARDWOOD LUMBER (2433)
UNITED STATES

TRADE STRUCTURE-- % OF TOTAL TRADE (VALUE BASIS)

YEAR	CANADA	BRAZIL	OTH S AM	MALAYSIA	PHIL REP	OTHERS	TOTAL
1967	52.	5.	9.	6.	5.	25.	100.
1968	50.	6.	7.	9.	6.	23.	100.
1969	46.	7.	8.	12.	6.	20.	100.
1970	48.	11.	7.	10.	4.	21.	100.
1971	48.	10.	7.	12.	4.	20.	100.
1972	43.	11.	7.	12.	5.	22.	100.
1973	36.	16.	9.	9.	7.	22.	100.
1974	28.	14.	9.	15.	8.	27.	100.
1975	30.	16.	15.	9.	6.	25.	100.
1976	29.	15.	10.	12.	9.	24.	100.

TRADE TRENDS-- % OF 1967 TRADE (VALUE BASIS)

YEAR	CANADA	BRAZIL	OTH S AM	MALAYSIA	PHIL REP	OTHERS	TOTAL
1967	100.	100.	100.	100.	100.	100.	100.
1968	99.	126.	84.	158.	123.	96.	103.
1969	113.	186.	120.	271.	167.	101.	126.
1970	92.	227.	77.	168.	95.	85.	100.
1971	98.	218.	82.	221.	89.	84.	105.
1972	121.	335.	107.	293.	165.	128.	143.
1973	139.	654.	209.	305.	297.	180.	196.
1974	115.	612.	221.	564.	358.	232.	213.
1975	64.	374.	187.	168.	132.	112.	110.
1976	89.	511.	184.	330.	298.	157.	157.

IMPORTS
HARDWOOD LUMBER (2433)

VALUE--MILLION CURRENT DOLLARS

YEAR	N ATLANTC	S ATLANTC	GULF	S PACIFIC	PACIFC NW	GRT LAKES	ALASKA	N CENTRAL	S CENTRAL	OTHERS	TOTAL
1967	19.561	2.826	8.468	3.361	3.594	21.664	0.002	0.114	0.011	1.793	61.394
1968	19.254	2.570	9.667	4.047	4.594	20.809	0.001	0.120	0.001	1.969	63.031
1969	20.942	4.362	11.616	6.435	9.358	23.531	0.000	0.181	0.000	1.774	77.200
1970	20.418	3.772	9.924	4.033	3.228	17.767	0.014	0.167	0.000	2.004	61.227
1971	19.332	3.598	11.639	3.932	4.230	19.792	0.015	0.254	0.000	1.872	64.662
1972	22.590	6.369	16.154	8.281	7.235	24.601	0.000	1.023	0.000	1.752	88.006
1973	29.652	12.471	27.029	10.542	8.244	26.979	0.000	1.795	0.010	1.785	120.508
1974	32.615	9.864	30.546	15.813	15.279	23.646	0.003	0.607	0.002	2.647	131.020
1975	18.939	5.037	17.556	5.561	4.065	14.597	0.001	0.121	0.000	1.792	67.669
1976	20.575	9.996	24.743	6.502	11.953	19.624	0.023	0.197	0.001	0.969	96.484

QUANTITY--MILLION BF

YEAR	N ATLANTC	S ATLANTC	GULF	S PACIFIC	PACIFC NW	GRT LAKES	ALASKA	N CENTRAL	S CENTRAL	OTHERS	TOTAL
1967	96.417	23.907	45.018	17.103	24.746	98.701	0.003	1.282	0.085	8.074	315.335
1968	91.335	22.375	55.673	21.053	29.973	93.299	0.001	1.132	0.009	8.533	323.382
1969	100.539	33.009	67.829	36.482	56.609	111.437	0.000	1.460	0.000	8.283	415.650
1970	96.114	30.086	55.179	19.757	20.081	83.211	0.054	1.417	0.000	8.545	314.443
1971	87.106	29.350	64.093	20.774	26.905	85.650	0.092	2.530	0.000	8.886	325.385
1972	85.902	44.490	94.106	39.576	39.644	89.907	0.000	10.015	0.000	7.187	410.770
1973	86.594	76.678	136.366	37.293	37.612	95.200	0.000	16.677	0.009	6.020	492.450
1974	75.701	41.979	115.242	37.775	56.862	72.426	0.018	7.053	0.012	7.609	414.677
1975	40.838	14.181	51.607	11.466	13.881	45.122	0.001	0.642	0.000	4.636	192.373
1976	45.548	25.134	76.202	16.330	33.996	58.328	0.063	1.106	0.005	2.883	259.602

IMPORTS
HARDWOOD LUMBER (2433)

TRADE STRUCTURE-- % OF TOTAL TRADE (VALUE BASIS)

YEAR	N ATLANTC	S ATLANTC	GULF	S PACIFIC	PACIFC NW	GRT LAKES	ALASKA	N CENTRAL	S CENTRAL	OTHERS	TOTAL
1967	32.	5.	14.	5.	6.	35.	0.	0.	0.	3.	100.
1968	31.	4.	15.	6.	7.	33.	0.	0.	0.	3.	100.
1969	27.	6.	15.	8.	11.	30.	0.	0.	0.	2.	100.
1970	33.	6.	16.	7.	5.	29.	0.	0.	0.	3.	100.
1971	30.	6.	18.	6.	7.	31.	0.	0.	0.	3.	100.
1972	26.	7.	18.	9.	8.	28.	0.	1.	0.	2.	100.
1973	25.	10.	22.	9.	7.	24.	0.	1.	0.	1.	100.
1974	25.	8.	23.	12.	12.	18.	0.	0.	0.	2.	100.
1975	28.	7.	26.	8.	6.	22.	0.	0.	0.	3.	100.
1976	21.	10.	26.	9.	12.	20.	0.	0.	0.	1.	100.

TRADE TRENDS-- % OF 1967 TRADE (VALUE BASIS)

YEAR	N ATLANTC	S ATLANTC	GULF	S PACIFIC	PACIFC NW	GRT LAKES	ALASKA	N CENTRAL	S CENTRAL	OTHERS	TOTAL
1967	100.	100.	100.	100.	100.	100.	100.	100.	100.	100.	100.
1968	98.	91.	114.	120.	128.	96.	49.	105.	8.	110.	103.
1969	127.	154.	137.	192.	233.	109.	0.	159.	0.	99.	126.
1970	104.	133.	116.	129.	90.	92.	824.	146.	0.	112.	100.
1971	99.	127.	137.	117.	113.	91.	881.	222.	0.	104.	105.
1972	115.	225.	191.	246.	201.	114.	0.	894.	0.	98.	143.
1973	152.	441.	319.	314.	229.	134.	0.	1569.	95.	100.	196.
1974	167.	349.	361.	471.	425.	109.	168.	530.	22.	148.	213.
1975	97.	178.	207.	166.	113.	67.	85.	106.	0.	100.	110.
1976	105.	354.	292.	253.	339.	91.	1497.	172.	11.	54.	157.

IMPORTS
WOOD VENEER (6311)
UNITED STATES

VALUE--MILLION CURRENT DOLLARS

YEAR	CANADA	BRAZIL	PHIL REP	OTH ASIA	AFRICA	OTHERS	TOTAL
1967	26.383	1.364	7.424	2.042	3.188	2.305	42.706
1968	27.210	1.725	11.013	3.370	3.758	3.569	50.645
1969	23.504	2.025	13.030	3.047	2.514	2.864	46.983
1970	21.127	3.080	6.531	2.286	2.718	3.204	38.945
1971	26.027	2.858	7.876	3.857	2.636	3.164	46.419
1972	37.227	3.919	12.957	6.901	2.742	4.718	68.464
1973	43.472	4.015	19.686	5.740	3.698	6.713	83.325
1974	39.512	3.465	19.882	6.704	2.759	7.109	79.422
1975	32.300	3.085	5.930	2.388	1.884	5.353	50.942
1976	45.307	5.564	11.604	4.541	0.692	6.910	74.619

QUANTITY--MIL SQ FT

YEAR	CANADA	BRAZIL	PHIL REP	OTH ASIA	AFRICA	OTHERS	TOTAL
1967	968.606	18.504	89.102	22.264	28.002	10.876	1137.354
1968	983.733	28.009	134.439	41.969	33.106	18.387	1239.642
1969	884.321	29.975	140.421	34.101	19.884	15.993	1124.695
1970	930.821	26.252	87.611	27.996	24.910	15.448	1113.038
1971	1100.105	29.217	109.179	52.033	20.171	26.327	1337.084
1972	1402.775	38.940	161.237	78.483	25.204	30.035	1736.675
1973	1310.376	37.879	195.324	40.657	24.518	34.333	1643.087
1974	1004.076	30.085	126.831	34.662	11.804	29.253	1235.711
1975	918.307	27.793	56.719	6.673	11.496	10.751	1031.740
1976	1184.528	41.559	85.895	11.801	2.606	16.548	1342.938

IMPORTS
WOOD VENEER (6311)
UNITED STATES

TRADE STRUCTURE-- % OF TOTAL TRADE (VALUE BASIS)

YEAR	CANADA	BRAZIL	PHIL REP	OTH ASIA	AFRICA	OTHERS	TOTAL
1967	62.	3.	17.	5.	7.	5.	100.
1968	54.	3.	22.	7.	7.	7.	100.
1969	50.	4.	28.	6.	5.	6.	100.
1970	54.	8.	17.	6.	7.	8.	100.
1971	56.	6.	17.	8.	6.	7.	100.
1972	54.	6.	19.	10.	4.	7.	100.
1973	52.	5.	24.	7.	4.	8.	100.
1974	50.	4.	25.	8.	3.	9.	100.
1975	63.	6.	12.	5.	4.	11.	100.
1976	61.	7.	16.	6.	1.	9.	100.

TRADE TRENDS-- % OF 1967 TRADE (VALUE BASIS)

YEAR	CANADA	BRAZIL	PHIL REP	OTH ASIA	AFRICA	OTHERS	TOTAL
1967	100.	100.	100.	100.	100.	100.	100.
1968	103.	126.	148.	165.	118.	155.	119.
1969	89.	148.	176.	149.	79.	124.	110.
1970	80.	226.	88.	112.	85.	139.	91.
1971	99.	210.	106.	189.	83.	137.	109.
1972	141.	287.	175.	338.	86.	205.	160.
1973	165.	294.	265.	281.	116.	291.	195.
1974	150.	254.	268.	328.	86.	308.	186.
1975	122.	226.	80.	117.	59.	232.	119.
1976	172.	408.	156.	222.	22.	300.	175.

IMPORTS
WOOD VENEER (6311)

VALUE--MILLION CURRENT DOLLARS

YEAR	N ATLANTC	S ATLANTC	GULF	S PACIFIC	PACIFC NW	GRT LAKES	ALASKA	N CENTRAL	S CENTRAL	OTHERS	TOTAL
1967	3.899	4.948	1.939	0.451	5.861	25.539	0.000	0.062	0.000	0.007	42.706
1968	6.242	8.660	2.001	0.776	6.863	26.084	0.000	0.000	0.000	0.018	50.645
1969	6.405	7.625	1.966	0.708	10.301	19.980	0.000	0.091	0.000	0.007	46.983
1970	5.035	7.033	1.014	0.323	6.790	18.707	0.000	0.010	0.016	0.016	38.945
1971	6.350	7.685	2.003	0.286	8.025	21.658	0.000	0.381	0.007	0.022	46.419
1972	7.432	13.182	4.201	1.304	13.653	28.679	0.000	0.006	0.000	0.008	68.464
1973	10.747	17.534	6.375	1.340	13.835	33.491	0.000	0.000	0.002	0.002	83.325
1974	11.227	18.998	3.723	1.636	12.280	31.454	0.007	0.098	0.000	0.000	79.422
1975	7.084	7.128	0.561	1.426	9.030	25.712	0.000	0.000	0.000	0.000	50.942
1976	10.791	13.913	1.442	2.726	12.840	32.906	0.000	0.000	0.000	0.002	74.619

QUANTITY--MIL SQ FT

YEAR	N ATLANTC	S ATLANTC	GULF	S PACIFIC	PACIFC NW	GRT LAKES	ALASKA	N CENTRAL	S CENTRAL	OTHERS	TOTAL
1967	71.840	57.947	19.661	4.241	247.933	734.260	0.000	1.470	0.000	0.003	1137.354
1968	115.921	102.717	20.268	6.364	224.608	769.848	0.000	0.000	0.000	0.016	1239.642
1969	105.394	86.333	16.047	6.381	265.916	640.852	0.000	3.770	0.000	0.003	1124.695
1970	90.834	78.594	11.048	2.440	319.660	610.004	0.000	0.263	0.174	0.021	1113.038
1971	147.582	89.737	29.284	2.351	311.128	727.117	0.000	29.656	0.061	0.168	1337.084
1972	151.114	150.954	52.538	10.958	507.789	862.846	0.000	0.465	0.000	0.011	1736.675
1973	148.526	158.858	60.449	5.354	441.418	828.468	0.000	0.000	0.011	0.003	1643.087
1974	117.033	125.264	25.026	5.155	307.965	651.279	0.492	4.497	0.000	0.000	1236.711
1975	81.207	57.961	4.308	3.718	354.493	530.051	0.000	0.000	0.000	0.001	1031.740
1976	145.815	99.829	6.612	6.224	368.420	716.029	0.000	0.000	0.000	0.008	1342.938

IMPORTS
WOOD VENEER (6311)

TRADE STRUCTURE-- % OF TOTAL TRADE (VALUE BASIS)

YEAR	N ATLANTC	S ATLANTC	GULF	S PACIFIC	PACIFC NW	GRT LAKES	ALASKA	N CENTRAL	S CENTRAL	OTHERS	TOTAL
1967	9.	12.	5.	1.	14.	60.	0.	0.	0.	0.	100.
1968	12.	17.	4.	2.	14.	52.	0.	0.	0.	0.	100.
1969	14.	16.	4.	2.	22.	43.	0.	0.	0.	0.	100.
1970	13.	18.	3.	1.	17.	48.	0.	0.	0.	0.	100.
1971	14.	17.	4.	1.	17.	47.	0.	1.	0.	0.	100.
1972	11.	19.	6.	2.	20.	42.	0.	0.	0.	0.	100.
1973	13.	21.	8.	2.	17.	40.	0.	0.	0.	0.	100.
1974	14.	24.	5.	2.	15.	40.	0.	0.	0.	0.	100.
1975	14.	14.	1.	3.	18.	50.	0.	0.	0.	0.	100.
1976	14.	19.	2.	4.	17.	44.	0.	0.	0.	0.	100.

TRADE TRENDS-- % OF 1967 TRADE (VALUE BASIS)

YEAR	N ATLANTC	S ATLANTC	GULF	S PACIFIC	PACIFC NW	GRT LAKES	ALASKA	N CENTRAL	S CENTRAL	OTHERS	TOTAL
1967	100.	100.	100.	100.	100.	100.	******	100.	******	100.	100.
1968	160.	175.	103.	172.	117.	102.	******	0.	******	255.	119.
1969	164.	154.	95.	157.	176.	78.	******	146.	******	92.	110.
1970	129.	142.	52.	72.	116.	73.	******	16.	******	225.	91.
1971	163.	155.	103.	64.	137.	85.	******	612.	******	298.	109.
1972	191.	266.	217.	290.	233.	112.	******	9.	******	109.	160.
1973	276.	354.	329.	298.	236.	131.	******	0.	******	21.	195.
1974	288.	334.	132.	364.	210.	123.	******	157.	******	0.	186.
1975	182.	144.	29.	317.	154.	101.	******	0.	******	4.	119.
1976	277.	281.	74.	606.	219.	129.	******	0.	******	23.	175.

IMPORTS
PLYWOOD (6312)
UNITED STATES

VALUE--MILLION CURRENT DOLLARS

YEAR	CANADA	PHIL REP	KOR REP	JAPAN	OTH ASIA	OTHERS	TOTAL
1967	5.199	23.711	30.140	47.469	21.841	13.942	142.302
1968	6.452	29.858	53.438	72.065	38.335	17.535	217.683
1969	6.116	30.124	72.836	69.453	47.101	24.537	250.166
1970	2.950	23.953	74.249	50.114	40.672	17.041	208.979
1971	5.230	24.509	96.886	53.521	62.583	16.879	259.608
1972	8.028	27.951	126.966	62.542	89.771	21.635	336.893
1973	12.088	40.506	165.962	55.690	92.298	25.000	391.545
1974	8.304	21.977	118.205	48.608	74.306	19.039	290.440
1975	8.324	14.734	132.690	35.832	58.824	11.593	261.997
1976	9.505	29.375	188.435	57.566	82.811	14.809	382.500

QUANTITY--MIL SQ FT

YEAR	CANADA	PHIL REP	KOR REP	JAPAN	OTH ASIA	OTHERS	TOTAL
1967	48.416	244.185	286.477	295.672	233.066	113.060	1220.875
1968	61.417	267.017	438.891	370.880	341.821	101.840	1581.866
1969	51.163	257.702	594.194	366.134	429.269	173.719	1872.180
1970	25.268	237.896	633.678	285.952	363.047	116.874	1662.715
1971	46.334	241.636	801.312	290.782	546.673	110.745	2037.482
1972	580.257	272.831	1047.989	283.967	786.011	113.714	3084.769
1973	78.287	296.959	941.566	173.634	531.508	100.200	2122.155
1974	48.390	127.358	620.812	131.431	378.650	63.480	1370.121
1975	51.168	101.763	802.975	126.796	373.710	41.949	1498.361
1976	54.091	160.849	990.381	166.876	423.977	65.484	1881.658

IMPORTS
PLYWOOD (6312)
UNITED STATES

TRADE STRUCTURE-- % OF TOTAL TRADE (VALUE BASIS)

YEAR	CANADA	PHIL REP	KOR REP	JAPAN	OTH ASIA	OTHERS	TOTAL
1967	4.	17.	21.	33.	15.	10.	100.
1968	3.	14.	25.	33.	18.	8.	100.
1969	2.	12.	29.	28.	19.	10.	100.
1970	1.	11.	36.	24.	19.	8.	100.
1971	2.	9.	37.	21.	24.	7.	100.
1972	2.	8.	38.	19.	27.	6.	100.
1973	3.	10.	42.	14.	24.	6.	100.
1974	3.	8.	41.	17.	26.	7.	100.
1975	3.	6.	51.	14.	22.	4.	100.
1976	2.	8.	49.	15.	22.	4.	100.

TRADE TRENDS-- % OF 1967 TRADE (VALUE BASIS)

YEAR	CANADA	PHIL REP	KOR REP	JAPAN	OTH ASIA	OTHERS	TOTAL
1967	100.	100.	100.	100.	100.	100.	100.
1968	124.	126.	177.	152.	176.	126.	153.
1969	118.	127.	242.	146.	216.	176.	176.
1970	57.	101.	246.	106.	186.	122.	147.
1971	101.	103.	321.	113.	287.	121.	182.
1972	154.	118.	421.	132.	411.	155.	237.
1973	233.	171.	551.	117.	423.	179.	275.
1974	160.	93.	392.	102.	340.	137.	204.
1975	160.	62.	440.	75.	269.	83.	184.
1976	183.	124.	625.	121.	379.	106.	269.

IMPORTS
PLYWOOD (6312)

VALUE--MILLION CURRENT DOLLARS

YEAR	N ATLANTC	S ATLANTC	GULF	S PACIFIC	PACIFC NW	GRT LAKES	ALASKA	N CENTRAL	S CENTRAL	OTHERS	TOTAL
1967	36.857	12.652	29.954	29.592	21.601	8.515	0.049	0.000	0.271	2.812	142.302
1968	55.962	21.686	49.316	45.195	31.420	10.437	0.021	0.065	0.217	3.365	217.683
1969	60.073	27.526	51.587	54.259	37.002	14.614	0.052	0.237	0.202	4.614	250.166
1970	54.555	28.307	46.371	37.204	25.969	11.947	0.065	0.002	0.118	3.941	208.979
1971	63.031	40.901	67.338	35.756	29.508	17.867	0.012	0.001	0.042	5.153	259.608
1972	85.188	61.660	85.471	46.728	30.582	20.080	0.035	0.211	0.080	6.858	336.893
1973	114.478	69.073	86.053	57.016	42.873	13.900	0.010	0.728	0.104	7.308	391.545
1974	75.555	63.538	66.055	38.880	27.341	10.561	0.144	0.002	0.064	8.299	290.440
1975	80.464	50.587	49.173	36.423	24.321	14.130	0.002	0.002	0.001	4.894	261.997
1976	108.676	75.711	91.019	55.399	33.955	8.524	0.001	0.012	0.202	9.001	382.500

QUANTITY--MIL SQ FT

YEAR	N ATLANTC	S ATLANTC	GULF	S PACIFIC	PACIFC NW	GRT LAKES	ALASKA	N CENTRAL	S CENTRAL	OTHERS	TOTAL
1967	328.655	111.658	242.467	247.369	197.541	71.876	0.283	0.006	1.676	19.325	1220.875
1968	390.923	175.690	348.541	326.049	241.417	79.610	0.074	0.411	1.482	17.669	1581.966
1969	446.389	221.081	371.054	402.534	288.215	111.057	0.271	0.985	2.130	28.465	1872.180
1970	429.677	243.476	365.131	286.641	208.337	104.289	0.226	0.099	0.789	23.550	1662.715
1971	482.943	342.302	503.822	275.886	246.268	153.478	0.080	0.001	0.234	32.468	2037.482
1972	644.080	512.192	623.097	332.847	247.219	680.797	0.145	1.009	0.439	42.946	3084.769
1973	652.264	389.133	456.300	278.231	228.236	81.159	0.036	2.290	0.563	33.445	2122.155
1974	368.543	321.613	287.537	166.333	133.193	65.915	0.408	0.011	0.250	31.317	1370.121
1975	463.912	307.120	267.709	206.525	139.859	88.988	0.010	0.010	0.009	24.219	1498.361
1976	556.568	399.326	409.863	255.327	172.391	48.096	0.002	0.049	0.604	39.432	1881.658

IMPORTS
PLYWOOD (6312)

TRADE STRUCTURE-- % OF TOTAL TRADE (VALUE BASIS)

YEAR	N ATLANTC	S ATLANTC	GULF	S PACIFIC	PACIFC NW	GRT LAKES	ALASKA	N CENTRAL	S CENTRAL	OTHERS	TOTAL
1967	26.	9.	21.	21.	15.	6.	0.	0.	0.	2.	100.
1968	26.	10.	23.	21.	14.	5.	0.	0.	0.	2.	100.
1969	24.	11.	21.	22.	15.	6.	0.	0.	0.	2.	100.
1970	26.	14.	22.	19.	12.	6.	0.	0.	0.	2.	100.
1971	24.	16.	26.	14.	11.	7.	0.	0.	0.	2.	100.
1972	25.	19.	25.	14.	9.	6.	0.	0.	0.	2.	100.
1973	29.	18.	22.	15.	11.	4.	0.	0.	0.	2.	100.
1974	26.	22.	23.	13.	9.	4.	0.	0.	0.	3.	100.
1975	31.	19.	19.	15.	9.	5.	0.	0.	0.	2.	100.
1976	28.	20.	24.	14.	9.	2.	0.	0.	0.	2.	100.

TRADE TRENDS-- % OF 1967 TRADE (VALUE BASIS)

YEAR	N ATLANTC	S ATLANTC	GULF	S PACIFIC	PACIFC NW	GRT LAKES	ALASKA	N CENTRAL	S CENTRAL	OTHERS	TOTAL
1967	100.	100.	100.	100.	100.	100.	100.	100.	100.	100.	100.
1968	152.	171.	165.	153.	145.	123.	42.	21714.	80.	120.	153.
1969	163.	218.	172.	183.	171.	172.	107.	79037.	75.	164.	176.
1970	148.	224.	156.	126.	120.	140.	133.	598.	43.	140.	147.
1971	171.	323.	225.	121.	137.	210.	24.	430.	16.	183.	182.
1972	231.	487.	285.	158.	142.	236.	73.	70254.	30.	244.	237.
1973	311.	546.	287.	193.	198.	163.	21.	******	38.	260.	275.
1974	205.	502.	221.	131.	127.	124.	295.	733.	24.	295.	204.
1975	218.	400.	164.	130.	113.	166.	5.	546.	0.	174.	184.
1976	295.	598.	304.	187.	157.	100.	1.	4033.	75.	320.	269.

IMPORTS
RECONSTITUTED WOOD (6314)
UNITED STATES

VALUE--MILLION CURRENT DOLLARS

YEAR	CANADA	S AMERIC	EUROPE	ASIA	AFRICA	OTHERS	TOTAL
1967	0.069	0.118	0.360	0.002	0.000	0.000	0.549
1968	0.020	0.103	0.378	0.008	0.000	0.000	0.509
1969	0.036	0.347	0.995	0.253	0.000	0.309	1.941
1970	0.282	0.206	0.425	0.388	0.000	0.009	1.310
1971	1.212	0.180	0.323	0.473	0.000	0.051	2.239
1972	1.499	0.117	0.686	0.063	0.000	0.014	2.379
1973	1.404	0.277	0.647	0.033	0.000	0.009	2.370
1974	0.709	0.205	0.982	0.026	0.000	0.000	1.922
1975	2.956	0.005	1.254	0.034	0.000	0.001	4.250
1976	11.328	0.013	1.219	0.084	0.000	0.006	12.651

QUANTITY--MIL SQ FT

YEAR	CANADA	S AMERIC	EUROPE	ASIA	AFRICA	OTHERS	TOTAL
1967	0.137	1.155	0.554	0.003	0.000	0.000	1.848
1968	0.125	0.992	0.549	0.014	0.000	0.000	1.680
1969	0.177	2.843	6.949	1.317	0.000	2.970	14.256
1970	1.667	1.770	1.108	0.248	0.000	0.080	4.874
1971	7.700	1.194	1.214	0.343	0.000	0.484	10.935
1972	8.609	0.811	4.439	0.101	0.000	0.041	14.001
1973	8.503	1.724	0.842	0.065	0.000	0.072	11.206
1974	5.176	0.998	1.490	0.072	0.000	0.000	7.736
1975	15.577	0.015	0.576	0.030	0.000	0.000	16.198
1976	60.091	0.122	0.571	0.089	0.000	0.033	60.906

IMPORTS
RECONSTITUTED WOOD (6314)
UNITED STATES

TRADE STRUCTURE-- % OF TOTAL TRADE (VALUE BASIS)

YEAR	CANADA	S AMERIC	EUROPE	ASIA	AFRICA	OTHERS	TOTAL
1967	13.	21.	66.	0.	0.	0.	100.
1968	4.	20.	74.	2.	0.	0.	100.
1969	2.	18.	51.	13.	0.	16.	100.
1970	22.	16.	32.	30.	0.	1.	100.
1971	54.	8.	14.	21.	0.	2.	100.
1972	63.	5.	29.	3.	0.	1.	100.
1973	59.	12.	27.	1.	0.	0.	100.
1974	37.	11.	51.	1.	0.	0.	100.
1975	70.	0.	30.	1.	0.	0.	100.
1976	90.	0.	10.	1.	0.	0.	100.

TRADE TRENDS-- % OF 1967 TRADE (VALUE BASIS)

YEAR	CANADA	S AMERIC	EUROPE	ASIA	AFRICA	OTHERS	TOTAL
1967	100.	100.	100.	100.	******	******	100.
1968	28.	88.	105.	434.	******	******	93.
1969	53.	295.	276.	13275.	******	******	354.
1970	410.	175.	118.	20391.	******	******	239.
1971	1764.	153.	90.	24834.	******	******	408.
1972	2181.	100.	191.	3312.	******	******	434.
1973	2043.	235.	180.	1728.	******	******	432.
1974	1031.	174.	273.	1369.	******	******	350.
1975	4302.	4.	346.	1804.	******	******	774.
1976	16485.	11.	338.	4404.	******	******	2305.

IMPORTS
RECONSTITUTED WOOD (6314)

VALUE--MILLION CURRENT DOLLARS

YEAR	N ATLANTC	S ATLANTC	GULF	S PACIFIC	PACIFC NW	GRT LAKES	ALASKA	N CENTRAL	S CENTRAL	OTHERS	TOTAL
1967	0.287	0.112	0.000	0.000	0.000	0.053	0.000	0.000	0.000	0.098	0.549
1968	0.235	0.149	0.000	0.008	0.000	0.014	0.000	0.002	0.000	0.102	0.509
1969	0.498	0.575	0.091	0.135	0.001	0.084	0.000	0.007	0.264	0.285	1.941
1970	0.482	0.173	0.017	0.110	0.020	0.122	0.000	0.153	0.000	0.232	1.310
1971	0.580	0.104	0.026	0.116	0.026	0.185	0.000	0.981	0.000	0.221	2.239
1972	0.468	0.360	0.001	0.027	0.001	0.166	0.000	1.241	0.011	0.104	2.379
1973	0.909	0.156	0.016	0.012	0.000	0.436	0.000	0.553	0.000	0.287	2.370
1974	1.361	0.211	0.001	0.000	0.009	0.102	0.000	0.000	0.000	0.238	1.922
1975	1.498	0.133	0.001	0.005	0.007	1.415	0.000	1.190	0.000	0.001	4.250
1976	1.510	0.156	0.011	0.031	0.212	9.210	0.000	1.511	0.000	0.009	12.651

QUANTITY--MIL SQ FT

YEAR	N ATLANTC	S ATLANTC	GULF	S PACIFIC	PACIFC NW	GRT LAKES	ALASKA	N CENTRAL	S CENTRAL	OTHERS	TOTAL
1967	0.558	0.289	0.000	0.000	0.000	0.011	0.000	0.000	0.000	0.990	1.848
1968	0.291	0.322	0.000	0.016	0.002	0.067	0.000	0.002	0.000	0.981	1.680
1969	2.918	4.673	0.787	0.333	0.001	0.568	0.000	0.043	2.581	2.352	14.256
1970	0.546	0.588	0.014	0.082	0.014	0.666	0.000	0.919	0.000	2.045	4.874
1971	0.868	0.792	0.018	0.120	0.021	1.062	0.000	6.357	0.000	1.697	10.935
1972	0.950	4.101	0.001	0.043	0.000	0.865	0.000	7.204	0.036	0.800	14.001
1973	3.661	0.245	0.055	0.060	0.001	2.252	0.000	3.090	0.000	1.842	11.206
1974	5.884	0.212	0.000	0.000	0.027	0.309	0.000	0.001	0.000	1.302	7.736
1975	2.747	0.131	0.000	0.015	0.035	7.484	0.000	5.780	0.000	0.005	16.138
1976	2.413	0.169	0.012	0.050	0.914	51.332	0.000	6.002	0.000	0.014	60.936

IMPORTS
RECONSTITUTED WOOD (6314)

TRADE STRUCTURE-- % OF TOTAL TRADE (VALUE BASIS)

YEAR	N ATLANTC	S ATLANTC	GULF	S PACIFIC	PACIFC NW	GPT LAKES	ALASKA	N CENTRAL	S CENTRAL	OTHERS	TOTAL
1967	52.	20.	0.	0.	0.	10.	0.	0.	0.	18.	100.
1968	46.	29.	0.	2.	0.	3.	0.	0.	0.	20.	100.
1969	26.	30.	5.	7.	0.	4.	0.	0.	14.	15.	100.
1970	37.	13.	1.	8.	2.	9.	0.	12.	0.	18.	100.
1971	26.	5.	1.	5.	1.	8.	0.	44.	0.	10.	100.
1972	20.	15.	0.	1.	0.	7.	0.	52.	0.	4.	100.
1973	38.	7.	1.	0.	0.	18.	0.	23.	0.	12.	100.
1974	71.	11.	0.	0.	0.	5.	0.	0.	0.	12.	100.
1975	35.	3.	0.	0.	0.	33.	0.	28.	0.	0.	100.
1976	12.	1.	0.	0.	2.	73.	0.	12.	0.	0.	100.

TRADE TRENDS-- % OF 1967 TRADE (VALUE BASIS)

YEAR	N ATLANTC	S ATLANTC	GULF	S PACIFIC	PACIFC NW	GPT LAKES	ALASKA	N CENTRAL	S CENTRAL	OTHERS	TOTAL
1967	100.	100.	******	******	******	100.	******	******	******	100.	100.
1968	82.	133.	******	******	******	26.	******	******	******	104.	93.
1969	174.	516.	******	******	******	159.	******	******	******	291.	354.
1970	168.	155.	******	******	******	232.	******	******	******	237.	239.
1971	202.	94.	******	******	******	350.	******	******	******	226.	408.
1972	163.	323.	******	******	******	316.	******	******	******	107.	434.
1973	317.	140.	******	******	******	828.	******	******	******	293.	432.
1974	475.	189.	******	******	******	193.	******	******	******	243.	350.
1975	523.	119.	******	******	******	2684.	******	******	******	1.	774.
1976	527.	140.	******	******	******	17467.	******	******	******	9.	2305.

IMPORTS
PULPWOOD CHIPS (631.8320)
UNITED STATES

VALUE--MILLION CURRENT DOLLARS

YEAR	CANADA	OTH N AM	S AMERIC	EUROPE	OCEANIA	OTHERS	TOTAL
1967	9.210	0.000	0.000	0.000	0.000	0.000	9.210
1968	8.375	0.000	0.000	0.000	0.000	0.000	8.375
1969	5.072	0.000	0.000	0.000	0.000	0.000	5.072
1970	7.476	0.000	0.000	0.001	0.001	0.000	7.478
1971	12.004	0.002	0.000	0.000	0.000	0.000	12.006
1972	10.416	0.000	0.000	0.000	0.000	0.000	10.416
1973	13.927	0.001	0.002	0.000	0.001	0.000	13.932
1974	13.639	0.000	0.000	0.000	0.000	0.000	13.639
1975	14.875	0.011	0.001	0.001	0.000	0.000	14.887
1976	23.892	0.000	0.000	0.001	0.100	0.000	23.994

QUANTITY--1000 STN

YEAR	CANADA	OTH N AM	S AMERIC	EUROPE	OCEANIA	OTHERS	TOTAL
1967	1237.524	0.000	0.000	0.000	0.000	0.000	1237.524
1968	1087.653	0.000	0.000	0.000	0.000	0.000	1087.653
1969	619.795	0.000	0.000	0.000	0.000	0.000	619.795
1970	861.994	0.000	0.000	0.004	0.080	0.000	862.078
1971	1201.118	0.071	0.000	0.000	0.000	0.000	1201.189
1972	977.929	0.000	0.029	0.000	0.000	0.000	977.958
1973	1251.372	0.002	2.105	0.000	0.055	0.000	1253.534
1974	818.419	0.000	0.000	0.003	0.000	0.000	818.422
1975	625.787	0.051	0.004	0.001	0.000	0.000	625.843
1976	1045.796	0.000	0.000	0.014	4.731	0.000	1050.541

IMPORTS
PULPWOOD CHIPS (631.8320)
UNITED STATES

TRADE STRUCTURE-- % OF TOTAL TRADE (VALUE BASIS)

YEAR	CANADA	OTH N AM	S AMERIC	EUROPE	OCEANIA	OTHERS	TOTAL
1967	100.	0.	0.	0.	0.	0.	100.
1968	100.	0.	0.	0.	0.	0.	100.
1969	100.	0.	0.	0.	0.	0.	100.
1970	100.	0.	0.	0.	0.	0.	100.
1971	100.	0.	0.	0.	0.	0.	100.
1972	100.	0.	0.	0.	0.	0.	100.
1973	100.	0.	0.	0.	0.	0.	100.
1974	100.	0.	0.	0.	0.	0.	100.
1975	100.	0.	0.	0.	0.	0.	100.
1976	100.	0.	0.	0.	0.	0.	100.

TRADE TRENDS-- % OF 1967 TRADE (VALUE BASIS)

YEAR	CANADA	OTH N AM	S AMERIC	EUROPE	OCEANIA	OTHERS	TOTAL
1967	100.	******	******	******	******	******	100.
1968	91.	******	******	******	******	******	91.
1969	55.	******	******	******	******	******	55.
1970	81.	******	******	******	******	******	81.
1971	130.	******	******	******	******	******	130.
1972	113.	******	******	******	******	******	113.
1973	151.	******	******	******	******	******	151.
1974	148.	******	******	******	******	******	148.
1975	161.	******	******	******	******	******	162.
1976	259.	******	******	******	******	******	261.

IMPORTS
PULPWOOD CHIPS (631.8320)

VALUE--MILLION CURRENT DOLLARS

YEAR	N ATLANTC	S ATLANTC	GULF	S PACIFIC	PACIFC NW	GRT LAKES	ALASKA	N CENTRAL	S CENTRAL	OTHERS	TOTAL
1967	0.149	0.000	0.000	0.000	8.650	0.273	0.138	0.000	0.000	0.000	9.210
1968	0.202	0.000	0.000	0.020	7.968	0.119	0.067	0.000	0.000	0.000	8.375
1969	0.301	0.000	0.000	0.000	4.519	0.252	0.000	0.000	0.000	0.000	5.072
1970	0.027	0.000	0.000	0.001	6.645	0.522	0.283	0.000	0.000	0.000	7.478
1971	0.128	0.000	0.000	0.000	11.075	0.731	0.071	0.000	0.000	0.000	12.006
1972	0.120	0.000	0.000	0.000	8.980	1.317	0.000	0.000	0.000	0.000	10.416
1973	0.567	0.002	0.000	0.000	11.148	2.097	0.000	0.116	0.000	0.001	13.932
1974	0.880	0.000	0.000	0.000	9.701	3.000	0.000	0.059	0.000	0.000	13.639
1975	1.469	0.000	0.000	0.000	11.109	2.298	0.000	0.000	0.011	0.000	14.887
1976	0.996	0.000	0.000	0.000	18.532	4.466	0.000	0.000	0.000	0.000	23.994

QUANTITY--1000 STN

YEAR	N ATLANTC	S ATLANTC	GULF	S PACIFIC	PACIFC NW	GRT LAKES	ALASKA	N CENTRAL	S CENTRAL	OTHERS	TOTAL
1967	11.359	0.000	0.000	0.000	1183.841	23.924	18.400	0.000	0.000	0.000	1237.524
1968	15.062	0.000	0.000	2.898	1049.259	12.234	8.200	0.000	0.000	0.000	1087.653
1969	16.412	0.000	0.000	0.000	581.167	22.216	0.000	0.000	0.000	0.000	619.795
1970	4.500	0.000	0.000	0.004	790.544	31.498	35.532	0.000	0.000	0.000	862.078
1971	8.570	0.000	0.000	0.000	1157.444	32.122	3.053	0.000	0.000	0.000	1201.189
1972	7.411	0.000	0.000	0.000	909.926	60.621	0.000	0.000	0.000	0.000	977.958
1973	39.885	2.105	0.000	0.000	1085.124	102.225	0.000	24.193	0.000	0.002	1253.534
1974	57.186	0.000	0.000	0.000	623.830	124.557	0.000	12.849	0.000	0.000	818.422
1975	66.975	0.000	0.000	0.000	493.762	65.055	0.000	0.000	0.051	0.000	625.843
1976	52.573	0.000	0.000	0.000	882.281	115.687	0.000	0.000	0.000	0.000	1050.541

IMPORTS
PULPWOOD CHIPS (631.8320)

TRADE STRUCTURE-- % OF TOTAL TRADE (VALUE BASIS)

YEAR	N ATLANTC	S ATLANTC	GULF	S PACIFIC	PACIFC NW	GRT LAKES	ALASKA	N CENTRAL	S CENTRAL	OTHERS	TOTAL
1967	2.	0.	0.	0.	94.	3.	1.	0.	0.	0.	100.
1968	2.	0.	0.	0.	95.	1.	1.	0.	0.	0.	100.
1969	6.	0.	0.	0.	89.	5.	0.	0.	0.	0.	100.
1970	0.	0.	0.	0.	89.	7.	4.	0.	0.	0.	100.
1971	1.	0.	0.	0.	92.	6.	1.	0.	0.	0.	100.
1972	1.	0.	0.	0.	86.	13.	0.	0.	0.	0.	100.
1973	4.	0.	0.	0.	80.	15.	0.	1.	0.	0.	100.
1974	6.	0.	0.	0.	71.	22.	0.	0.	0.	0.	100.
1975	10.	0.	0.	0.	75.	15.	0.	0.	0.	0.	100.
1976	4.	0.	0.	0.	77.	19.	0.	0.	0.	0.	100.

TRADE TRENDS-- % OF 1967 TRADE (VALUE BASIS)

YEAR	N ATLANTC	S ATLANTC	GULF	S PACIFIC	PACIFC NW	GRT LAKES	ALASKA	N CENTRAL	S CENTRAL	OTHERS	TOTAL
1967	100.	******	******	******	100.	100.	100.	******	******	******	100.
1968	136.	******	******	******	92.	43.	49.	******	******	******	91.
1969	202.	******	******	******	52.	92.	0.	******	******	******	55.
1970	18.	******	******	******	77.	191.	295.	******	******	******	81.
1971	86.	******	******	******	128.	267.	51.	******	******	******	130.
1972	81.	******	******	******	104.	482.	0.	******	******	******	113.
1973	381.	******	******	******	129.	767.	0.	******	******	******	151.
1974	591.	******	******	******	112.	1097.	0.	******	******	******	148.
1975	986.	******	******	******	128.	841.	0.	******	******	******	162.
1976	669.	******	******	******	214.	1634.	0.	******	******	******	261.

IMPORTS
PULPWOOD (EXC CHIPS)(2421)
UNITED STATES

VALUE--MILLION CURRENT DOLLARS

YEAR	CANADA	BAHAMAS	OTH N AM	S AMERIC	EUROPE	OTHERS	TOTAL
1967	12.591	3.957	0.042	0.000	0.000	0.000	16.590
1968	11.124	5.526	0.000	0.000	0.000	0.000	16.650
1969	8.816	5.510	0.001	0.000	0.000	0.000	14.326
1970	8.523	6.044	0.000	0.000	0.000	0.000	14.567
1971	4.747	6.177	0.000	0.000	0.000	0.000	10.924
1972	3.610	6.346	0.000	0.000	0.000	0.000	9.956
1973	3.749	4.257	0.017	0.000	0.000	0.002	8.026
1974	8.560	2.478	0.000	0.000	0.000	0.000	11.038
1975	8.079	1.160	0.000	0.058	0.000	0.000	9.297
1976	10.071	1.621	0.000	0.010	0.000	0.000	11.702

QUANTITY--1000 CORDS

YEAR	CANADA	BAHAMAS	OTH N AM	S AMERIC	EUROPE	OTHERS	TOTAL
1967	544.701	282.547	2.954	0.000	0.000	0.000	830.202
1968	476.819	312.148	0.000	0.000	0.000	0.000	788.967
1969	374.395	301.496	0.040	0.000	0.000	0.000	675.931
1970	322.314	316.187	0.000	0.000	0.000	0.000	638.501
1971	188.768	267.497	0.000	0.000	0.000	0.000	456.265
1972	126.861	276.422	0.000	0.000	0.000	0.000	403.283
1973	129.864	206.588	0.107	0.000	0.000	0.054	336.612
1974	263.625	134.777	0.000	0.000	0.000	0.000	398.402
1975	236.971	72.761	0.000	2.620	0.000	0.000	312.352
1976	227.538	101.344	0.000	0.471	0.000	0.000	329.353

IMPORTS
PULPWOOD (EXC CHIPS)(2421)
UNITED STATES

TRADE STRUCTURE-- % OF TOTAL TRADE (VALUE BASIS)

YEAR	CANADA	BAHAMAS	OTH N AM	S AMERIC	EUROPE	OTHERS	TOTAL
1967	76.	24.	0.	0.	0.	0.	100.
1968	67.	33.	0.	0.	0.	0.	100.
1969	62.	38.	0.	0.	0.	0.	100.
1970	59.	41.	0.	0.	0.	0.	100.
1971	43.	57.	0.	0.	0.	0.	100.
1972	36.	64.	0.	0.	0.	0.	100.
1973	47.	53.	0.	0.	0.	0.	100.
1974	78.	22.	0.	0.	0.	0.	100.
1975	87.	12.	0.	1.	0.	0.	100.
1976	86.	14.	0.	0.	0.	0.	100.

TRADE TRENDS-- % OF 1967 TRADE (VALUE BASIS)

YEAR	CANADA	BAHAMAS	OTH N AM	S AMERIC	EUROPE	OTHERS	TOTAL
1967	100.	100.	100.	******	******	******	100.
1968	88.	140.	0.	******	******	******	100.
1969	70.	139.	2.	******	******	******	86.
1970	68.	153.	0.	******	******	******	88.
1971	38.	156.	0.	******	******	******	66.
1972	29.	160.	0.	******	******	******	60.
1973	30.	108.	41.	******	******	******	48.
1974	68.	63.	0.	******	******	******	67.
1975	64.	29.	0.	******	******	******	56.
1976	80.	41.	0.	******	******	******	71.

IMPORTS
PULPWOOD (EXC CHIPS)(2421)

VALUE--MILLION CURRENT DOLLARS

YEAR	N ATLANTC	S ATLANTC	GULF	S PACIFIC	PACIFC NW	GRT LAKES	ALASKA	N CENTRAL	S CENTRAL	OTHERS	TOTAL
1967	2.478	3.998	0.000	0.000	0.193	9.791	0.000	0.121	0.008	0.000	16.590
1968	1.896	5.528	0.003	0.000	0.251	8.911	0.000	0.060	0.002	0.000	16.650
1969	1.849	5.510	0.002	0.000	0.308	6.600	0.000	0.057	0.000	0.000	14.326
1970	1.976	6.044	0.000	0.000	1.687	4.815	0.000	0.044	0.000	0.000	14.567
1971	0.753	6.177	0.000	0.000	0.045	3.930	0.000	0.020	0.000	0.000	10.924
1972	0.639	6.346	0.000	0.000	0.109	2.854	0.000	0.008	0.000	0.000	9.956
1973	0.766	4.257	0.017	0.000	0.002	2.975	0.000	0.009	0.000	0.000	8.026
1974	3.677	2.478	0.000	0.000	1.922	2.946	0.000	0.012	0.004	0.000	11.038
1975	6.090	1.160	0.000	0.000	0.552	1.415	0.000	0.079	0.000	0.000	9.297
1976	7.811	1.621	0.000	0.000	0.064	2.175	0.000	0.031	0.000	0.000	11.702

QUANTITY--1000 CORDS

YEAR	N ATLANTC	S ATLANTC	GULF	S PACIFIC	PACIFC NW	GRT LAKES	ALASKA	N CENTRAL	S CENTRAL	OTHERS	TOTAL
1967	141.110	285.500	0.000	0.000	3.529	392.054	0.000	7.650	0.360	0.000	830.202
1968	109.044	312.222	0.165	0.000	3.720	359.990	0.000	3.766	0.060	0.000	788.967
1969	101.257	301.496	0.069	0.000	3.174	266.689	0.000	3.246	0.000	0.000	675.931
1970	107.679	316.187	0.000	0.016	17.501	194.451	0.000	2.667	0.000	0.000	638.501
1971	39.884	267.497	0.000	0.000	2.330	145.229	0.000	1.325	0.000	0.000	456.265
1972	35.363	276.422	0.000	0.000	2.300	88.648	0.000	0.550	0.000	0.000	403.283
1973	34.918	206.588	0.107	0.000	0.054	94.396	0.000	0.550	0.000	0.000	336.612
1974	132.884	134.777	0.000	0.000	31.998	98.226	0.000	0.375	0.142	0.000	398.402
1975	181.320	72.761	0.000	0.000	14.137	41.834	0.000	2.300	0.000	0.000	312.352
1976	187.552	101.344	0.000	0.000	2.170	37.766	0.000	0.521	0.000	0.000	329.353

IMPORTS
PULPWOOD (EXC CHIPS)(2421)

TRADE STRUCTURE-- % OF TOTAL TRADE (VALUE BASIS)

YEAR	N ATLANTC	S ATLANTC	GULF	S PACIFIC	PACIFC NW	GRT LAKES	ALASKA	N CENTRAL	S CENTRAL	OTHERS	TOTAL
1967	15.	24.	0.	0.	1.	59.	0.	1.	0.	0.	100.
1968	11.	33.	0.	0.	2.	54.	0.	0.	0.	0.	100.
1969	13.	38.	0.	0.	2.	46.	0.	0.	0.	0.	100.
1970	14.	41.	0.	0.	12.	33.	0.	0.	0.	0.	100.
1971	7.	57.	0.	0.	0.	36.	0.	0.	0.	0.	100.
1972	6.	64.	0.	0.	1.	29.	0.	0.	0.	0.	100.
1973	10.	53.	0.	0.	0.	37.	0.	0.	0.	0.	100.
1974	33.	22.	0.	0.	17.	27.	0.	0.	0.	0.	100.
1975	66.	12.	0.	0.	6.	15.	0.	1.	0.	0.	100.
1976	67.	14.	0.	0.	1.	19.	0.	0.	0.	0.	100.

TRADE TRENDS-- % OF 1967 TRADE (VALUE BASIS)

YEAR	N ATLANTC	S ATLANTC	GULF	S PACIFIC	PACIFC NW	GRT LAKES	ALASKA	N CENTRAL	S CENTRAL	OTHERS	TOTAL
1967	100.	100.	******	******	100.	100.	******	100.	100.	******	100.
1968	77.	138.	******	******	130.	91.	******	49.	18.	******	100.
1969	75.	138.	******	******	159.	67.	******	47.	0.	******	86.
1970	80.	151.	******	******	872.	49.	******	37.	0.	******	88.
1971	30.	154.	******	******	23.	40.	******	16.	0.	******	66.
1972	26.	159.	******	******	57.	29.	******	7.	0.	******	60.
1973	31.	106.	******	******	1.	30.	******	7.	0.	******	48.
1974	148.	62.	******	******	994.	30.	******	10.	42.	******	67.
1975	246.	29.	******	******	285.	14.	******	66.	0.	******	56.
1976	315.	41.	******	******	33.	22.	******	26.	0.	******	71.

IMPORTS
WOOD PULP (2510)
UNITED STATES

VALUE--MILLION CURRENT DOLLARS

YEAR	CANADA	OTH N AM	SWEDEN	OTH EURO	REP S AF	OTHERS	TOTAL
1967	328.410	0.057	17.081	7.934	4.386	0.436	358.303
1968	363.112	0.000	10.793	9.986	5.897	0.519	390.307
1969	440.070	0.050	6.056	6.520	5.982	0.192	458.870
1970	427.515	0.000	3.562	5.161	7.938	0.184	444.359
1971	410.488	0.402	2.316	4.751	7.261	0.370	425.588
1972	477.433	0.000	3.497	2.887	7.711	0.613	492.141
1973	623.092	0.020	14.238	4.755	11.626	1.126	654.858
1974	1054.621	0.019	14.774	1.745	21.032	0.830	1093.021
1975	980.267	0.258	9.756	2.042	13.774	0.541	1006.638
1976	1201.241	0.000	16.172	4.622	21.215	1.038	1244.289

QUANTITY--1000 STN

YEAR	CANADA	OTH N AM	SWEDEN	OTH EURO	REP S AF	OTHERS	TOTAL
1967	2677.824	0.441	149.941	84.101	37.212	2.452	2951.971
1968	3032.106	0.000	95.552	113.271	51.064	2.289	3294.283
1969	3628.643	0.353	49.322	68.231	50.722	0.899	3798.169
1970	3171.275	0.000	26.256	51.391	67.291	0.675	3316.888
1971	3381.898	2.543	16.157	42.201	62.316	1.651	3506.766
1972	3594.567	0.000	26.869	27.743	62.290	2.023	3713.492
1973	3795.569	0.104	68.281	36.346	64.731	3.957	3968.988
1974	4002.406	0.063	47.380	4.416	82.502	19.767	4156.534
1975	2991.153	0.757	27.611	5.494	49.776	1.045	3075.836
1976	3583.887	0.000	51.373	15.292	78.120	3.818	3732.490

IMPORTS
WOOD PULP (2510)
UNITED STATES

TRADE STRUCTURE-- % OF TOTAL TRADE (VALUE BASIS)

YEAR	CANADA	OTH N AM	SWEDEN	OTH EURO	REP S AF	OTHERS	TOTAL
1967	92.	0.	5.	2.	1.	0.	100.
1968	93.	0.	3.	3.	2.	0.	100.
1969	96.	0.	1.	1.	1.	0.	100.
1970	96.	0.	1.	1.	2.	0.	100.
1971	96.	0.	1.	1.	2.	0.	100.
1972	97.	0.	1.	1.	2.	0.	100.
1973	95.	0.	2.	1.	2.	0.	100.
1974	96.	0.	1.	0.	2.	0.	100.
1975	97.	0.	1.	0.	1.	0.	100.
1976	97.	0.	1.	0.	2.	0.	100.

TRADE TRENDS-- % OF 1967 TRADE (VALUE BASIS)

YEAR	CANADA	OTH N AM	SWEDEN	OTH EURO	REP S AF	OTHERS	TOTAL
1967	100.	100.	100.	100.	100.	100.	100.
1968	111.	0.	63.	126.	134.	119.	109.
1969	134.	88.	35.	82.	136.	44.	128.
1970	130.	0.	21.	65.	181.	42.	124.
1971	125.	708.	14.	60.	166.	85.	119.
1972	145.	0.	20.	36.	176.	141.	137.
1973	190.	36.	83.	60.	265.	259.	183.
1974	321.	34.	86.	22.	480.	190.	305.
1975	298.	455.	57.	26.	314.	124.	281.
1976	366.	0.	95.	58.	484.	238.	347.

IMPORTS
WOOD PULP (2510)

VALUE--MILLION CURRENT DOLLARS

YEAR	N ATLANTC	S ATLANTC	GULF	S PACIFIC	PACIFC NW	GRT LAKES	ALASKA	N CENTRAL	S CENTRAL	OTHERS	TOTAL
1967	112.987	0.148	5.001	9.124	27.287	175.938	0.000	27.437	0.052	0.327	358.303
1968	111.486	0.097	6.469	8.681	22.550	211.046	0.008	29.674	0.006	0.290	390.307
1969	116.934	0.007	6.142	11.432	33.871	249.213	0.000	40.824	0.022	0.426	458.870
1970	111.705	0.047	8.398	9.936	21.854	252.331	0.000	39.743	0.000	0.345	444.359
1971	115.442	0.000	7.907	8.257	28.029	222.714	0.000	42.998	0.000	0.240	425.588
1972	119.645	0.003	8.352	10.671	61.990	241.833	0.000	49.122	0.000	0.525	492.141
1973	173.406	0.805	11.892	11.878	69.089	316.806	0.000	70.816	0.000	0.167	654.858
1974	262.023	0.913	24.313	14.686	106.507	546.569	0.000	138.010	0.000	0.000	1093.021
1975	224.140	3.296	13.880	6.548	124.088	515.522	0.000	119.164	0.000	0.000	1006.638
1976	269.743	11.061	22.605	8.802	138.097	640.504	0.000	153.476	0.000	0.000	1244.289

QUANTITY--1000 STN

YEAR	N ATLANTC	S ATLANTC	GULF	S PACIFIC	PACIFC NW	GRT LAKES	ALASKA	N CENTRAL	S CENTRAL	OTHERS	TOTAL
1967	911.178	1.082	48.799	116.073	238.197	1417.872	0.000	215.447	0.430	2.893	2951.971
1968	906.297	0.770	60.559	126.552	199.806	1731.479	0.058	266.141	0.059	2.561	3294.283
1969	934.351	0.063	53.515	161.456	287.306	2008.270	0.000	349.233	0.171	3.805	3798.169
1970	825.132	0.239	74.245	106.694	161.581	1852.432	0.000	294.032	0.000	2.533	3316.887
1971	815.911	0.000	69.033	81.379	429.821	1787.014	0.000	322.043	0.000	1.565	3506.766
1972	888.178	0.020	69.146	143.837	398.773	1817.348	0.000	392.690	0.000	3.501	3713.492
1973	1047.263	3.705	66.316	106.804	392.795	1922.912	0.000	427.888	0.000	1.304	3968.988
1974	1050.244	3.030	92.193	85.659	396.912	2030.390	0.000	498.107	0.001	0.000	4156.534
1975	709.906	13.803	50.108	22.862	352.291	1580.357	0.000	346.508	0.001	0.000	3075.836
1976	865.236	41.483	83.411	35.763	390.354	1885.511	0.000	430.733	0.000	0.000	3732.490

IMPORTS
WOOD PULP (2510)

TRADE STRUCTURE-- % OF TOTAL TRADE (VALUE BASIS)

YEAR	N ATLANTC	S ATLANTC	GULF	S PACIFIC	PACIFC NW	GRT LAKES	ALASKA	N CENTRAL	S CENTRAL	OTHERS	TOTAL
1967	32.	0.	1.	3.	8.	49.	0.	8.	0.	0.	100.
1968	29.	0.	2.	2.	6.	54.	0.	8.	0.	0.	100.
1969	25.	0.	1.	2.	7.	54.	0.	9.	0.	0.	100.
1970	25.	0.	2.	2.	5.	57.	0.	9.	0.	0.	100.
1971	27.	0.	2.	2.	7.	52.	0.	10.	0.	0.	100.
1972	24.	0.	2.	2.	13.	49.	0.	10.	0.	0.	100.
1973	26.	0.	2.	2.	11.	48.	0.	11.	0.	0.	100.
1974	24.	0.	2.	1.	10.	50.	0.	13.	0.	0.	100.
1975	22.	0.	1.	1.	12.	51.	0.	12.	0.	0.	100.
1976	22.	1.	2.	1.	11.	51.	0.	12.	0.	0.	100.

TRADE TRENDS-- % OF 1967 TRADE (VALUE BASIS)

YEAR	N ATLANTC	S ATLANTC	GULF	S PACIFIC	PACIFC NW	GRT LAKES	ALASKA	N CENTRAL	S CENTRAL	OTHERS	TOTAL
1967	100.	100.	100.	100.	100.	100.	******	100.	100.	100.	100.
1968	99.	66.	129.	95.	83.	120.	******	108.	11.	89.	109.
1969	103.	5.	123.	125.	124.	142.	******	149.	41.	130.	128.
1970	99.	31.	168.	109.	80.	143.	******	145.	0.	106.	124.
1971	102.	0.	158.	90.	103.	127.	******	157.	0.	73.	119.
1972	106.	2.	167.	117.	227.	137.	******	179.	0.	161.	137.
1973	153.	542.	238.	130.	253.	180.	******	258.	0.	51.	183.
1974	232.	615.	486.	161.	390.	311.	******	503.	1.	0.	305.
1975	198.	2220.	278.	72.	455.	293.	******	434.	1.	0.	281.
1976	239.	7450.	452.	96.	506.	364.	******	559.	0.	0.	347.

IMPORTS
NEWSPRINT (6411)
UNITED STATES

VALUE--MILLION CURRENT DOLLARS

YEAR	CANADA	FINLAND	OTH EURO	ASIA	AFRICA	OTHERS	TOTAL
1967	860.042	32.030	0.017	0.000	0.000	0.000	892.090
1968	829.378	33.211	0.206	0.000	0.000	0.000	862.795
1969	903.524	34.752	0.349	0.000	0.000	0.001	938.626
1970	890.763	38.058	0.805	0.000	0.000	0.000	929.626
1971	953.980	32.205	1.635	0.000	0.000	0.000	987.819
1972	1014.458	39.289	0.000	0.000	0.000	0.000	1053.747
1973	1132.589	52.390	0.000	0.000	0.000	0.000	1184.979
1974	1451.014	33.039	0.000	0.000	0.000	0.042	1484.095
1975	1411.111	7.009	0.005	0.002	0.000	0.000	1418.127
1976	1726.548	1.394	14.198	0.004	0.000	0.000	1742.144

QUANTITY--1000 STN

YEAR	CANADA	FINLAND	OTH EURO	ASIA	AFRICA	OTHERS	TOTAL
1967	6315.032	281.121	0.178	0.000	0.000	0.000	6596.331
1968	6178.179	284.107	1.839	0.000	0.000	0.000	6464.125
1969	6496.632	287.461	3.014	0.000	0.000	0.008	6787.115
1970	6320.353	309.169	6.906	0.000	0.000	0.000	6636.428
1971	6566.633	299.770	13.599	0.000	0.000	0.000	6880.002
1972	6757.801	325.320	0.000	0.000	0.000	0.000	7083.121
1973	7029.365	384.207	0.000	0.000	0.000	0.000	7413.572
1974	7185.046	222.255	0.000	0.000	0.000	0.017	7407.317
1975	5812.460	34.852	0.003	0.010	0.000	0.000	5847.325
1976	6509.886	5.839	48.652	0.020	0.000	0.001	6564.398

IMPORTS
NEWSPRINT (6411)
UNITED STATES

TRADE STRUCTURE-- % OF TOTAL TRADE (VALUE BASIS)

YEAR	CANADA	FINLAND	OTH EURO	ASIA	AFRICA	OTHERS	TOTAL
1967	96.	4.	0.	0.	0.	0.	100.
1968	96.	4.	0.	0.	0.	0.	100.
1969	96.	4.	0.	0.	0.	0.	100.
1970	96.	4.	0.	0.	0.	0.	100.
1971	97.	3.	0.	0.	0.	0.	100.
1972	96.	4.	0.	0.	0.	0.	100.
1973	96.	4.	0.	0.	0.	0.	100.
1974	98.	2.	0.	0.	0.	0.	100.
1975	100.	0.	0.	0.	0.	0.	100.
1976	99.	0.	1.	0.	0.	0.	100.

TRADE TRENDS-- % OF 1967 TRADE (VALUE BASIS)

YEAR	CANADA	FINLAND	OTH EURO	ASIA	AFRICA	OTHERS	TOTAL
1967	100.	100.	100.	******	******	******	100.
1968	96.	104.	1183.	******	******	******	97.
1969	105.	108.	2005.	******	******	******	105.
1970	104.	119.	4625.	******	******	******	104.
1971	111.	101.	9396.	******	******	******	111.
1972	118.	123.	0.	******	******	******	118.
1973	132.	164.	0.	******	******	******	133.
1974	169.	103.	0.	******	******	******	166.
1975	164.	22.	28.	******	******	******	159.
1976	201.	4.	81609.	******	******	******	195.

IMPORTS
NEWSPRINT (6411)

VALUE--MILLION CURRENT DOLLARS

YEAR	N ATLANTC	S ATLANTC	GULF	S PACIFIC	PACIFC NW	GRT LAKES	ALASKA	N CENTRAL	S CENTRAL	OTHERS	TOTAL
1967	179.473	7.019	42.655	79.822	44.854	449.717	0.006	54.178	0.098	34.267	892.090
1968	190.577	7.221	43.503	65.020	49.334	449.262	0.088	51.971	0.075	5.745	862.795
1969	190.805	8.199	36.005	69.218	60.792	516.739	0.313	49.308	0.070	7.179	938.626
1970	194.307	9.006	41.102	68.866	55.626	502.354	0.517	48.887	0.046	8.915	929.626
1971	184.038	11.151	38.009	56.383	86.452	549.355	0.614	51.424	0.000	10.393	987.819
1972	206.866	10.301	49.412	68.861	77.028	577.218	0.741	53.868	0.000	9.453	1053.747
1973	218.769	13.404	60.916	94.332	70.942	663.370	0.683	52.277	0.000	10.286	1184.979
1974	263.755	6.756	68.749	111.793	79.084	864.249	1.053	76.567	0.000	12.088	1484.095
1975	232.376	2.816	63.961	107.644	100.997	834.423	1.312	63.909	0.000	10.689	1418.127
1976	308.500	2.704	66.709	131.073	146.520	1000.461	0.928	66.245	0.035	18.969	1742.144

QUANTITY--1000 STN

YEAR	N ATLANTC	S ATLANTC	GULF	S PACIFIC	PACIFC NW	GRT LAKES	ALASKA	N CENTRAL	S CENTRAL	OTHERS	TOTAL
1967	1362.396	54.499	345.855	681.448	366.745	3318.593	0.050	415.682	0.725	50.337	6596.331
1968	1424.344	55.347	348.109	537.067	392.417	3257.139	0.810	400.952	0.591	47.349	6464.125
1969	1358.879	70.603	286.638	542.631	458.634	3632.710	2.739	376.398	0.519	57.364	6787.115
1970	1351.303	69.984	304.078	544.564	408.705	3516.726	4.567	366.463	0.344	69.694	6636.428
1971	1291.197	96.324	322.705	457.935	639.470	3618.866	5.350	369.442	0.000	78.713	6880.002
1972	1353.176	96.392	342.592	606.378	514.965	3721.346	4.822	374.759	0.000	68.691	7083.121
1973	1339.409	94.004	401.937	630.801	448.011	4064.261	4.144	342.668	0.000	88.337	7413.572
1974	1300.496	41.554	378.119	606.324	386.208	4222.174	5.302	407.012	0.000	60.127	7407.317
1975	953.768	12.718	277.450	451.985	395.079	3432.714	5.536	276.352	0.000	41.723	5847.325
1976	1156.635	11.028	301.220	561.002	541.124	3672.241	3.527	251.940	0.146	65.535	6564.398

IMPORTS
NEWSPRINT (6411)

TRADE STRUCTURE-- % OF TOTAL TRADE (VALUE BASIS)

YEAR	N ATLANTC	S ATLANTC	GULF	S PACIFIC	PACIFC NW	GRT LAKES	ALASKA	N CENTRAL	S CENTRAL	OTHERS	TOTAL
1967	20.	1.	5.	9.	5.	50.	0.	6.	0.	4.	100.
1968	22.	1.	5.	8.	6.	52.	0.	6.	0.	1.	100.
1969	20.	1.	4.	7.	6.	55.	0.	5.	0.	1.	100.
1970	21.	1.	4.	7.	6.	54.	0.	5.	0.	1.	100.
1971	19.	1.	4.	6.	9.	56.	0.	5.	0.	1.	100.
1972	20.	1.	5.	7.	7.	55.	0.	5.	0.	1.	100.
1973	18.	1.	5.	8.	6.	56.	0.	4.	0.	1.	100.
1974	18.	0.	5.	8.	5.	58.	0.	5.	0.	1.	100.
1975	16.	0.	5.	8.	7.	59.	0.	5.	0.	1.	100.
1976	18.	0.	4.	8.	8.	57.	0.	4.	0.	1.	100.

TRADE TRENDS-- % OF 1967 TRADE (VALUE BASIS)

YEAR	N ATLANTC	S ATLANTC	GULF	S PACIFIC	PACIFC NW	GRT LAKES	ALASKA	N CENTRAL	S CENTRAL	OTHERS	TOTAL
1967	100.	100.	100.	100.	100.	100.	100.	100.	100.	100.	100.
1968	106.	103.	102.	81.	110.	100.	1508.	96.	76.	17.	97.
1969	106.	117.	84.	87.	136.	115.	5347.	91.	71.	21.	105.
1970	108.	128.	96.	86.	124.	112.	8830.	90.	46.	26.	104.
1971	103.	159.	89.	71.	193.	122.	10473.	95.	0.	30.	111.
1972	115.	147.	116.	86.	172.	128.	12637.	99.	0.	28.	118.
1973	122.	191.	143.	118.	158.	148.	11652.	96.	0.	30.	133.
1974	147.	96.	161.	140.	176.	192.	17973.	141.	0.	35.	166.
1975	129.	40.	150.	135.	225.	186.	22382.	118.	0.	31.	159.
1976	172.	39.	156.	164.	327.	222.	15842.	122.	35.	55.	195.

IMPORTS
BUILDING BOARD (6416)
UNITED STATES

VALUE--MILLION CURRENT DOLLARS

YEAR	CANADA	BRAZIL	OTH S AM	SCANDINA	OTH EURO	OTHERS	TOTAL
1967	3.850	1.798	0.031	7.837	0.881	0.757	15.153
1968	5.855	1.936	0.089	8.974	1.001	2.457	20.312
1969	7.038	1.394	0.153	8.627	1.398	3.582	22.192
1970	6.839	1.898	0.146	5.638	0.892	1.565	16.979
1971	7.746	2.843	0.134	6.986	0.887	2.012	20.608
1972	32.772	4.423	0.442	10.764	3.022	4.423	55.846
1973	16.540	5.968	1.108	10.449	3.933	6.475	44.474
1974	12.004	8.773	1.779	8.768	3.874	4.644	39.843
1975	5.455	5.253	0.663	1.297	0.779	1.187	14.634
1976	6.231	8.677	1.159	3.899	3.084	3.003	26.052

QUANTITY--1000 STN

YEAR	CANADA	BRAZIL	OTH S AM	SCANDINA	OTH EURO	OTHERS	TOTAL
1967	37.945	20.624	0.530	106.608	15.992	9.768	191.465
1968	58.916	22.836	1.577	126.591	17.757	31.675	259.352
1969	68.936	17.349	2.999	121.062	22.998	45.211	278.555
1970	62.542	24.845	2.657	71.682	13.663	41.626	217.016
1971	74.592	36.007	2.234	94.913	13.621	28.465	249.832
1972	110.333	51.767	7.221	132.973	43.951	60.998	407.242
1973	131.214	56.463	12.251	89.800	45.084	72.003	406.815
1974	72.651	61.648	16.734	59.403	28.734	44.999	284.170
1975	36.840	37.697	6.746	9.318	8.861	10.594	110.056
1976	38.162	56.495	9.856	31.287	33.519	22.717	192.037

IMPORTS
BUILDING BOARD (6416)
UNITED STATES

TRADE STRUCTURE-- % OF TOTAL TRADE (VALUE BASIS)

YEAR	CANADA	BRAZIL	OTH S AM	SCANDINA	OTH EURO	OTHERS	TOTAL
1967	25.	12.	0.	52.	6.	5.	100.
1968	29.	10.	0.	44.	5.	12.	100.
1969	32.	6.	1.	39.	6.	16.	100.
1970	40.	11.	1.	33.	5.	9.	100.
1971	38.	14.	1.	34.	4.	10.	100.
1972	59.	8.	1.	19.	5.	8.	100.
1973	37.	13.	2.	23.	9.	15.	100.
1974	30.	22.	4.	22.	10.	12.	100.
1975	37.	36.	5.	9.	5.	8.	100.
1976	24.	33.	4.	15.	12.	12.	100.

TRADE TRENDS-- % OF 1967 TRADE (VALUE BASIS)

YEAR	CANADA	BRAZIL	OTH S AM	SCANDINA	OTH EURO	OTHERS	TOTAL
1967	100.	100.	100.	100.	100.	100.	100.
1968	152.	108.	290.	115.	114.	325.	134.
1969	183.	78.	496.	110.	159.	473.	146.
1970	178.	106.	474.	72.	101.	207.	112.
1971	201.	158.	434.	89.	101.	266.	136.
1972	851.	246.	1437.	137.	343.	584.	369.
1973	430.	332.	3599.	133.	447.	856.	294.
1974	312.	488.	5776.	112.	440.	614.	263.
1975	142.	292.	2154.	17.	88.	157.	97.
1976	162.	483.	3762.	50.	350.	397.	172.

IMPORTS
BUILDING BOARD (6416)

VALUE--MILLION CURRENT DOLLARS

YEAR	N ATLANTC	S ATLANTC	GULF	S PACIFIC	PACIFC NW	GRT LAKES	ALASKA	N CENTRAL	S CENTRAL	OTHERS	TOTAL
1967	7.612	0.693	1.975	1.094	0.830	2.535	0.002	0.200	0.000	0.211	15.153
1968	8.879	1.223	3.181	2.023	0.570	3.886	0.000	0.266	0.000	0.284	20.312
1969	6.812	1.243	5.509	2.307	0.593	5.176	0.000	0.269	0.000	0.282	22.192
1970	5.373	0.964	2.987	2.413	0.429	4.258	0.000	0.263	0.000	0.292	16.979
1971	6.659	1.356	4.439	1.846	0.779	4.930	0.000	0.312	0.000	0.286	20.608
1972	30.336	3.557	5.849	4.401	1.128	9.547	0.000	0.497	0.000	0.531	55.846
1973	11.076	4.162	9.057	5.447	1.784	11.404	0.000	1.154	0.000	0.391	44.474
1974	9.648	4.236	8.714	6.682	1.533	7.860	0.000	0.877	0.000	0.292	39.843
1975	3.680	0.462	2.588	2.400	0.692	3.748	0.000	0.633	0.000	0.432	14.634
1976	5.617	2.697	5.837	5.070	1.436	4.019	0.000	1.147	0.000	0.229	26.052

QUANTITY--1000 STN

YEAR	N ATLANTC	S ATLANTC	GULF	S PACIFIC	PACIFC NW	GRT LAKES	ALASKA	N CENTRAL	S CENTRAL	OTHERS	TOTAL
1967	99.872	10.306	26.516	15.151	10.099	23.512	0.011	2.464	0.000	3.534	191.465
1968	114.507	19.251	43.253	30.047	6.851	37.467	0.000	3.267	0.000	4.709	259.352
1969	87.022	19.225	73.531	33.938	7.196	49.769	0.000	3.585	0.000	4.290	278.555
1970	64.566	12.592	60.014	32.240	4.685	35.168	0.000	3.390	0.000	4.361	217.016
1971	84.042	20.936	58.872	24.025	9.123	45.211	0.000	3.789	0.000	3.834	249.832
1972	120.066	46.838	73.700	56.251	11.219	86.164	0.000	5.171	0.000	7.833	407.242
1973	102.138	41.261	87.174	52.647	15.985	91.861	0.000	11.339	0.000	4.410	406.815
1974	68.882	33.053	66.371	50.875	10.549	45.210	0.000	7.199	0.000	2.031	284.170
1975	29.902	3.930	20.586	18.580	4.474	23.716	0.000	5.175	0.000	3.693	110.056
1976	44.754	23.734	45.057	36.396	10.074	22.277	0.000	8.064	0.000	1.680	192.037

IMPORTS
BUILDING BOARD (6416)

TRADE STRUCTURE-- % OF TOTAL TRADE (VALUE BASIS)

YEAR	N ATLANTC	S ATLANTC	GULF	S PACIFIC	PACIFC NW	GRT LAKES	ALASKA	N CENTRAL	S CENTRAL	OTHERS	TOTAL
1967	50.	5.	13.	7.	5.	17.	0.	1.	0.	1.	100.
1968	44.	6.	16.	10.	3.	19.	0.	1.	0.	1.	100.
1969	31.	6.	25.	10.	3.	23.	0.	1.	0.	1.	100.
1970	32.	6.	18.	14.	3.	25.	0.	2.	0.	2.	100.
1971	32.	7.	22.	9.	4.	24.	0.	2.	0.	1.	100.
1972	54.	6.	10.	8.	2.	17.	0.	1.	0.	1.	100.
1973	25.	9.	20.	12.	4.	26.	0.	3.	0.	1.	100.
1974	24.	11.	22.	17.	4.	20.	0.	2.	0.	1.	100.
1975	25.	3.	18.	16.	5.	26.	0.	4.	0.	3.	100.
1976	22.	10.	22.	19.	6.	15.	0.	4.	0.	1.	100.

TRADE TRENDS-- % OF 1967 TRADE (VALUE BASIS)

YEAR	N ATLANTC	S ATLANTC	GULF	S PACIFIC	PACIFC NW	GRT LAKES	ALASKA	N CENTRAL	S CENTRAL	OTHERS	TOTAL
1967	100.	100.	100.	100.	100.	100.	100.	100.	******	100.	100.
1968	117.	177.	161.	185.	69.	153.	0.	133.	******	135.	134.
1969	89.	179.	279.	211.	71.	204.	0.	134.	******	134.	146.
1970	71.	139.	151.	220.	52.	168.	0.	132.	******	138.	112.
1971	87.	196.	225.	169.	94.	194.	0.	156.	******	136.	136.
1972	399.	513.	296.	402.	136.	377.	0.	248.	******	252.	369.
1973	146.	601.	459.	498.	215.	450.	0.	576.	******	185.	294.
1974	127.	611.	441.	611.	185.	310.	0.	438.	******	139.	263.
1975	48.	67.	131.	219.	83.	148.	0.	316.	******	205.	97.
1976	74.	389.	296.	463.	173.	159.	0.	573.	******	109.	172.

IMPORTS
PAPER & PAPERBOARD (6410)
UNITED STATES

VALUE--MILLION CURRENT DOLLARS

YEAR	CANADA	SWEDEN	FINLAND	OTH EURO	JAPAN	OTHERS	TOTAL
1967	33.419	6.221	7.040	9.824	2.426	0.520	59.450
1968	34.988	5.708	8.324	10.277	2.514	1.064	62.874
1969	48.364	5.189	9.071	10.329	2.938	1.649	77.539
1970	61.541	5.136	10.149	10.458	3.808	1.339	92.432
1971	67.460	3.949	8.285	9.886	2.885	0.987	93.452
1972	66.666	3.566	8.890	14.991	3.402	1.660	99.174
1973	102.826	7.037	16.588	19.132	4.403	2.691	152.678
1974	144.542	12.875	25.479	26.403	8.257	3.423	220.979
1975	85.642	5.804	19.383	25.605	5.280	3.433	145.147
1976	140.295	9.548	21.283	32.498	8.302	7.040	218.966

QUANTITY--1000 STN

YEAR	CANADA	SWEDEN	FINLAND	OTH EURO	JAPAN	OTHERS	TOTAL
1967	188.702	36.750	42.740	16.596	8.037	0.244	293.067
1968	187.105	32.526	56.217	20.568	2.316	2.062	300.794
1969	253.446	27.959	58.782	19.478	5.987	7.145	372.795
1970	315.129	22.403	64.822	16.412	7.176	1.399	427.341
1971	373.463	16.239	53.444	14.055	4.396	0.974	462.570
1972	357.207	12.764	54.645	18.155	1.852	5.228	449.852
1973	504.923	20.217	88.070	23.320	1.909	6.129	644.569
1974	482.618	24.110	81.200	24.155	4.025	2.810	618.919
1975	280.892	9.186	60.311	18.055	1.820	4.830	375.093
1976	411.333	15.310	76.340	19.158	2.419	9.790	534.350

IMPORTS
PAPER & PAPERBOARD (6410)
UNITED STATES

TRADE STRUCTURE-- % OF TOTAL TRADE (VALUE BASIS)

YEAR	CANADA	SWEDEN	FINLAND	OTH EURO	JAPAN	OTHERS	TOTAL
1967	56.	10.	12.	17.	4.	1.	100.
1968	56.	9.	13.	16.	4.	2.	100.
1969	62.	7.	12.	13.	4.	2.	100.
1970	67.	6.	11.	11.	4.	1.	100.
1971	72.	4.	9.	11.	3.	1.	100.
1972	67.	4.	9.	15.	3.	2.	100.
1973	67.	5.	11.	13.	3.	2.	100.
1974	65.	6.	12.	12.	4.	2.	100.
1975	59.	4.	13.	18.	4.	2.	100.
1976	64.	4.	10.	15.	4.	3.	100.

TRADE TRENDS-- % OF 1967 TRADE (VALUE BASIS)

YEAR	CANADA	SWEDEN	FINLAND	OTH EURO	JAPAN	OTHERS	TOTAL
1967	100.	100.	100.	100.	100.	100.	100.
1968	105.	92.	118.	105.	104.	204.	106.
1969	145.	83.	129.	105.	121.	317.	130.
1970	184.	83.	144.	106.	157.	257.	155.
1971	202.	63.	118.	101.	119.	190.	157.
1972	199.	57.	126.	153.	140.	319.	167.
1973	308.	113.	236.	195.	182.	517.	257.
1974	433.	207.	362.	269.	340.	658.	372.
1975	256.	93.	275.	261.	218.	660.	244.
1976	420.	153.	302.	331.	342.	1353.	368.

IMPORTS
PAPER & PAPERBOARD (6410)

VALUE--MILLION CURRENT DOLLARS

YEAR	N ATLANTC	S ATLANTC	GULF	S PACIFIC	PACIFC NW	GRT LAKES	ALASKA	N CENTRAL	S CENTRAL	OTHERS	TOTAL
1967	21.558	1.100	1.026	4.700	0.953	29.029	0.000	0.575	0.000	0.510	59.450
1968	21.982	1.631	1.882	5.499	1.470	29.054	0.006	0.719	0.012	0.619	62.874
1969	22.408	1.106	2.453	6.003	3.295	40.118	0.002	1.597	0.001	0.557	77.539
1970	27.384	1.017	3.638	7.212	3.947	46.854	0.000	1.835	0.001	0.544	92.432
1971	26.015	0.448	2.390	6.315	5.955	50.362	0.001	1.244	0.000	0.722	93.452
1972	32.721	0.689	1.425	6.614	3.281	51.084	0.002	1.643	0.013	1.703	99.174
1973	53.761	0.829	1.962	9.993	2.702	75.911	0.014	4.570	0.005	2.932	152.678
1974	76.250	1.684	4.275	9.632	5.709	114.213	0.000	5.989	0.106	3.120	220.979
1975	50.863	1.052	2.179	6.472	1.664	74.199	0.000	3.537	0.284	4.898	145.147
1976	71.400	4.202	3.368	14.273	3.047	115.304	0.000	4.322	0.419	2.631	218.966

QUANTITY--1000 STN

YEAR	N ATLANTC	S ATLANTC	GULF	S PACIFIC	PACIFC NW	GRT LAKES	ALASKA	N CENTRAL	S CENTRAL	OTHERS	TOTAL
1967	85.594	6.161	4.738	24.199	7.509	158.895	0.000	3.302	0.000	2.670	293.067
1968	85.880	11.369	11.825	26.886	10.673	146.593	0.057	4.239	0.062	3.210	300.794
1969	87.420	5.884	15.448	28.144	23.129	201.441	0.041	8.688	0.000	2.600	372.795
1970	99.448	4.183	24.072	32.632	24.377	230.471	0.000	10.102	0.007	2.049	427.341
1971	100.565	1.443	14.986	31.506	38.327	265.761	0.004	6.990	0.000	2.987	462.570
1972	119.421	1.852	7.014	28.384	15.952	258.434	0.010	7.749	0.005	11.030	449.852
1973	202.226	3.371	6.352	35.959	16.019	344.378	0.008	19.068	0.032	17.155	644.569
1974	170.529	2.621	10.414	15.835	23.778	370.110	0.000	17.928	0.775	6.929	618.919
1975	98.417	1.176	3.745	5.805	5.647	239.003	0.000	11.307	1.997	7.996	375.093
1976	137.736	8.519	6.724	20.034	11.052	328.041	0.000	13.889	3.392	4.961	534.350

IMPORTS
PAPER & PAPERBOARD (6410)

TRADE STRUCTURE-- % OF TOTAL TRADE (VALUE BASIS)

YEAR	N ATLANTC	S ATLANTC	GULF	S PACIFIC	PACIFC NW	GRT LAKES	ALASKA	N CENTRAL	S CENTRAL	OTHERS	TOTAL
1967	36.	2.	2.	8.	2.	49.	0.	1.	0.	1.	100.
1968	35.	3.	3.	9.	2.	46.	0.	1.	0.	1.	100.
1969	29.	1.	3.	8.	4.	52.	0.	2.	0.	1.	100.
1970	30.	1.	4.	8.	4.	51.	0.	2.	0.	1.	100.
1971	28.	0.	3.	7.	6.	54.	0.	1.	0.	1.	100.
1972	33.	1.	1.	7.	3.	52.	0.	2.	0.	2.	100.
1973	35.	1.	1.	7.	2.	50.	0.	3.	0.	2.	100.
1974	35.	1.	2.	4.	3.	52.	0.	3.	0.	1.	100.
1975	35.	1.	2.	4.	1.	51.	0.	2.	0.	3.	100.
1976	33.	2.	2.	7.	1.	53.	0.	2.	0.	1.	100.

TRADE TRENDS-- % OF 1967 TRADE (VALUE BASIS)

YEAR	N ATLANTC	S ATLANTC	GULF	S PACIFIC	PACIFC NW	GRT LAKES	ALASKA	N CENTRAL	S CENTRAL	OTHERS	TOTAL
1967	100.	100.	100.	100.	100.	100.	******	100.	******	100.	100.
1968	102.	148.	184.	117.	154.	100.	******	125.	******	121.	106.
1969	104.	101.	239.	128.	346.	138.	******	278.	******	109.	130.
1970	127.	92.	355.	153.	414.	161.	******	319.	******	107.	155.
1971	121.	41.	233.	134.	625.	173.	******	217.	******	142.	157.
1972	152.	63.	139.	141.	344.	176.	******	286.	******	334.	167.
1973	249.	75.	191.	213.	284.	262.	******	795.	******	575.	257.
1974	354.	153.	417.	205.	599.	393.	******	1042.	******	612.	372.
1975	236.	96.	212.	138.	175.	256.	******	615.	******	961.	244.
1976	331.	382.	328.	304.	320.	397.	******	752.	******	516.	368.

Appendix D

THE PACIFIC NORTHWEST REGION

Exports

Regional Overview

Throughout the 1967-76 period, the Pacific Northwest was the largest forest-product-exporting region in the United States. In 1976 the region exported products valued at $1.57 billion, which represented 39 percent of the total U.S. exports of forest products. The 1976 value was more than five times the value of 1967 exports, which totaled $311.9 million.

Since 1967, the region's export trade has been on a persistent upward trend which mirrors the growth in forest product exports for the United States as a whole. The region's percentage share of U.S. exports has grown slightly but, in general, has been fairly stable. One exceptional year was 1973, in which the Pacific Northwest exported 47 percent of the value of U.S. forest product exports.

The region's export trade is heavily concentrated in solid-wood commodities (non-pulp and paper), especially softwood logs. In 1976, 79 percent of the value of the region's forest product exports was in softwood logs, softwood lumber, veneer, plywood, and pulpwood chips. Softwood logs constituted 63 percent of the $1.23 billion in exports of the five solid-wood commodities.

Wood pulp accounted for $173.5 million, or 11 percent of the value of the region's exports in 1976, and paper and paperboard other than newsprint and building board accounted for another 9 percent ($135.3 million).

Throughout the period 1967-76, pulp, paper, and paperboard have steadily assumed less significance as a Pacific Northwest export; despite nearly tripling in value since 1967, pulp, paper, and paperboard have decreased their percentage share of total forest product exports from 35 percent in 1967 to 21 percent in 1976. The nominal value of the solid wood products during the same period increased by more than 500 percent, from $200.5 million in 1967 to $1.25 billion in 1976.

The commodity structure of the region's trade is quite different from the commodity structure of exports from the United States as a whole. As noted above, the Pacific Northwest is a heavy exporter of solid wood products; these products in 1976 represented 46 percent of the value of total U.S. forest product exports. Paper and paperboard other than newsprint and building board was the largest single component of U.S. export trade, accounting for 30 percent of the value in 1976, while constituting only 9 percent of the Pacific Northwest exports. Similarly, wood pulp was 22 percent of the value of national exports, but only 11 percent of regional exports.

<u>Softwood Sawlogs</u>

The United States is the world's largest exporter of softwood sawlogs, and the Pacific Northwest region in 1976 accounted for over 90 percent of the value of U.S. softwood log exports. Log exports from the Pacific Northwest in 1976 totaled nearly 3.5 billion board-feet and were valued at $778.4 million. That value represented an increase of 456 percent over the 1967 value.

Japan has been far and away the largest importer of sawlogs from this region, accounting for at least 93 percent of the value of the region's log exports every year during the 1967-76 period. Although Japan's percentage share remained fairly constant during the study period, its absolute consumption of logs from this area drastically increased. In value terms, Japanese imports of logs in 1976 were 545 percent of its 1967 imports, but the volume of logs increased by only 85 percent during the same period.

Korea has rapidly expanded its import of logs from the Pacific Northwest, but in 1976 consumed only 4 percent of the region's exports. Nevertheless, that 4 percent represented exports of 197 million board-feet, valued at $28.6 million. In value terms, Korea's 1976 consumption of logs is more than ten times the 1967 amount; in volume terms, exports to Korea increased by 433 percent.

Throughout 1967-76, the residual of logs not shipped to Japan or Korea was exported to Canada. This amount represented only 1 percent of the region's log exports, but in 1976 totaled 48 million board-feet and was valued at nearly $8 million.

Softwood Lumber

The Pacific Northwest accounted for nearly half the U.S. exports of softwood lumber in 1976, shipping out 800 million board-feet (valued at $224.5 million). The 1976 value represented an increase of 470 percent over the 1967 value for the Pacific Northwest, while the United States as a whole increased the value of softwood lumber exports by 350 percent. The U.S. peak export year, in terms of quantity of softwood lumber, was 1973, and in that year the Pacific Northwest accounted for 51 percent of the value of total U.S. softwood lumber exports and 45 percent of the volume.

Nearly three-fourths of the Pacific Northwest's softwood lumber exports are shipped to only four countries--Canada, Italy, Japan, and Australia. In 1976, in value terms, Australia and Italy each took 22 percent of the total exports, while Japan took 18 percent and Canada 11 percent.

Most (21 percent) of the remainder was shipped to European countries other than Italy.

During 1967-76, Canada, Australia, Italy, and the other European countries maintained relatively stable market share levels of exports from the Pacific Northwest. Japan, however, increased its share from only 8 percent of the value in 1967 to 18 percent in 1976. "Other" countries declined in importance, from 14 percent in 1967 to only 6 percent in 1976.

Throughout 1967-76, relatively low unit values prevailed for Japan--in 1967 Japan imported 8 percent of the value of exports from this region, but 15 percent of the volume. In 1976 the figures were 18 percent of value and 26 percent of volume.

During the same period, Italy faced relatively high unit values, as, for example, in 1976, when it consumed 22 percent of the value of the region's exports, but only 15 percent of the volume.

Wood Pulp

The Pacific Northwest region in 1976 was the origin of one-fifth of the value of U.S. exports of wood pulp. The 1976 export value ($173.5 million) increased 165 percent over the 1967 value ($65.5 million); but in volume terms, exports declined by 2 percent from 1967 to 1976.

Japan is the major importer of Pacific Northwest pulp, in 1976 consuming 155,000 tons. Valued at $52.5 million, this amounted to 30 percent of the value of the region's wood pulp exports in that year.

European nations consumed 47 percent of the value of the region's pulp exports in 1976. The major importing countries were the United Kingdom ($21.8 million, 13 percent), Italy ($16.5 million, 10 percent), and Belgium ($14.7 million, 8 percent). Other significant importers include

France, West Germany, and the Netherlands. Asian countries other than Japan took 11 percent of the value of the region's exports in 1976, a total of 175,000 tons (valued at $18.3 million).

Since 1967 the country structure of the region's wood-pulp export trade has changed only slightly; the European countries accounted for about 7 percent more of the value in 1976 than in 1967, and Asian countries other than Japan accounted for about 7 percent less than in 1967.

Paper and Paperboard

The Pacific Northwest throughout 1967-76 exported 8 to 12 percent of the value of U.S. exports of paper and paperboard (excluding newsprint and building board). The region's percentage share of U.S. exports has remained fairly stable over the period. In 1976 the region exported 410,000 tons valued at $135.3 million, which represented 11 percent of the value of U.S. exports of paper and paperboard.

From 1967 to 1976, the value of Pacific Northwest exports of paper and paperboard increased at an average annual rate of 13 percent, and the volume increased at a 4 percent rate.

As with other forest products in the Pacific Northwest region, paper and paperboard exports are shipped principally to Canada and Japan. Each of these countries, in 1976, imported 27 percent of the value of the region's paper and paperboard exports. Japan, however, imported 13 percent less volume than did Canada.

Australia is the third largest single importer of the region's paper and paperboard, accounting in 1976 for 24,000 tons, valued at $11.7 million, which was 9 percent of the value of the region's exports.

Asian countries other than Japan in 1976 imported 92,000 tons valued at $22.5 million, which represented 17 percent of the value of the region's exports of paper and paperboard.

European countries took 15 percent of the 1976 value of regional exports, consuming 88,000 tons valued at $20.2 million.

The country structure of the paper and paperboard export trade has undergone some significant changes since 1967. Japan, in particular, has increased its share of the value of regional exports from 1 percent in 1967 to 27 percent in 1976. Canada has also experienced a drastic increase, from 15 percent in 1967 to 27 percent in 1976. Australia has increased its share by about 6 percent, while Europe and Asian countries other than Japan have declined 5 and 9 percent, respectively. The role of all other countries has steadily declined from 1967, when this residual group captured 35 percent of the market, to 1976, when it accounted for only 5 percent of the value of regional exports.

Japan's rate of growth in paper and paperboard imports from the Pacific Northwest has been phenomenal. During the 1967-76 period, Japan's imports increased in value at an annual rate of 58 percent and in volume at 43 percent per year. Canada also experienced substantial growth, increasing the value of its imports, on the average, by 22 percent annually, and the volume by 16 percent per year. During this period, all countries importing Pacific Northwest paper and paperboard had average annual compounded increases of 13 percent in value and 4 percent in volume.

Pulpwood Chips

Although pulpwood roundwood was the least valuable forest product export from the Pacific Northwest in 1976, pulpwood in chip form was very valuable, having a 1976 export volume of nearly 4 million tons (valued at $130.5 million).

The Pacific Northwest led the nation in the export of pulpwood chips during 1967-76, always accounting for more than 85 percent of the value of U.S. pulpwood chip exports. During that period the value of Pacific Northwest chip exports rose from $11.9 million to $130.5 million, an average annual increase of 30 percent, while volume increased from 765,000 to 3.3 million tons, or an annual growth rate of 18 percent. In 1976 pulpwood chips constituted 8 percent of the value of all forest product exports from the Pacific Northwest.

In 1976 and throughout the 1967-76 period, Japan consumed greater than 99 percent of the value of pulpwood chip exports. In 1967 Japan imported 765,000 tons (valued at $11.9 million). The 1976 value of $129.1 million and volume of 3.2 million tons represented average annual growth rates of 30 percent in value, and 17 percent in volume.

Plywood

The Pacific Northwest, in 1976, exported 63 percent of the value of U.S. plywood exports, more than three times the amount exported by any other U.S. region. The $93.4 million of Pacific Northwest plywood exports amounted to some 698 million square feet, which was 70 percent of the volume exported by the entire United States.

Consumption of exported plywood has been heavily concentrated in Western European countries. Denmark (30 percent), the United Kingdom (24 percent), Belgium (12 percent), and the Netherlands (10 percent) accounted for 76 percent of the region's exports in 1976, and West Germany, France, and Italy together took another 10 percent.

Other than European nations, Canada has been the largest importer of the region's plywood, taking, in 1976, 51 million square feet, valued at nearly $7 million. This amount represented 7 percent of the value of the region's total plywood exports.

Pacific Northwest exports of plywood grew in value at an average annual rate of 36 percent, although the value dropped in 1970, 1971, and 1974. The largest percentage gain in value (312 percent) was recorded in 1969.

During 1967-76, Belgium was the fastest growing importer of plywood, gaining, on average, 73 percent in value per year. Exports to Canada and the Netherlands each grew at 52 percent per year, and exports to Denmark grew at 40 percent per year.

Average growth rates can give a misleading impression of steady year-to-year growth. For example, although Denmark averaged a 40 percent increase in value of Pacific Northwest plywood exports, it experienced declines in four of the ten years, and year-to-year increases of over 100 percent in three years. Denmark's percentage share of the region's plywood market also varied tremendously over the period, with peaks in 1969 (48 percent), 1972 (56 percent), and 1973 (47 percent), and with low points in 1967 (23 percent), 1970 (36 percent), and 1974 (23 percent).

Other Forest Products

The remaining forest product exports from the Pacific Northwest totaled $34.1 million, which was only 2 percent of the value of all forest products exported from the region.

The most valuable commodity in this group was newsprint, which had an export value of $11.7 million in 1976. In that year, the region exported nearly one-fourth of the value of all U.S. exports of newsprint. Very nearly all of the region's exports of 37,000 tons were shipped to Asia, principally to Hong Kong and Taiwan.

Wood veneer, valued at $8.1 million, was exported from the Pacific Northwest in 1976. That value represented nearly one-fifth of all U.S. exports of veneer and was an increase of 272 percent over the 1967 regional export value. Canada has been the importer of virtually all wood veneer exported from the region during 1967-76.

In 1976 the Pacific Northwest exported $5.6 million of reconstituted wood, which was 30 percent of the total value of U.S. exports of this commodity. During 1967-76, exports increased at a rapid rate, both in value and volume. The average annual growth rate, in value, was 24 percent and, in volume, was 54 percent. Canada was the major importer of reconstituted wood from the Pacific Northwest, taking on average 94 percent of the value of the region's exports.

Imports

Regional Overview

The Pacific Northwest region in 1976 was, in value terms, the third largest forest-product-importing region in the United States, accounting for 13 percent of the total U.S. forest product imports. This percentage share had grown steadily since 1967, when the Pacific Northwest imported 8 percent of the country's forest product imports. In current dollars, the region's imports grew from $153.3 million in 1967 to $692.6 million in 1976. This growth rate was substantially above the national growth rate in forest product imports for the 1967-76 period, which averaged 12 percent annually.

Softwood lumber was the region's major forest product import throughout 1967-76, constituting nearly half (in value terms) of the region's total forest product imports in 1976. The remainder of the Pacific Northwest imports was composed largely of wood pulp (20 percent) and newsprint (21 percent). The trend during the period was toward the import of more solid wood products and less pulp and paper products; while the composition of the region's imports was split evenly between these two product groups in 1967, the solid wood products comprised nearly 60 percent of the total regional imports in 1976. A similar trend, although not as pronounced, can be observed in the national statistics. This trend was not, however, reflected in every commodity within the solid wood group. Plywood, for example, declined in importance, from 14 percent of the value of the region's imports in 1967 to 5 percent in 1976.

Softwood lumber was not only the largest import in terms of absolute value, but was also the fastest-growing forest product import in the region during 1967-76; the value of softwood lumber imports increased from $46.2 million in 1967 to $338.8 million in 1976, an average annual gain of 25 percent. Above average gains were also recorded for wood pulp imports, which increased in value by 19 percent annually.

Softwood Lumber

In 1976 the Pacific Northwest imported one-fourth (in value terms) of all of the softwood lumber imported to the United States. This represented a substantial increase over the 1967 share of regional softwood lumber imports, when the Pacific Northwest accounted for only 14 percent of the U.S. total. The nearly 2 billion board-feet imported to the region in 1976 was valued at $338.8 million.

Throughout 1967-76, Canada was the only major supplier of softwood lumber to the region; only once during the ten-year period did Canada's percentage share of the region's imports drop below 100 percent. Regional imports of softwood lumber from Canada increased in value at an annual rate of 25 percent, with the largest increase coming in 1976, when the value of imports jumped $153.2 million, a gain of 83 percent over the 1975 value.

Newsprint

In 1976 the Pacific Northwest imported 8 percent of the total U.S. imports of newsprint. More than one-fifth of the region's total forest product imports consisted of newsprint. Regional imports of newsprint totaled 541,000 tons valued at $146.5 million in 1976.

Canada is the principal origin of many forest product imports to the region, and newsprint is probably the most extreme example of a commodity for which Canada is the only source. Of the approximate $4.6 billion of newsprint imported into the Pacific Northwest during 1967-76, only $1.1 million, or two one-hundredths of 1 percent, originated in countries other than Canada.

The value of regional imports of newsprint tripled over the period, from $44.9 million in 1967 to $146.5 million in 1976, while the volume of imports increased from 368,000 to 541,000 tons. The regional rate of growth in imports during the period was higher than that for the United States, which nearly doubled the value of newsprint imports from 1967 to 1976.

Wood Pulp

In 1976, 20 percent of the value of Pacific Northwest forest product imports consisted of wood pulp. Regional imports of 390,000 tons (valued at $138.1 million) represented 11 percent of the total value of U.S. imports of wood pulp. The value of Pacific Northwest imports of pulp during 1967-76 increased at an average annual rate of 20 percent. Canada is the major country of origin for wood pulp imports to the Pacific Northwest. The only other country of any significance at all during the period was Finland, which, in 1969 and 1973, supplied 1 percent of the value of the region's imports of wood pulp.

Canada was a steady supplier of wood pulp to the region during the period, sending 237,000 tons in 1967, peaking at 430,000 tons in 1971, and averaging 385,000 tons over the period 1972-76. In contrast, Finland was an erratic supplier which shipped 2,800 tons in 1969, 5,800 tons in 1973, but did not export at all to the region in 1971-72 and 1974-76.

Plywood

During the 1967-76 period, the Pacific Northwest declined in relative importance as a plywood-importing region. In 1967 the region imported $21.6 million worth of plywood, which represented 15 percent of the value of total U.S. plywood imports. By 1976 the percentage share had dropped to 9 percent, when the region imported 258.2 million square feet valued at $34 million. Plywood was less important in relation to the total imports of the Pacific Northwest region, where it comprised 5 percent of the value of all forest product imports in 1976.

Asian countries were the origins of nearly all of the region's plywood imports, accounting for 97 to 100 percent of the total value throughout 1967-76. The major plywood exporters to the Pacific Northwest in 1976 were Korea (46 percent of the total value), Taiwan (19 percent), Japan (15 percent), the Philippines (14 percent), and Malaysia (4 percent).

Korea showed phenomenal growth as an exporter during the period, tripling the volume of plywood shipped to the Pacific Northwest, while increasing the value from $2.8 million in 1967 to $15.7 million in 1976. This growth rate allowed Korea to increase its share of the value of plywood imports by the Pacific Northwest from 13 percent in 1967 to 46 percent in 1976.

Taiwan also increased its share of the region's imports during the period, from 10 percent of the total value in 1967 to 19 percent in 1976. Although the 1976 volume of Taiwanese plywood imports was 11 percent higher than the 1967 volume, it represented a greater than 100 percent decline from the quantities imported during 1968–73, when the average volume was 101.6 million square feet. During the 1967–76 period the value of imports from Taiwan increased, on the average, at an annual rate of 14 percent.

Japan and the Philippines both substantially reduced their plywood exports to the region from their 1967 levels. The volume shipped by Japan in 1976 (19.2 million square feet) was less than one-third the 1967 volume (74.9 million square feet), and value was reduced by more than one-third. The reduction in imports from the Philippines was 62 percent in volume terms, and 33 percent in value. In 1967 the two countries supplied 72 percent of the value of the region's imports, but in 1976 contributed only 29 percent.

Malaysia slightly decreased its volume of plywood exports to the region, and increased the nominal value by only 50 percent. Malaysia's share of Pacific Northwest plywood imports was fairly stable throughout the period, averaging 4 percent of the total value.

Wood Veneer

Although in 1976 the value of wood veneer imports was only 2 percent of the value of all Pacific Northwest forest product imports, the region's veneer imports did account for 17 percent of the value of the total U.S. veneer imports.

During the 1967-76 period, the value of the region's imports of veneer rose from $5.9 million to $12.8 million, and volume increased from 376 million to 421 million square feet.

For most of the period, Canada was the leading supplier of wood veneer to the region; imports from Canada averaged 46 percent of the value of total veneer imports over the ten years. While the value of Canadian imports more than quadrupled from 1967 to 1976, the volume increased by almost 75 percent. The 1976 volume of veneer imported to the Pacific Northwest from Canada was 349 million square feet (valued at $9.1 million). In 1976 Canada captured 71 percent of the region's wood veneer-import market.

The remainder of the region's veneer imports was supplied largely by three southeast Asian countries: the Philippines (14 percent of the 1976 value), Singapore (8 percent), and Malaysia (1 percent). During the period, the share of the imports accounted for by these three nations dropped sharply from 61 percent in 1967 to 23 percent in 1976.

The Philippines was the principal foreign supplier of wood veneer to the region in 1967-69, each year taking 55 percent of the value of the region's imports. Since 1969 the decline has been precipitous; the 1976 volume was only 20 percent of the quantity imported in 1969.

Although the volume imported from Singapore in 1976 was greater than that in 1967, the 1973-76 average volume is only 21 percent of the 1968-72 average import volume. A similar trend can be seen in Malaysia's import volumes, which dropped from 14 million square feet in 1967 to 4.7 million square feet in 1976.

Hardwood Lumber

The Pacific Northwest in 1976 imported 12 percent of the value of total U.S. hardwood lumber imports. The $11.9 million of hardwood lumber imported by the region was, however, only 2 percent of the region's total forest product imports.

Southeast Asian countries were the principal sources of hardwood lumber imports to the region throughout 1967-76. Malaysia and the Philippines were the most important, together providing 70 percent of the value and 77 percent of the volume of all hardwood lumber imports to the Pacific Northwest in 1976. Both countries increased their exports to the region at about the same rate during the period, and the values and volumes were nearly identical for the two countries in both 1967 and 1976.

Other Asian countries supplied 21 percent of the value and 13 percent of the volume of the region's 1976 imports. Thailand and Singapore accounted for nearly 80 percent of the value of imports from Asian countries other than Malaysia and the Philippines. In 1976 Singapore shipped 2.3 million board-feet (valued at $974,000) while Thailand sent 796,000 board feet (valued at $1 million).

Canada's shipment of hardwood lumber to the Pacific Northwest declined slightly during the period, from $405,000 to $371,000. In 1967 Canada supplied 11 percent of the value of the region's imports, but by 1976 Canada's share was only 3 percent.

Other Forest Products

The remaining forest products (pulpwood, softwood and hardwood logs, reconstituted wood, building board, and paper and paperboard) imports to the region in 1976 were valued at $10.5 million, which represented less than 2 percent of the region's total forest products imports. Nearly 60 percent of the value of imports in this residual group was in solid wood products, principally softwood sawlogs.

Canada in 1976 provided 90 percent of the value of the region's imports of products in this group. Hardwood logs were the only commodity in which Canada was not the dominant foreign supplier throughout the period. The Philippine Republic was the major supplier of hardwood logs for much of the period, although Canada provided a greater percentage of the total value in the later years.

The Pacific Northwest imported 73 percent of the value of the total U.S. softwood sawlog imports in 1976. However, imports of softwood sawlogs were insignificant in regard to total regional trade in forest products. Nationally, softwood sawlogs account for less than 1 percent of the value of all forest product imports.

EXPORTS
SOFTWOOD LOGS (2422)
PACIFIC NORTHWEST

VALUE--MILLION CURRENT DOLLARS

YEAR	CANADA	EUROPE	KOR REP	JAPAN	OTH ASIA	OTHERS	TOTAL
1967	1.582	0.405	2.451	135.495	0.000	0.022	139.956
1968	3.423	0.347	5.901	192.844	0.132	0.004	202.651
1969	3.535	0.594	3.199	211.581	0.000	0.141	219.050
1970	2.023	0.324	3.658	274.669	0.721	0.197	281.592
1971	4.892	0.514	7.951	220.428	0.021	0.385	234.190
1972	14.327	0.659	5.094	337.253	0.005	0.471	357.808
1973	9.594	0.534	19.045	668.149	0.017	1.391	698.730
1974	13.842	0.559	27.434	573.900	0.106	0.398	616.239
1975	8.403	0.734	14.870	597.810	0.518	0.668	623.002
1976	7.908	0.924	28.626	737.921	2.773	0.242	778.394

QUANTITY--MILLION BF

YEAR	CANADA	EUROPE	KOR REP	JAPAN	OTH ASIA	OTHERS	TOTAL
1967	45.383	1.168	36.981	1752.865	0.000	0.061	1836.459
1968	47.524	2.004	76.814	2304.507	1.237	0.025	2432.112
1969	44.480	7.058	36.455	2248.106	0.000	1.031	2337.191
1970	27.997	1.163	44.048	2761.839	8.713	2.088	2845.849
1971	55.712	2.362	77.990	2265.155	0.331	3.912	2405.463
1972	173.494	1.521	54.734	3065.886	0.031	2.376	3298.042
1973	72.164	0.852	129.259	3092.752	0.022	4.094	3299.143
1974	73.740	0.600	173.983	2577.057	0.168	0.512	2826.060
1975	58.506	1.218	115.717	2716.879	2.347	1.240	2895.906
1976	48.289	1.029	197.225	3241.650	3.836	0.471	3492.506

EXPORTS
HARDWOOD LOGS (2423)
PACIFIC NORTHWEST

VALUE--MILLION CURRENT DOLLARS

YEAR	CANADA	W GERMNY	ITALY	OTH EURO	JAPAN	OTHERS	TOTAL
1967	0.035	0.000	0.007	0.006	0.067	0.000	0.115
1968	0.028	0.005	0.001	0.009	0.212	0.000	0.254
1969	0.000	0.001	0.005	0.002	0.879	0.001	0.888
1970	0.032	0.000	0.000	0.000	0.966	0.002	1.000
1971	0.013	0.000	0.000	0.000	0.846	0.000	0.859
1972	0.221	0.000	0.005	0.004	0.840	0.000	1.070
1973	0.133	0.005	0.000	0.019	1.819	0.000	1.976
1974	0.048	0.000	0.000	0.000	5.293	0.000	5.341
1975	0.075	0.007	0.003	0.004	0.942	0.000	1.033
1976	0.097	0.001	0.007	0.003	1.605	0.001	1.715

QUANTITY--MILLION BF

YEAR	CANADA	W GERMNY	ITALY	OTH EURO	JAPAN	OTHERS	TOTAL
1967	0.054	0.000	0.013	0.012	0.101	0.000	0.180
1968	0.100	0.015	0.003	0.010	1.296	0.000	1.424
1969	0.000	0.003	0.017	0.004	2.019	0.003	2.044
1970	0.380	0.000	0.000	0.000	2.207	0.002	2.589
1971	0.024	0.000	0.000	0.000	1.206	0.000	1.230
1972	3.726	0.000	0.022	0.005	1.864	0.000	5.618
1973	2.089	0.011	0.000	0.011	4.124	0.000	6.235
1974	0.261	0.000	0.000	0.000	36.409	0.000	36.670
1975	0.404	0.015	0.004	0.007	3.964	0.000	4.394
1976	0.360	0.002	0.011	0.005	4.001	0.001	4.379

EXPORTS
SOFTWOOD LUMBER (2432)
PACIFIC NW

VALUE--MILLION CURRENT DOLLARS

YEAR	CANADA	ITALY	OTH EURO	JAPAN	AUSTRLIA	OTHERS	TOTAL
1967	4.436	9.137	9.662	3.207	7.469	5.505	39.437
1968	5.024	14.759	11.436	2.642	9.871	5.817	49.549
1969	4.810	25.137	13.326	0.758	10.959	7.921	62.911
1970	5.230	23.708	18.186	2.833	12.205	9.617	70.778
1971	5.822	18.159	14.684	2.062	12.802	5.574	59.104
1972	10.090	27.912	21.357	3.438	14.879	6.321	83.998
1973	15.940	62.433	57.265	31.352	38.473	11.023	216.506
1974	22.731	45.464	33.605	45.700	35.603	17.754	200.857
1975	20.072	35.388	26.686	39.720	26.745	12.847	161.457
1976	24.476	50.277	47.649	41.200	48.502	12.347	224.451

QUANTITY--MILLION BF

YEAR	CANADA	ITALY	OTH EURO	JAPAN	AUSTRLIA	OTHERS	TOTAL
1967	49.798	67.341	74.313	59.833	82.803	56.533	390.621
1968	45.365	106.788	78.279	41.084	94.685	49.954	416.155
1969	33.320	122.454	70.415	9.430	99.482	59.294	394.395
1970	42.607	107.880	95.514	38.091	89.798	60.820	434.710
1971	45.485	100.389	73.542	25.359	96.126	41.537	382.437
1972	71.424	116.739	94.640	25.098	104.389	38.074	450.364
1973	198.835	147.860	193.769	190.993	162.533	45.787	939.796
1974	106.296	103.434	121.931	259.939	139.659	64.924	796.182
1975	102.329	99.205	83.607	279.767	121.739	54.661	741.308
1976	100.981	116.698	136.801	213.928	191.192	40.454	800.053

EXPORTS
HARDWOOD LUMBER (2433)
PACIFIC NORTHWEST

VALUE--MILLION CURRENT DOLLARS

YEAR	CANADA	S AMER	EUROPE	JAPAN	OTH ASIA	OTHERS	TOTAL
1967	0.562	0.000	0.001	0.075	0.006	0.003	0.648
1968	0.714	0.003	0.032	0.065	0.007	0.002	0.823
1969	0.569	0.002	0.010	0.070	0.006	0.005	0.662
1970	0.777	0.000	0.003	0.107	0.010	0.216	1.112
1971	0.691	0.001	0.010	1.533	0.003	0.001	2.239
1972	1.224	0.026	0.020	2.466	0.136	0.007	3.880
1973	1.422	0.011	0.007	0.585	0.028	0.019	2.072
1974	2.586	0.063	0.005	0.568	0.067	0.044	3.352
1975	2.466	0.078	0.007	0.273	0.031	0.265	3.119
1976	2.851	0.057	0.069	0.196	0.163	0.064	3.399

QUANTITY--MILLION BF

YEAR	CANADA	S AMER	EUROPE	JAPAN	OTH ASIA	OTHERS	TOTAL
1967	3.060	0.000	0.004	0.683	0.009	0.022	3.777
1968	4.272	0.015	0.173	0.232	0.058	0.011	4.761
1969	2.821	0.010	0.029	0.113	0.053	0.042	3.068
1970	4.991	0.000	0.012	0.164	0.019	0.293	5.478
1971	4.156	0.001	0.040	4.102	0.010	0.005	8.314
1972	4.813	0.091	0.045	4.360	0.485	0.036	9.831
1973	4.857	0.029	0.028	0.854	0.065	0.057	5.891
1974	7.847	0.284	0.018	0.967	0.219	0.110	9.445
1975	8.080	0.215	0.013	0.250	0.033	0.762	9.354
1976	7.470	0.172	0.127	0.317	1.900	0.219	10.204

EXPORTS
WOOD VENEER (6311)
PACIFIC NORTHWEST

VALUE--MILLION CURRENT DOLLARS

YEAR	CANADA	W GERMNY	SWITZLND	OTH EURO	ASIA	OTHERS	TOTAL
1967	2.150	0.000	0.000	0.007	0.014	0.000	2.170
1968	3.170	0.000	0.000	0.003	0.000	0.004	3.178
1969	4.228	0.000	0.000	0.003	0.010	0.000	4.241
1970	3.590	0.000	0.000	0.012	0.000	0.000	3.603
1971	4.704	0.000	0.000	0.007	0.004	0.000	4.714
1972	8.885	0.000	0.000	0.005	0.056	0.000	8.946
1973	13.097	0.005	0.000	0.032	0.110	0.000	13.244
1974	9.861	0.000	0.000	0.000	0.000	0.000	9.861
1975	10.093	0.000	0.000	0.000	0.011	0.000	10.104
1976	8.068	0.002	0.000	0.000	0.010	0.000	8.080

QUANTITY--MIL SQ FT

YEAR	CANADA	W GERMNY	SWITZLND	OTH EURO	ASIA	OTHERS	TOTAL
1967	94.892	0.000	0.000	0.007	0.204	0.000	95.103
1968	133.811	0.000	0.000	0.005	0.000	0.012	133.828
1969	165.256	0.000	0.000	0.010	0.022	0.000	165.288
1970	141.231	0.000	0.000	0.044	0.000	0.000	141.276
1971	388.727	0.000	0.000	0.046	0.014	0.000	388.787
1972	276.165	0.000	0.000	0.015	0.033	0.000	276.213
1973	312.674	0.008	0.000	0.032	0.071	0.000	312.787
1974	202.382	0.000	0.000	0.000	0.000	0.000	202.382
1975	250.909	0.000	0.000	0.000	0.026	0.000	250.935
1976	211.956	0.001	0.000	0.000	0.010	0.000	211.967

EXPORTS
PLYWOOD (6312)
PACIFIC NORTHWEST

VALUE--MILLION CURRENT DOLLARS

YEAR	CANADA	DENMARK	U KINGDM	NETHLNDS	BELGIUM	OTHERS	TOTAL
1967	0.163	1.369	1.678	0.232	0.078	2.370	5.889
1968	0.453	1.438	0.661	0.229	0.215	2.048	5.044
1969	1.178	7.492	1.724	1.249	0.574	3.544	15.761
1970	0.967	3.687	1.921	0.486	0.257	3.012	10.331
1971	1.113	3.666	1.724	0.341	0.343	2.231	9.418
1972	2.589	11.236	2.459	0.370	0.123	3.123	19.901
1973	3.741	21.089	6.033	2.937	1.089	10.000	44.889
1974	11.674	9.240	5.605	0.742	3.358	9.371	39.990
1975	10.722	18.503	10.944	3.644	5.242	7.810	56.866
1976	6.946	27.707	22.785	9.806	10.837	15.326	93.407

QUANTITY--MIL SQ FT

YEAR	CANADA	DENMARK	U KINGDM	NETHLNDS	BELGIUM	OTHERS	TOTAL
1967	0.732	12.044	16.709	2.501	0.675	24.599	57.259
1968	3.075	11.079	4.322	0.753	0.873	14.337	34.438
1969	9.541	63.484	16.376	10.599	5.597	25.252	130.850
1970	6.772	28.259	19.506	4.936	2.449	25.828	87.748
1971	8.448	23.437	14.730	2.702	3.089	13.999	66.405
1972	20.115	93.928	18.989	1.473	0.683	15.189	150.377
1973	30.112	110.618	36.421	13.456	6.987	98.887	296.481
1974	76.343	48.121	32.262	4.225	21.535	47.384	229.869
1975	70.075	103.450	67.967	22.504	34.551	41.021	339.568
1976	51.198	121.876	129.701	50.784	60.270	68.840	482.670

EXPORTS
RECONSTITUTED WOOD (6314)
PACIFIC NORTHWEST

VALUE--MILLION CURRENT DOLLARS

YEAR	CANADA	MEXICO	EUROPE	JAPAN	OTH ASIA	OTHERS	TOTAL
1967	0.360	0.000	0.004	0.000	0.018	0.020	0.402
1968	0.513	0.000	0.000	0.000	0.008	0.067	0.587
1969	0.460	0.000	0.000	0.000	0.000	0.025	0.485
1970	0.990	0.000	0.002	0.000	0.000	0.008	1.000
1971	1.958	0.000	0.001	0.019	0.000	0.007	1.986
1972	2.712	0.000	0.000	0.002	0.000	0.005	2.718
1973	5.491	0.000	0.000	0.258	0.000	0.015	5.765
1974	5.877	0.000	0.000	0.342	0.000	0.815	7.034
1975	5.137	0.000	0.000	0.008	0.000	0.066	5.211
1976	5.244	0.000	0.008	0.282	0.045	0.039	5.619

QUANTITY--MIL SQ FT

YEAR	CANADA	MEXICO	EUROPE	JAPAN	OTH ASIA	OTHERS	TOTAL
1967	0.722	0.000	0.008	0.000	0.113	0.079	0.922
1968	4.397	0.000	0.000	0.000	0.002	0.120	4.520
1969	3.614	0.000	0.000	0.000	0.000	0.147	3.761
1970	4.985	0.000	0.004	0.000	0.000	0.050	5.039
1971	13.912	0.000	0.011	0.083	0.000	0.068	14.074
1972	23.621	0.000	0.000	0.002	0.000	0.063	23.686
1973	42.072	0.000	0.000	1.548	0.000	0.138	43.757
1974	46.126	0.000	0.000	1.572	0.000	4.942	52.640
1975	47.639	0.000	0.000	0.015	0.000	0.327	47.981
1976	42.296	0.000	0.021	1.516	0.442	0.205	44.481

EXPORTS
PULPWOOD CHIPS (631.8320)
PACIFIC NORTHWEST

VALUE--MILLION CURRENT DOLLARS

YEAR	CANADA	OTH N AM	EUROPE	JAPAN	OTH ASIA	OTHERS	TOTAL
1967	0.000	0.000	0.000	11.916	0.000	0.000	11.916
1968	0.016	0.000	0.001	24.978	0.168	0.000	25.163
1969	0.067	0.000	0.000	34.752	0.000	0.000	34.819
1970	0.000	0.000	0.000	35.650	0.000	0.000	35.650
1971	0.026	0.000	0.000	35.574	0.000	0.001	35.601
1972	0.230	0.000	0.000	50.004	0.015	0.000	50.249
1973	0.586	0.000	0.000	75.537	0.000	0.000	76.123
1974	0.668	0.000	0.000	104.038	0.000	0.000	104.706
1975	0.317	0.000	0.000	99.905	0.000	0.000	100.222
1976	0.031	0.000	0.610	129.069	0.800	0.000	130.510

QUANTITY--1000 STN

YEAR	CANADA	OTH N AM	EUROPE	JAPAN	OTH ASIA	OTHERS	TOTAL
1967	0.000	0.000	0.000	765.204	0.000	0.000	765.204
1968	1.516	0.000	0.043	1539.625	10.654	0.000	1551.838
1969	5.150	0.000	0.000	2284.735	0.000	0.000	2289.885
1970	0.000	0.000	0.000	2349.039	0.000	0.000	2349.039
1971	1.641	0.000	0.000	1714.392	0.000	0.075	1716.108
1972	13.211	0.000	0.000	2235.850	0.696	0.000	2249.757
1973	38.707	0.000	0.000	3052.830	0.000	0.000	3091.537
1974	49.640	0.000	0.000	3518.195	0.000	0.000	3567.835
1975	16.764	0.000	0.000	2746.149	0.000	0.000	2762.913
1976	0.301	0.000	32.033	3289.044	18.000	0.000	3339.378

EXPORTS
PULPWOOD (EXC CHIPS)(2421)
PACIFIC NORTHWEST

VALUE--MILLION CURRENT DOLLARS

YEAR	CANADA	OTH N AM	EUROPE	JAPAN	OTH ASIA	OTHERS	TOTAL
1967	0.004	0.000	0.000	0.000	0.000	0.000	0.004
1968	0.000	0.000	0.000	0.005	0.000	0.000	0.005
1969	0.000	0.000	0.000	0.000	0.000	0.000	0.000
1970	0.005	0.000	0.000	0.315	0.000	0.000	0.320
1971	0.000	0.000	0.000	0.477	0.000	0.000	0.477
1972	0.012	0.000	0.000	0.383	0.005	0.000	0.399
1973	0.009	0.000	0.000	0.221	0.000	0.000	0.230
1974	0.057	0.000	0.000	0.601	0.000	0.000	0.658
1975	0.018	0.000	0.000	0.121	0.000	0.000	0.139
1976	0.000	0.000	0.000	0.113	0.000	0.000	0.113

QUANTITY--1000 CORDS

YEAR	CANADA	OTH N AM	EUROPE	JAPAN	OTH ASIA	OTHERS	TOTAL
1967	0.420	0.000	0.000	0.000	0.000	0.000	0.420
1968	0.000	0.000	0.000	0.314	0.000	0.000	0.314
1969	0.000	0.000	0.000	0.000	0.000	0.000	0.000
1970	0.220	0.000	0.000	39.224	0.000	0.000	39.444
1971	0.000	0.000	0.000	47.609	0.000	0.000	47.609
1972	0.494	0.000	0.000	33.535	0.137	0.000	34.166
1973	0.160	0.000	0.000	25.960	0.000	0.000	26.120
1974	1.198	0.000	0.000	47.335	0.000	0.000	48.533
1975	0.813	0.000	0.000	10.929	0.000	0.000	11.742
1976	0.000	0.000	0.000	5.583	0.000	0.000	5.583

EXPORTS
WOOD PULP (2510)
PACIFIC NORTHWEST

VALUE--MILLION CURRENT DOLLARS

YEAR	U KINGDM	ITALY	OTH EURO	JAPAN	OTH ASIA	OTHERS	TOTAL
1967	7.316	8.660	10.519	18.148	11.920	8.968	65.532
1968	8.964	7.963	11.315	14.995	9.842	12.096	55.194
1969	9.859	10.336	13.569	18.607	14.876	13.994	81.242
1970	12.416	12.802	23.743	20.720	14.785	24.504	108.971
1971	11.346	7.423	18.778	17.125	8.914	13.951	77.436
1972	9.844	12.893	21.176	22.501	7.475	15.557	89.446
1973	8.736	12.354	29.476	28.436	10.876	17.624	107.501
1974	17.338	24.062	41.549	55.367	11.901	30.057	180.275
1975	24.811	25.047	69.515	54.570	10.229	24.066	208.255
1976	21.810	16.506	42.316	52.512	18.313	22.085	173.542

QUANTITY--1000 STN

YEAR	U KINGDM	ITALY	OTH EURO	JAPAN	OTH ASIA	OTHERS	TOTAL
1967	50.806	69.022	82.470	157.729	107.143	74.399	541.570
1968	64.704	62.593	95.315	130.019	97.114	108.423	558.167
1969	69.961	79.396	105.670	153.616	132.278	117.552	658.474
1970	80.415	91.770	177.145	141.625	112.037	183.928	786.920
1971	68.132	48.985	123.133	97.494	69.503	100.835	508.082
1972	58.443	83.214	141.656	135.088	65.765	121.994	606.160
1973	50.418	72.129	178.946	147.998	70.419	138.848	658.758
1974	56.680	85.174	150.425	188.920	42.095	106.701	629.995
1975	70.089	71.223	209.202	158.022	34.511	78.994	622.040
1976	54.962	45.942	125.771	155.194	74.725	73.404	529.998

EXPORTS
NEWSPRINT (6411)
PACIFIC NORTHWEST

VALUE--MILLION CURRENT DOLLARS

YEAR	MEXICO	VENZUELA	OTH S AM	EUROPE	ASIA	OTHERS	TOTAL
1967	0.000	0.000	0.104	0.000	1.321	0.008	1.433
1968	0.000	0.000	0.511	0.000	0.662	0.000	1.173
1969	0.000	0.000	0.637	0.000	1.431	0.062	2.129
1970	0.000	0.000	0.396	0.000	2.505	0.183	3.084
1971	0.000	0.000	0.000	0.000	5.064	0.354	5.418
1972	0.000	0.000	0.000	0.000	0.544	0.133	0.677
1973	0.000	0.011	0.000	0.000	0.000	0.057	0.068
1974	0.000	0.000	0.258	0.002	0.444	0.024	0.728
1975	0.000	0.000	0.000	0.000	1.380	0.442	1.822
1976	0.000	0.000	0.000	0.000	11.448	0.273	11.721

QUANTITY--1000 STN

YEAR	MEXICO	VENZUELA	OTH S AM	EUROPE	ASIA	OTHERS	TOTAL
1967	0.000	0.000	0.926	0.000	11.290	0.016	12.232
1968	0.000	0.000	4.501	0.000	6.306	0.000	10.807
1969	0.000	0.000	5.041	0.000	13.766	0.298	19.104
1970	0.000	0.000	2.812	0.000	22.582	0.939	26.333
1971	0.000	0.000	0.000	0.000	47.150	2.676	49.826
1972	0.000	0.000	0.000	0.000	4.052	0.980	5.032
1973	0.000	0.077	0.000	0.000	0.000	0.450	0.527
1974	0.000	0.000	1.077	0.002	1.609	0.133	2.822
1975	0.000	0.000	0.000	0.000	4.366	1.468	5.835
1976	0.000	0.000	0.000	0.000	35.810	0.847	36.657

EXPORTS
BUILDING BOARD (6416)
PACIFIC NORTHWEST

VALUE--MILLION CURRENT DOLLARS

YEAR	CANADA	OTH N AM	SOUTH AM	EUROPE	ASIA	OTHERS	TOTAL
1967	0.237	0.013	0.000	0.221	0.025	0.061	0.556
1968	0.319	0.004	0.000	0.225	0.072	0.166	0.786
1969	0.581	0.007	0.004	0.427	0.014	0.072	1.104
1970	0.756	0.004	0.001	0.553	0.022	0.056	1.392
1971	0.745	0.008	0.006	0.576	0.023	0.003	1.360
1972	1.137	0.002	0.058	0.451	0.002	0.047	1.697
1973	3.010	0.000	0.000	0.618	0.009	0.017	3.654
1974	3.974	0.006	0.000	0.310	0.048	0.032	4.370
1975	2.283	0.000	0.000	0.025	0.105	0.041	2.453
1976	3.367	0.000	0.000	0.005	0.094	0.021	3.487

QUANTITY--1000 STN

YEAR	CANADA	OTH N AM	SOUTH AM	EUROPE	ASIA	OTHERS	TOTAL
1967	0.824	0.091	0.000	0.941	0.117	0.262	2.236
1968	1.462	0.032	0.000	0.988	0.184	0.790	3.456
1969	2.802	0.026	0.014	1.854	0.047	0.263	5.006
1970	2.932	0.013	0.002	2.459	0.117	0.227	5.751
1971	3.477	0.043	0.020	2.106	0.107	0.011	5.764
1972	5.757	0.014	0.230	1.625	0.008	0.170	7.804
1973	14.077	0.000	0.000	2.252	0.019	0.053	16.402
1974	16.419	0.007	0.000	0.848	0.127	0.144	17.544
1975	8.246	0.000	0.000	0.081	0.253	0.115	8.695
1976	10.433	0.000	0.000	0.020	0.164	0.058	10.675

EXPORTS
PAPER & PAPERBOARD (6410)
PACIFIC NORTHWEST

VALUE--MILLION CURRENT DOLLARS

YEAR	CANADA	EUROPE	JAPAN	OTH ASIA	AUSTRLIA	OTHERS	TOTAL
1967	6.405	8.891	0.584	11.544	1.226	15.200	43.851
1968	6.283	11.272	1.832	14.068	1.710	10.926	46.091
1969	9.628	14.761	1.562	15.291	2.073	11.558	54.872
1970	9.137	12.035	0.888	14.452	1.779	11.150	49.440
1971	10.996	5.251	4.568	11.978	1.897	7.152	41.843
1972	12.419	12.349	2.587	17.082	2.898	11.880	59.215
1973	17.121	14.880	9.372	20.965	4.147	14.137	80.621
1974	26.487	26.340	18.069	30.387	6.275	16.579	124.138
1975	32.400	9.769	19.007	16.258	3.156	8.434	89.023
1976	37.098	20.203	36.687	22.453	11.766	7.137	135.345

QUANTITY--1000 STN

YEAR	CANADA	EUROPE	JAPAN	OTH ASIA	AUSTRLIA	OTHERS	TOTAL
1967	26.666	73.764	3.520	92.780	6.806	96.589	300.125
1968	24.074	87.636	13.886	114.071	7.387	72.833	319.887
1969	33.199	121.602	12.595	133.957	10.782	69.482	381.617
1970	35.633	96.378	4.595	108.126	8.445	66.752	319.928
1971	45.913	55.273	27.811	109.096	8.776	42.591	289.461
1972	50.396	102.267	15.042	135.264	12.384	75.114	390.468
1973	60.858	96.286	44.691	128.969	17.346	69.680	417.829
1974	81.047	97.602	64.593	103.654	18.203	45.238	410.336
1975	94.875	32.896	53.079	64.352	6.979	25.321	277.502
1976	99.533	87.916	86.579	91.874	24.244	20.395	410.540

IMPORTS
SOFTWOOD LOGS (2422)
PACIFIC NORTHWEST

VALUE--MILLION CURRENT DOLLARS

YEAR	CANADA	OTH N AM	S AMERIC	EUROPE	OCEANIA	OTHERS	TOTAL
1967	1.635	0.000	0.000	0.000	0.000	0.000	1.635
1968	1.325	0.000	0.000	0.000	0.000	0.000	1.325
1969	2.981	0.000	0.000	0.000	0.000	0.000	2.981
1970	8.212	0.000	0.000	0.000	0.000	0.000	8.212
1971	3.904	0.000	0.000	0.000	0.000	0.000	3.904
1972	0.680	0.000	0.000	0.000	0.000	0.000	0.680
1973	0.262	0.000	0.000	0.000	0.000	0.000	0.262
1974	7.865	0.000	0.000	0.000	0.000	0.000	7.865
1975	11.494	0.000	0.000	0.000	0.000	0.000	11.494
1976	5.447	0.000	0.000	0.000	0.000	0.000	5.447

QUANTITY--MILLION BF

YEAR	CANADA	OTH N AM	S AMERIC	EUROPE	OCEANIA	OTHERS	TOTAL
1967	32.567	0.000	0.000	0.000	0.000	0.000	32.567
1968	38.660	0.000	0.000	0.000	0.000	0.000	38.660
1969	40.445	0.000	0.000	0.000	0.000	0.000	40.445
1970	99.462	0.000	0.000	0.000	0.000	0.000	99.462
1971	54.689	0.000	0.000	0.000	0.000	0.000	54.689
1972	8.451	0.000	0.000	0.000	0.000	0.000	8.451
1973	2.102	0.000	0.000	0.000	0.000	0.000	2.102
1974	31.625	0.000	0.000	0.000	0.000	0.000	31.625
1975	55.494	0.000	0.000	0.000	0.000	0.000	55.494
1976	44.434	0.000	0.000	0.000	0.000	0.003	44.437

IMPORTS
HARDWOOD LOGS (2423)
PACIFIC NORTHWEST

VALUE--MILLION CURRENT DOLLARS

YEAR	CANADA	S AMERIC	PHIL REP	OTH ASIA	AFRICA	OTHERS	TOTAL
1967	0.008	0.000	0.272	0.019	0.000	0.008	0.306
1968	0.005	0.000	0.147	0.013	0.000	0.005	0.170
1969	0.002	0.000	0.229	0.000	0.000	0.003	0.233
1970	0.002	0.050	0.045	0.001	0.000	0.010	0.107
1971	0.014	0.000	0.008	0.000	0.000	0.014	0.036
1972	0.004	0.000	0.047	0.001	0.000	0.001	0.054
1973	0.043	0.000	0.213	0.214	0.000	0.001	0.470
1974	0.044	0.002	0.119	0.382	0.000	0.004	0.550
1975	1.183	0.000	0.027	0.027	0.000	0.002	1.239
1976	0.172	0.000	0.101	0.000	0.000	0.002	0.276

QUANTITY--MILLION BF

YEAR	CANADA	S AMERIC	PHIL REP	OTH ASIA	AFRICA	OTHERS	TOTAL
1967	0.096	0.000	2.560	0.149	0.000	0.028	2.832
1968	0.117	0.000	1.169	0.006	0.000	0.003	1.294
1969	0.021	0.000	1.817	0.000	0.000	0.015	1.853
1970	0.055	1.360	0.388	0.001	0.000	0.017	1.821
1971	0.283	0.000	0.057	0.000	0.000	0.019	0.359
1972	0.011	0.000	0.385	0.002	0.000	0.001	0.399
1973	1.645	0.009	1.693	0.770	0.000	0.015	4.131
1974	1.202	0.002	0.529	2.170	0.000	0.009	3.911
1975	4.514	0.000	0.128	0.020	0.000	0.001	4.663
1976	1.391	0.000	0.385	0.000	0.000	0.003	1.778

IMPORTS
SOFTWOOD LUMBER (2432)
PACIFIC NORTHWEST

VALUE--MILLION CURRENT DOLLARS

YEAR	CANADA	OTH N AM	S AMERIC	EUROPE	OCEANIA	OTHERS	TOTAL
1967	46.101	0.001	0.000	0.000	0.000	0.063	46.165
1968	70.635	0.006	0.000	0.000	0.000	0.097	70.738
1969	76.891	0.002	0.000	0.002	0.011	0.066	76.971
1970	58.788	0.002	0.007	0.000	0.000	0.035	58.833
1971	85.087	0.000	0.000	0.000	0.000	0.034	85.121
1972	163.336	0.000	0.000	0.000	0.000	0.083	163.419
1973	215.649	0.000	0.003	0.000	0.005	0.769	216.426
1974	207.336	0.000	0.000	0.014	0.000	1.438	208.788
1975	185.370	0.000	0.005	0.001	0.000	0.223	185.598
1976	338.578	0.101	0.021	0.000	0.000	0.110	338.810

QUANTITY--MILLION BF

YEAR	CANADA	OTH N AM	S AMERIC	EUROPE	OCEANIA	OTHERS	TOTAL
1967	612.024	0.000	0.000	0.000	0.000	0.455	612.479
1968	778.689	0.080	0.000	0.000	0.000	0.776	779.545
1969	789.413	0.019	0.000	0.006	0.132	0.392	789.963
1970	709.526	0.026	0.031	0.002	0.000	0.167	709.752
1971	834.617	0.000	0.000	0.000	0.000	0.164	834.781
1972	1295.263	0.000	0.000	0.000	0.000	0.512	1295.775
1973	1382.819	0.000	0.020	0.000	0.009	2.301	1385.149
1974	1534.742	0.000	0.000	0.094	0.000	3.527	1538.363
1975	1497.921	0.000	0.043	0.000	0.000	0.673	1498.637
1976	1962.168	0.686	0.045	0.000	0.000	0.372	1963.271

IMPORTS
HARDWOOD LUMBER (2433)
PACIFIC NORTHWEST

VALUE--MILLION CURRENT DOLLARS

YEAR	CANADA	THAILAND	MALAYSIA	SINGAPOR	PHIL REP	OTHERS	TOTAL
1967	0.405	0.043	1.013	0.032	1.115	0.988	3.594
1968	0.481	0.058	1.030	0.082	1.437	1.506	4.594
1969	0.718	0.052	3.112	0.519	2.015	1.942	8.358
1970	0.247	0.069	1.206	0.185	1.023	0.498	3.228
1971	0.256	0.037	1.917	0.324	0.979	0.716	4.230
1972	0.425	0.126	3.229	0.558	2.136	0.761	7.235
1973	0.574	0.348	3.080	0.268	3.195	0.780	8.244
1974	0.431	0.678	7.671	1.266	4.062	1.170	15.279
1975	0.224	0.433	1.258	0.172	1.215	0.763	4.065
1976	0.371	1.023	4.313	0.974	4.016	1.156	11.853

QUANTITY--MILLION BF

YEAR	CANADA	THAILAND	MALAYSIA	SINGAPOR	PHIL REP	OTHERS	TOTAL
1967	2.085	0.071	8.934	0.299	8.576	4.781	24.746
1968	2.381	0.084	7.202	0.497	9.693	10.117	29.973
1969	3.309	0.079	23.095	4.535	15.782	9.810	56.609
1970	1.356	0.107	8.480	1.596	6.947	1.594	20.081
1971	1.073	0.071	13.210	2.173	6.744	3.634	26.905
1972	2.108	0.253	18.026	2.767	13.809	2.682	39.644
1973	2.508	0.511	13.231	1.111	18.163	2.088	37.612
1974	2.003	0.466	30.757	4.612	16.245	2.779	56.862
1975	0.800	0.440	5.006	0.737	4.042	2.855	13.881
1976	1.269	0.796	13.132	2.331	13.155	3.313	33.996

IMPORTS
WOOD VENEER (6311)
PACIFIC NORTHWEST

VALUE--MILLION CURRENT DOLLARS

YEAR	CANADA	EUROPE	MALAYSIA	SINGAPOR	PHIL REP	OTHERS	TOTAL
1967	2.072	0.082	0.124	0.223	3.241	0.119	5.861
1968	1.830	0.058	0.152	1.022	3.752	0.048	6.863
1969	2.759	0.052	0.226	1.505	5.731	0.028	10.301
1970	3.038	0.002	0.390	0.996	2.350	0.015	6.790
1971	3.221	0.056	0.505	1.313	2.926	0.003	8.025
1972	7.288	0.136	1.006	1.734	3.416	0.072	13.653
1973	7.240	0.149	0.931	0.727	4.204	0.586	13.835
1974	4.915	0.098	1.160	0.822	4.789	0.496	12.280
1975	6.317	0.000	0.230	0.460	1.717	0.305	9.030
1976	9.105	0.132	0.193	1.037	1.781	0.592	12.840

QUANTITY--MIL SQ FT

YEAR	CANADA	EUROPE	MALAYSIA	SINGAPOR	PHIL REP	OTHERS	TOTAL
1967	201.267	0.070	3.731	2.526	39.685	0.653	247.933
1968	152.345	0.061	3.504	14.381	53.952	0.364	224.608
1969	177.264	0.109	3.138	20.047	65.301	0.057	265.916
1970	266.523	0.002	4.711	12.812	35.586	0.025	319.660
1971	241.055	0.079	6.124	17.094	46.770	0.006	311.128
1972	432.202	0.209	12.177	19.285	43.029	0.888	507.789
1973	371.584	0.099	6.323	4.530	54.925	3.956	441.418
1974	258.369	0.052	6.257	3.697	37.959	1.632	307.965
1975	333.397	0.000	1.309	1.714	17.403	0.671	354.493
1976	349.240	0.066	1.275	4.353	12.883	0.603	368.420

IMPORTS
PLYWOOD (6312)
PACIFIC NORTHWEST

VALUE--MILLION CURRENT DOLLARS

YEAR	MALAYSIA	PHIL REP	KOR REP	TAIWAN	JAPAN	OTHERS	TOTAL
1967	1.021	7.277	2.794	2.081	8.162	0.266	21.601
1968	0.702	9.284	2.151	7.584	11.836	0.862	31.420
1969	1.606	8.813	3.011	11.291	10.693	1.589	37.002
1970	1.005	4.978	5.190	6.316	8.161	0.319	25.969
1971	1.029	4.763	9.966	6.623	6.828	0.299	29.508
1972	1.283	5.952	8.200	7.446	7.267	0.434	30.582
1973	1.481	7.153	15.122	10.927	7.148	1.042	42.873
1974	1.916	3.690	8.592	6.908	5.885	0.350	27.341
1975	1.141	2.993	10.816	4.600	4.571	0.200	24.321
1976	1.500	4.872	15.739	6.604	5.084	0.156	33.955

QUANTITY--MIL SQ FT

YEAR	MALAYSIA	PHIL REP	KOR REP	TAIWAN	JAPAN	OTHERS	TOTAL
1967	13.232	79.067	26.952	25.987	50.005	2.298	197.541
1968	8.899	78.786	18.570	66.820	61.145	7.197	241.417
1969	17.249	76.715	24.208	101.848	54.198	13.997	288.215
1970	10.894	49.332	46.157	55.195	45.255	2.004	208.837
1971	11.376	50.361	86.902	57.848	36.727	3.053	246.268
1972	14.032	60.236	69.820	69.762	30.912	2.456	247.218
1973	10.452	52.065	86.838	55.160	19.837	3.884	228.236
1974	10.604	22.582	49.396	32.355	15.665	2.591	133.193
1975	8.324	20.541	68.307	27.017	14.763	0.906	139.859
1976	11.611	29.714	88.960	28.767	12.817	0.522	172.391

IMPORTS
RECONSTITUTED WOOD (6314)
PACIFIC NORTHWEST

VALUE--MILLION CURRENT DOLLARS

YEAR	CANADA	S AMERIC	EUROPE	ASIA	OCEANIA	OTHERS	TOTAL
1967	0.000	0.000	0.000	0.000	0.000	0.000	0.000
1968	0.000	0.000	0.000	0.000	0.000	0.000	0.000
1969	0.000	0.000	0.000	0.001	0.000	0.000	0.001
1970	0.000	0.000	0.000	0.020	0.000	0.000	0.020
1971	0.004	0.000	0.000	0.022	0.000	0.000	0.026
1972	0.000	0.000	0.000	0.000	0.001	0.000	0.001
1973	0.000	0.000	0.000	0.000	0.000	0.000	0.000
1974	0.001	0.000	0.000	0.008	0.000	0.000	0.009
1975	0.007	0.000	0.000	0.000	0.000	0.000	0.007
1976	0.195	0.012	0.000	0.005	0.000	0.000	0.212

QUANTITY--MIL SQ FT

YEAR	CANADA	S AMERIC	EUROPE	ASIA	OCEANIA	OTHERS	TOTAL
1967	0.000	0.000	0.000	0.000	0.000	0.000	0.000
1968	0.002	0.000	0.000	0.000	0.000	0.000	0.002
1969	0.000	0.000	0.000	0.001	0.000	0.000	0.001
1970	0.000	0.000	0.000	0.014	0.000	0.000	0.014
1971	0.004	0.000	0.000	0.016	0.000	0.000	0.021
1972	0.000	0.000	0.000	0.000	0.000	0.000	0.000
1973	0.001	0.000	0.000	0.000	0.000	0.000	0.001
1974	0.001	0.000	0.000	0.026	0.000	0.000	0.027
1975	0.035	0.000	0.000	0.000	0.000	0.000	0.035
1976	0.773	0.121	0.000	0.020	0.000	0.000	0.914

IMPORTS
PULPWOOD CHIPS (631.8320)
PACIFIC NORTHWEST

VALUE--MILLION CURRENT DOLLARS

YEAR	CANADA	S AMERIC	FRANCE	OTH EURO	AUSTRALA	OTHERS	TOTAL
1967	8.650	0.000	0.000	0.000	0.000	0.000	8.650
1968	7.968	0.000	0.000	0.000	0.000	0.000	7.968
1969	4.519	0.000	0.000	0.000	0.000	0.000	4.519
1970	6.645	0.000	0.000	0.000	0.000	0.000	6.645
1971	11.075	0.000	0.000	0.000	0.000	0.000	11.075
1972	8.980	0.000	0.000	0.000	0.000	0.000	8.980
1973	11.148	0.000	0.000	0.000	0.000	0.000	11.149
1974	9.701	0.000	0.000	0.000	0.000	0.000	9.701
1975	11.109	0.000	0.000	0.000	0.000	0.000	11.109
1976	18.431	0.000	0.000	0.000	0.100	0.000	18.532

QUANTITY--1000 STN

YEAR	CANADA	S AMERIC	FRANCE	OTH EURO	AUSTRALA	OTHERS	TOTAL
1967	1183.841	0.000	0.000	0.000	0.000	0.000	1183.841
1968	1049.259	0.000	0.000	0.000	0.000	0.000	1049.259
1969	581.167	0.000	0.000	0.000	0.000	0.000	581.167
1970	790.544	0.000	0.000	0.000	0.000	0.000	790.544
1971	1157.444	0.000	0.000	0.000	0.000	0.000	1157.444
1972	909.926	0.000	0.000	0.000	0.000	0.000	909.926
1973	1085.124	0.000	0.000	0.000	0.000	0.000	1085.124
1974	623.830	0.000	0.000	0.000	0.000	0.000	623.830
1975	493.761	0.000	0.001	0.000	0.000	0.000	493.762
1976	877.550	0.000	0.000	0.000	4.731	0.000	882.281

IMPORTS
PULPWOOD(EXC CHIPS) (2421)
PACIFIC NORTHWEST

VALUE--MILLION CURRENT DOLLARS

YEAR	CANADA	OTH N AM	S AMERIC	EUROPE	ASIA	OTHERS	TOTAL
1967	0.193	0.000	0.000	0.000	0.000	0.000	0.193
1968	0.251	0.000	0.000	0.000	0.000	0.000	0.251
1969	0.308	0.000	0.000	0.000	0.000	0.000	0.308
1970	1.687	0.000	0.000	0.000	0.000	0.000	1.687
1971	0.045	0.000	0.000	0.000	0.000	0.000	0.045
1972	0.109	0.000	0.000	0.000	0.000	0.000	0.109
1973	0.000	0.000	0.000	0.000	0.000	0.002	0.002
1974	1.922	0.000	0.000	0.000	0.000	0.000	1.922
1975	0.494	0.000	0.058	0.000	0.000	0.000	0.552
1976	0.054	0.000	0.010	0.000	0.000	0.000	0.064

QUANTITY--1000 CORDS

YEAR	CANADA	OTH N AM	S AMERIC	EUROPE	ASIA	OTHERS	TOTAL
1967	3.529	0.000	0.000	0.000	0.000	0.000	3.529
1968	3.720	0.000	0.000	0.000	0.000	0.000	3.720
1969	3.174	0.000	0.000	0.000	0.000	0.000	3.174
1970	17.501	0.000	0.000	0.000	0.000	0.000	17.501
1971	2.330	0.000	0.000	0.000	0.000	0.000	2.330
1972	2.300	0.000	0.000	0.000	0.000	0.000	2.300
1973	0.000	0.000	0.000	0.000	0.000	0.054	0.054
1974	31.998	0.000	0.000	0.000	0.000	0.000	31.998
1975	11.517	0.000	2.620	0.000	0.000	0.000	14.137
1976	1.699	0.000	0.471	0.000	0.000	0.000	2.170

IMPORTS
WOOD PULP (2510)
PACIFIC NORTHWEST

VALUE--MILLION CURRENT DOLLARS

YEAR	CANADA	SWEDEN	FINLAND	OTH EURO	REP S AF	OTHERS	TOTAL
1967	27.203	0.000	0.048	0.000	0.000	0.036	27.287
1968	22.479	0.000	0.070	0.000	0.000	0.000	22.550
1969	33.633	0.000	0.229	0.009	0.000	0.000	33.871
1970	21.806	0.000	0.049	0.000	0.000	0.000	21.854
1971	28.029	0.000	0.000	0.000	0.000	0.000	28.029
1972	61.989	0.000	0.000	0.000	0.000	0.001	61.990
1973	68.145	0.002	0.583	0.346	0.000	0.012	69.089
1974	106.468	0.029	0.000	0.006	0.000	0.003	106.507
1975	124.086	0.000	0.000	0.000	0.000	0.002	124.088
1976	138.061	0.000	0.000	0.002	0.000	0.034	138.097

QUANTITY--1000 STN

YEAR	CANADA	SWEDEN	FINLAND	OTH EURO	REP S AF	OTHERS	TOTAL
1967	237.350	0.000	0.524	0.000	0.000	0.323	238.197
1968	199.331	0.000	0.475	0.000	0.000	0.000	199.806
1969	284.419	0.000	2.771	0.116	0.000	0.000	287.306
1970	161.030	0.000	0.551	0.000	0.000	0.000	161.581
1971	429.821	0.000	0.000	0.000	0.000	0.000	429.821
1972	398.770	0.000	0.000	0.000	0.000	0.003	398.773
1973	384.902	0.008	5.787	1.984	0.000	0.115	392.795
1974	396.763	0.112	0.000	0.028	0.000	0.010	396.912
1975	352.280	0.000	0.000	0.000	0.000	0.011	352.291
1976	389.986	0.000	0.000	0.023	0.000	0.345	390.354

IMPORTS
NEWSPRINT (6411)
PACIFIC NORTHWEST

VALUE--MILLION CURRENT DOLLARS

YEAR	CANADA	S AMERIC	EUROPE	ASIA	AFRICA	OTHERS	TOTAL
1967	44.854	0.000	0.000	0.000	0.000	0.000	44.854
1968	49.334	0.000	0.000	0.000	0.000	0.000	49.334
1969	60.792	0.000	0.000	0.000	0.000	0.000	60.792
1970	55.623	0.000	0.002	0.000	0.000	0.000	55.626
1971	86.452	0.000	0.000	0.000	0.000	0.000	86.452
1972	77.028	0.000	0.000	0.000	0.000	0.000	77.028
1973	70.797	0.000	0.146	0.000	0.000	0.000	70.942
1974	79.084	0.000	0.000	0.000	0.000	0.000	79.084
1975	100.997	0.000	0.000	0.000	0.000	0.000	100.997
1976	146.520	0.000	0.000	0.000	0.000	0.000	146.520

QUANTITY--1000 STN

YEAR	CANADA	S AMERIC	EUROPE	ASIA	AFRICA	OTHERS	TOTAL
1967	366.745	0.000	0.000	0.000	0.000	0.000	366.745
1968	392.417	0.000	0.000	0.000	0.000	0.000	392.417
1969	458.634	0.000	0.000	0.000	0.000	0.000	458.534
1970	408.691	0.000	0.014	0.000	0.000	0.000	408.705
1971	639.470	0.000	0.000	0.000	0.000	0.000	639.470
1972	514.965	0.000	0.000	0.000	0.000	0.000	514.965
1973	446.943	0.000	1.068	0.000	0.000	0.000	448.011
1974	386.208	0.000	0.000	0.000	0.000	0.000	386.208
1975	395.079	0.000	0.000	0.000	0.000	0.000	395.079
1976	541.124	0.000	0.000	0.000	0.000	0.000	541.124

IMPORTS
BUILDING BOARD (6416)
PACIFIC NORTHWEST

VALUE--MILLION CURRENT DOLLARS

YEAR	CANADA	S AMERIC	EUROPE	ASIA	OCEANIA	OTHERS	TOTAL
1967	0.796	0.020	0.014	0.000	0.000	0.000	0.830
1968	0.446	0.081	0.033	0.009	0.000	0.000	0.570
1969	0.524	0.000	0.067	0.000	0.000	0.003	0.593
1970	0.397	0.006	0.025	0.000	0.000	0.000	0.429
1971	0.765	0.005	0.010	0.000	0.000	0.000	0.779
1972	1.090	0.009	0.021	0.000	0.008	0.000	1.128
1973	1.628	0.002	0.001	0.153	0.000	0.000	1.784
1974	1.260	0.000	0.163	0.110	0.000	0.000	1.533
1975	0.690	0.000	0.000	0.002	0.000	0.000	0.692
1976	1.336	0.076	0.000	0.024	0.000	0.000	1.436

QUANTITY--1000 STN

YEAR	CANADA	S AMERIC	EUROPE	ASIA	OCEANIA	OTHERS	TOTAL
1967	9.774	0.201	0.124	0.000	0.000	0.000	10.099
1968	5.687	0.670	0.454	0.040	0.000	0.000	6.851
1969	6.259	0.000	0.902	0.000	0.000	0.000	7.161
1970	4.346	0.057	0.283	0.000	0.000	0.000	4.685
1971	8.973	0.035	0.115	0.000	0.000	0.000	9.123
1972	10.811	0.103	0.240	0.000	0.064	0.000	11.219
1973	14.257	0.027	0.011	1.691	0.000	0.000	15.985
1974	8.460	0.000	1.117	0.973	0.000	0.000	10.549
1975	4.461	0.000	0.000	0.013	0.000	0.000	4.474
1976	9.417	0.486	0.000	0.171	0.000	0.000	10.074

IMPORTS
PAPER & PAPERBOARD (6410)
PACIFIC NORTHWEST

VALUE--MILLION CURRENT DOLLARS

YEAR	CANADA	S AMERIC	EUROPE	JAPAN	OTH ASIA	OTHERS	TOTAL
1967	0.899	0.000	0.018	0.036	0.000	0.000	0.953
1968	1.321	0.004	0.086	0.045	0.014	0.000	1.470
1969	3.048	0.016	0.160	0.049	0.019	0.001	3.295
1970	3.714	0.000	0.092	0.103	0.038	0.000	3.947
1971	5.635	0.004	0.048	0.248	0.019	0.003	5.955
1972	3.099	0.000	0.004	0.128	0.041	0.009	3.281
1973	2.563	0.000	0.016	0.093	0.023	0.007	2.702
1974	5.226	0.000	0.073	0.307	0.100	0.003	5.709
1975	1.113	0.000	0.095	0.402	0.053	0.000	1.664
1976	2.182	0.000	0.109	0.637	0.118	0.000	3.047

QUANTITY--1000 STN

YEAR	CANADA	S AMERIC	EUROPE	JAPAN	OTH ASIA	OTHERS	TOTAL
1967	7.387	0.000	0.093	0.028	0.000	0.000	7.509
1968	10.147	0.022	0.439	0.058	0.007	0.000	10.673
1969	22.187	0.086	0.808	0.033	0.009	0.005	23.129
1970	23.635	0.000	0.437	0.294	0.012	0.000	24.377
1971	37.194	0.022	0.210	0.894	0.005	0.003	38.327
1972	15.867	0.000	0.010	0.057	0.012	0.006	15.952
1973	15.910	0.000	0.034	0.064	0.007	0.004	16.019
1974	23.523	0.000	0.099	0.106	0.048	0.001	23.778
1975	5.417	0.000	0.048	0.142	0.039	0.000	5.647
1976	10.721	0.000	0.084	0.223	0.024	0.000	11.052

Appendix E

THE SOUTH ATLANTIC REGION

Exports

Regional Overview

The South Atlantic region, in 1976, exported $492.9 million worth of forest products, which represented 12 percent of the total U.S. exports of wood, pulp and paper goods. The North Atlantic, Gulf, and Great Lakes regions each exported roughly the same amount of forest products as did the South Atlantic in 1976.

The percentage share of the total U.S. exports shipped from the South Atlantic declined slightly, from 15 percent to 12 percent, during 1967-76. However, the nominal value of the region's exports increased over the decade at an average annual rate of 15 percent.

Pulp, paper and paperboard products constitute a very large share of the region's total forest products. The share captured by the non-solid-wood group was 91 percent in 1976, when the region exported $450.9 million worth of wood pulp, newsprint, building board, and other paper and paperboard. Of these commodities, wood pulp was the most important, as it accounted for 54 percent of the pulp and paper exports. Other paper and paperboard made up most of the remainder of the exports from that group; newsprint and building board exports were relatively insignificant.

Solid wood products constituted only 9 percent of the value of the South Atlantic region's 1976 forest product exports. Softwood lumber was the most important solid wood commodity. The $18 million of softwood lumber shipped in 1976 represented 43 percent of the value of the region's solid wood exports, but only 4 percent of the region's total forest product exports.

The commodity structure of the region's exports did not change appreciably from 1967 to 1976. Throughout the period pulp and paper products accounted for about 90 percent of the region's exports, and solid wood products accounted for the remaining 10 percent.

Wood Pulp

The South Atlantic is one of three major U.S. wood-pulp-exporting regions, the others being the Gulf and Pacific Northwest regions. In 1976 the South Atlantic was the leader in wood pulp exports, shipping 708,000 tons (valued at $242.2 million). That value represented 28 percent of all U.S. wood pulp exports. In the same year, the Gulf accounted for 26 percent of U.S. exports, and the Pacific Northwest accounted for 20 percent.

During 1967-76, the South Atlantic's exports increased in nominal value at an average annual rate of 16 percent, while volumes increased at an average rate of 5 percent per year. Those rates were only slightly higher than the growth rates recorded by the United States as a whole--15 percent in value and 4 percent in volume.

European nations were the biggest importers of South Atlantic wood pulp--in 1976 Europe was the destination of 534,000 tons, valued at $189.7 million, which represented more than three-fourths of the South Atlantic's total wood pulp exports.

Together, the United Kingdom, France, and West Germany imported an average of 43 percent of the value of the region's wood pulp exports during 1967-76. In 1976 the United Kingdom accounted for 7 percent of the value of the region's exports, while France accounted for 12 percent and West Germany took 23 percent.

During 1967-76, West Germany exhibited the largest average import value growth rates of the three major European countries, increasing its wood pulp imports from the South Atlantic by 20 percent annually. Although the nominal value of exports to the United Kingdom increased by an average of 3 percent per year, the volumes declined by a total of 47 percent, which is equivalent to an annual decrease of 7 percent.

European countries other than United Kingdom, France, and West Germany imported 36 percent (by value) of the region's wood pulp exports in 1976. The growth rate of the nominal value of exports to this group of European countries rivaled that of West Germany--the $87.4 million exported in 1976 represented a 500 percent gain over the 1967 value, or an equivalent average annual increase of 20 percent.

The 1976 volume of South Atlantic wood pulp imported by Asian countries was one-third less than the amount imported in 1967. However, the nominal value of wood pulp exports to Asia increased during the period at an annual average rate of 6 percent. Asian countries accounted for 6 percent of the 1976 value of the region's wood pulp exports.

Paper and Paperboard

More than 40 percent of the value of the South Atlantic region's forest product exports during 1967-76 consisted of paper and paperboard other than newsprint and building board. In 1976 the region shipped 738,000 tons (valued at $206.4 million), which represented 17 percent of the total U.S. exports of paper and paperboard.

During the study period, the region's paper and paperboard exports grew in volume at an average annual rate of 3 percent, and in nominal value at a rate of 13 percent per year.

Throughout most of the period, Europe was the destination of more than half of the region's paper and paperboard exports. The United Kingdom, West Germany, and Italy were the leading European importers. During 1967-76, West Germany on average imported about one-third of the region's exports to Europe, which in 1976 amounted to 311,000 tons (valued at $84.6 million). In that year, Italy and the United Kingdom together were responsible for another one-third of the South Atlantic paper and paperboard exports.

Ecuador was a steady market for the region's paper and paperboard throughout 1967-76. In 1976 the South American country imported 78,000 tons worth $22.5 million, which represented 11 percent of the South Atlantic's total paper and paperboard exports.

Exports to various countries in North America, Asia, and Africa represented about 30 to 40 percent of the region's paper and paperboard exports during 1967-76. North American countries imported 141,000 tons worth $41.6 million, or 20 percent of the value of the region's total paper and paperboard exports in 1976.

Asia and Africa imported approximately equal amounts of paper and paperboard from the region. In 1976 South Atlantic exports to Asia amounted to $24.7 million and those to Africa $22.3 million, which represented 12 and 11 percent, respectively, of the region's total paper and paperboard exports.

Softwood Lumber

The third largest export commodity from the South Atlantic region has been softwood lumber. However, 1976 exports of softwood lumber, which totaled 43.4 million board-feet (valued at $18 million), represented only 4 percent (by value) of the region's total forest product exports. In the same year, the region's softwood lumber exports also represented 4 percent of U.S. softwood lumber exports.

During 1967-76, the volumes and values of softwood lumber exported from the region fluctuated, but by 1976 they showed increases over the 1967 volumes and values of 29 and 187 percent, respectively.

West Germany is the largest single importer of softwood lumber from the South Atlantic region. In 1976 exports to West Germany amounted to 14 million board-feet (valued at $5.2 million), which represented 29 percent of the region's total softwood lumber exports.

The Bahamas, in 1967, imported 53 percent of the value of softwood lumber shipped from the region, but by 1976, that share had dipped to 12 percent. During that period the volumes of softwood lumber exported to the Bahamas from the South Atlantic dropped from 21 million to 6 million board-feet.

North American countries other than the Bahamas (largely Central American nations) imported 29 percent of the value of the region's softwood lumber exports. The value of exports to these countries had increased at an average annual rate of 23 percent during the 1967-76 period.

The group of European countries other than West Germany also showed a high growth rate for the period, with an average increase of 19 percent per year in the value of softwood lumber imported from the South Atlantic.

Wood Veneer

The South Atlantic region, in 1976, exported 12.3 million square feet of wood veneer, which was valued at $3.6 million. That value represented 8 percent of the total U.S. exports of veneer.

The region drastically increased its exports of veneer over the 1967-76 period--the average yearly growth in volume was 55 percent, and in value was 70 percent.

Less than 1 percent of the region's 1976 veneer exports were destined for countries outside of Europe. West Germany imported more than three-fourths of the region's exports in 1976, and was the major importer throughout the 1967-76 period.

Finland and Norway combined accounted for 11 percent, and all other European countries combined accounted for 12 percent of the region's 1976 veneer exports.

Hardwood Logs

The South Atlantic region, in 1976, exported 5 percent of the value of U.S. total hardwood log exports. The $3.3 million of logs exported by the region represented only 1 percent of the region's total forest product exports.

During 1967-76, hardwood log exports from the region grew in volume at an average annual rate of 11 percent, and in value at 7 percent per year.

Two countries--Brazil and West Germany--received an average of 88 percent of the value of the region's hardwood log exports during 1967-76. In 1976 Brazil imported 2.6 million board-feet (valued at $1.9 million), and West Germany imported 640,000 board-feet (valued at $1 million). In that year, Brazil and West Germany accounted for 56 percent and 31 percent, respectively, of the value of the region's total hardwood log exports.

Hardwood Lumber

The South Atlantic's 1976 hardwood lumber exports amounted to 8.2 million board-feet worth $5 million, which represented 6 percent of the total U.S. hardwood lumber exports. The 1976 export volume and nominal value were, respectively, 236 and 535 percent greater than the 1967 export figures.

European countries were the destinations of an average 73 percent of the region's hardwood lumber exports during 1967-76. The United Kingdom and West Germany were major European trading partners, but neither country held a dominant market position for the entire period. For example, in 1976 West Germany's share of the value of the region's exports was 19 percent, the United Kingdom's share was only 6 percent, and the combined share of other European countries was 33 percent. This is in contrast to 1973, when the shares captured by the United Kingdom, West Germany, and other European countries were 60 percent, 2 percent, and 18 percent, respectively.

North American countries (primarily nations in Central America, defined here as in the North American classification) in 1976 imported 2.9 million board-feet of South Atlantic hardwood lumber, which was valued at $1.6 million. The 1976 value represented 31 percent of the South Atlantic region's total hardwood lumber exports, by far the largest market share recorded by North America during 1967-76.

Other Forest Products

Together, the remaining forest products (pulpwood, pulpwood chips, softwood logs, plywood, reconstituted wood, newsprint, and building board) constituted only 3 percent of the South Atlantic region's total forest product exports in 1976.

Plywood exports of 32.4 million square feet (valued at $6.3 million) represented 44 percent of the 1976 regional exports of this residual commodity group. Plywood exports from the region increased in volume by 73 percent and in value by 293 percent from 1967 to 1976.

European countries, in 1976, imported half the region's plywood exports, with West Germany (28 percent) and the United Kingdom (15 percent) being the principal markets involved. During the period 1967-76, North American countries, principally the Bahamas, were importing more than 90 percent (by value) of the region's plywood exports.

The 1976 exports of softwood logs from the region totaled 3.2 million board-feet worth $2.1 million. The 1976 figures reflect the exceptional export growth rates achieved by the region during 1967-76--the value of softwood log exports increased in volume at an average annual rate of 56 percent, and in value at 67 percent per year.

The principal destination for South Atlantic softwood logs--West Germany--on the average imported 60 percent of the value of the region's softwood log exports.

The South Atlantic region in 1976 exported 4,036 tons of building board worth $1.9 million. North American countries other than Canada were the major destinations for the region's building board, providing on average, 59 percent (by value) of the South Atlantic export market during 1967-76. France was the most important European market during that same period and, in 1976, imported 11 percent (by value) of the region's building board exports.

Imports

Regional Overview

The South Atlantic region is not a major U.S. regional importer of forest products. The $124 million worth of forest products imported by the region in 1976 was only 2 percent of total U.S. forest product imports.

The nominal value of the region's imports grew at a faster rate than did most regions' imports during the 1967-76 period. The total increase in value over the period was 261 percent, which is equivalent to an annual increase of 15 percent.

The commodity mix of the region's imports is weighted heavily in favor of solid wood products, particularly wood veneer and plywood. In 1976, 84 percent of the region's imports was in solid wood products, and 15 percent was in pulp and paper. This situation contrasted with the commodity composition of imports to the entire United States, where, in 1976, 63 percent of total imports of forest products was in pulp and paper and 37 percent was in solid wood products.

Plywood constituted more than 60 percent ($75.7 million) of the region's forest product imports in 1976. Wood veneer ($13.9 million) and hardwood lumber ($10 million) were also significant solid wood imports.

Wood pulp is the region's major non-solid wood import. In 1976 the $11.1 million worth of pulp represented 9 percent of the region's total forest product imports, and 56 percent of the region's pulp and paper imports. The remainder of the region's pulp and paper imports was fairly evenly divided among newsprint ($2.7 million), building board ($2.7 million), and other paper and paperboard ($4.2 million).

The commodity structure of the region's imports changed somewhat from 1967 to 1976. Pulp and paper made up a larger percentage of the region's forest product imports in 1967--25 percent, as opposed to 16 percent in 1976. Moreover, newsprint imports represented 80 percent of the region's pulp and paper imports in 1967, but only 13 percent in 1976.

There was also some shifting among commodities in the solid wood group. Plywood was not as important in 1967 (37 percent of the region's total forest product imports) as it was in 1976 (61 percent). Pulpwood was an important commodity in 1967 (12 percent of the total forest product imports) but was not significant in 1976 (2 percent).

Plywood

Of the major plywood importing regions, the South Atlantic region was the most rapidly growing in imports of plywood for the period 1967-76. The volume imported to the region increased at an average annual rate of 15 percent, while the value grew at 22 percent per year. The 1976 import volume of 598.5 million square feet was valued at $75.5 million, which represented 20 percent of U.S. total plywood imports.

The region's chief foreign sources of plywood are eastern Asian nations. In particular, Japan, Korea, and Taiwan together provided 90 percent of the region's plywood imports in 1976.

Korea was the leading exporter to the region in 1976, when it shipped 347.2 million square feet worth $43.7 million. The 1976 volume and value represented tremendous increases over the 1967 figures--volume was up by more than 900 percent, and the value increased by more than 1,600 percent. The total increases were equivalent to annual gains of 29 percent in volume and 37 percent in nominal value.

Taiwan also showed spectacular growth in its export of plywood to the South Atlantic region. Volumes increased at an average annual rate of 13 percent, and values increased by an average of 22 percent per year during 1967-76. In 1976 Taiwan shipped 156.7 million square feet to the region, or 26 percent of the total plywood volume imported by the South Atlantic region.

Japan's shipments of plywood to the South Atlantic declined during the study period, from 35.6 million square feet in 1967 to 22.1 million square feet in 1976.

The plywood imports which were not supplied by Korea, Taiwan, or Japan originated in other Asian countries. On average, this residual group provided 15 percent of the region's plywood imports during the period.

Wood Veneer

The South Atlantic is the second largest veneer-importing region in the United States. In 1976 the region imported 374.1 million square feet worth $13.9 million, which represented nearly one-fifth of the total U.S. imports of wood veneer.

The region's veneer imports peaked during 1972-74, when an average of 544.4 million square feet was imported annually. For the remainder of the 1967-76 period, the average was 307.2 million square feet, but overall there was an upward trend in wood veneer imports to the region.

In an average year, over 70 percent (by value) of the South Atlantic region's veneer imports originated in just two countries--Brazil and the Philippines. The Philippines was the dominant supplier, providing an average of 44 percent of the region's imports. In 1976 veneer imports from the Philippines totaled 188.7 million square feet (worth $6.7 million), or 48 percent of the South Atlantic's total veneer imports.

Brazil's 1976 exports to the region, 150.7 million square feet (valued at $5.1 million), represented 36 percent of the South Atlantic's veneer imports.

Both countries' exports to the region grew faster than average during 1967-76. While the value of all countries' veneer exports to the South Atlantic increased at an average rate of 12 percent per year, Brazil and the Philippines recorded annual growth rates of 17 percent and 14 percent, respectively.

South American and Asian countries (other than Brazil and the Philippines) were significant veneer suppliers to the South Atlantic in the early years of the period, but their market shares had dropped off considerably by 1976. In 1967 these countries together provided 28 percent of the value of the region's veneer imports, but by 1976 that share was only 8 percent.

Wood Pulp

Wood pulp is the region's third most important forest product import, but in 1976 it accounted for only 9 percent (by value) of the region's total wood, pulp, and paper imports.

Wood pulp was the region's fastest growing forest product import during 1967-76, increasing from 1,082 tons (worth $148,000) in 1967 to 42,000 tons (worth $11.1 million) in 1976. Those figures represent average annual growth rates of 50 percent in volume and 61 percent in nominal value.

Canada and Sweden have been the dominant foreign suppliers of wood pulp to the South Atlantic. In fact, beginning in 1973, when the region first started to import significant quantities of wood pulp, Canada and Sweden have provided 100 percent of the region's pulp imports.

The 33,000 tons of pulp which were imported from Canada in 1976 were valued at $8.4 million, or 76 percent of the value of the region's total wood pulp imports. Sweden in that year exported to the region 8,500 tons (worth $2.7 million).

Hardwood Lumber

The South Atlantic region, in an average year, imported 7 percent (by value) of the total U.S. imports of hardwood lumber during 1967-76. In 1976 the region imported 25.1 million board-feet (worth $10 million).

The South Atlantic was one of the few regions to increase its imports of hardwood lumber over the 1967-76 period. The 1976 volume and value of South Atlantic hardwood lumber imports represented increases of 5 percent and 254 percent, respectively, over the 1967 volume of 23.9 million board-feet and nominal value of $2.8 million.

The South American countries of Colombia, Ecuador, and Brazil were responsible for most of the hardwood lumber imports to the South Atlantic region during 1967-76. Brazil was the most important supplier, accounting for an average of 55 percent of the region's imports each year. Brazil increased the volume of its exports to the region by an average of 8 percent per year from 1967 to 1976. In 1976 Brazilian exports of hardwood lumber to the South Atlantic amounted to 18.4 million board-feet worth $4.5 million, or 45 percent of the region's total hardwood lumber imports.

Colombia drastically declined as an exporter of lumber to the region, going from 10.1 million board-feet in 1967 to 745,000 board-feet in 1976.

While in 1967 Colombia accounted for 34 percent of the value of the region's hardwood lumber imports, by 1976 that share was reduced to only 1 percent.

Ecuador has partly filled the gap left by Colombia's decline in exports of hardwood lumber. Although Ecuador did not export to the region in 1967, in 1976 it shipped 2.8 million board-feet (valued at $4 million), which represented 40 percent of the value of the region's total hardwood lumber imports. Note, however, that the 2.8 million board-feet exported by Ecuador was only 11 percent of the volume of the region's total hardwood lumber imports; apparently, Ecuador is shipping a much higher valued product to the South Atlantic than are either Colombia or Brazil.

Asian exports of hardwood lumber to the South Atlantic region declined from 2.8 million board-feet in 1967 to 1.4 million board-feet in 1976. The 1976 value of imports from Asia was $812,000 or 8 percent of the region's total hardwood lumber imports.

Pulpwood (Except Chips)

The South Atlantic region is one of only two significant pulpwood-importing regions in the United States, the other being the North Atlantic. The South Atlantic region, in 1976, imported 14 percent of the value, and 30 percent of the volume of the total U.S. pulpwood imports.

As in the entire country, pulpwood imports to the South Atlantic region declined considerably from 1967 to 1976. Both the value and volume of pulpwood imported into the region in 1976 were less than half the 1967 value and volume. In 1976 the region imported 101,000 cords (valued at $1.6 million).

The Bahamas was practically the only exporter of pulpwood to the South Atlantic throughout 1967-76.

Other Forest Products

The remaining forest products (softwood and hardwood logs, softwood lumber, reconstituted wood, pulpwood chips, newsprint, building board, other paper and paperboard, and other industrial roundwood) together constituted only 9 percent (by value) of the region's 1976 total forest product imports. In 1967, that same group of products represented 28 percent of total imports.

Most outstanding has been the decline in imports of newsprint. In 1967 the $7 million worth of newsprint imports was 20 percent of the South Atlantic's total forest product imports, but by 1976 the nominal value of newsprint imports had dropped to $2.7 million, only 2 percent of total imports. Canada and Finland were the sole exporters of newsprint to the region, with Canada supplying, on average, 65 percent and Finland 35 percent of the region's newsprint imports.

Paper and paperboard in 1976 represented 3 percent of the value of the South Atlantic's total forest product imports. During the 1967-76 period, the total value of paper and paperboard imports fluctuated considerably, and no trend was discernible. The chief foreign sources of paper and paperboard were the Nordic countries of Sweden and Finland. In 1976 the $1.1 million worth imported from Sweden represented 27 percent of the region's total paper and paperboard imports, and the $1.8 million imported from Finland was 43 percent. European countries other than Sweden and Finland captured 20 percent of the value of the region's 1976 imports. The large disparities observable in unit values among the European countries--$733 per ton in Sweden, $289 per ton in Finland, and $2,225 per ton in other European nations--indicate that other European countries are exporting a more highly processed product than are the Nordic nations.

The South Atlantic region, in 1976, imported 24,000 tons of building board (valued at $2.7 million), which represented 10 percent of the total U.S. imports of building board. Building board imports to the region increased in value during 1967-76 at an average annual rate of 16 percent. Sweden was the most important supplier of building board, providing nearly 60 percent (by value) of the region's imports in 1976. South American countries were the sources of 17 percent of the region's 1976 building board imports, and European countries other than Sweden were the sources of another 21 percent.

EXPORTS
SOFTWOOD LOGS (2422)
SOUTH ATLANTIC

VALUE--MILLION CURRENT DOLLARS

YEAR	N AMERIC	BRAZIL	W GERMNY	OTH EURO	AFRICA	OTHERS	TOTAL
1967	0.003	0.000	0.018	0.000	0.000	0.000	0.021
1968	0.000	0.000	0.044	0.000	0.000	0.000	0.044
1969	0.003	0.000	0.006	0.000	0.000	0.000	0.008
1970	0.002	0.002	0.015	0.002	0.000	0.000	0.021
1971	0.036	0.002	0.019	0.012	0.000	0.002	0.072
1972	0.031	0.015	0.062	0.033	0.000	0.000	0.140
1973	0.082	0.015	0.407	0.183	0.000	0.008	0.695
1974	0.321	0.007	1.734	0.429	0.082	0.000	2.573
1975	0.419	0.076	1.667	0.252	0.245	0.000	2.660
1976	0.405	0.158	0.807	0.495	0.105	0.125	2.095

QUANTITY--MILLION BF

YEAR	N AMERIC	BRAZIL	W GERMNY	OTH EURO	AFRICA	OTHERS	TOTAL
1967	0.014	0.000	0.044	0.000	0.000	0.000	0.058
1968	0.000	0.000	0.174	0.000	0.000	0.000	0.174
1969	0.025	0.000	0.013	0.000	0.000	0.000	0.038
1970	0.002	0.020	0.068	0.018	0.000	0.000	0.107
1971	0.100	0.060	0.103	0.050	0.000	0.013	0.327
1972	0.141	0.130	0.156	0.096	0.000	0.000	0.524
1973	0.130	0.153	1.064	0.420	0.000	0.012	1.779
1974	0.531	0.175	2.479	0.813	0.110	0.000	4.109
1975	0.763	0.252	1.656	1.520	0.202	0.000	4.392
1976	0.513	0.522	1.125	0.799	0.120	0.134	3.213

EXPORTS
HARDWOOD LOGS (2423)
SOUTH ATLANTIC

VALUE--MILLION CURRENT DOLLARS

YEAR	N AMERIC	BRAZIL	W GERMNY	SWITZLND	OTH EURO	OTHERS	TOTAL
1967	0.001	0.000	1.703	0.053	0.007	0.062	1.826
1968	0.002	0.000	2.786	0.315	0.115	0.003	3.221
1969	0.002	0.000	1.164	0.230	0.444	0.014	1.854
1970	0.003	1.747	1.159	0.126	0.215	0.000	3.249
1971	0.004	1.150	0.455	0.000	0.150	0.000	1.759
1972	0.033	3.317	0.510	0.026	0.250	0.003	4.139
1973	0.039	3.738	0.227	0.033	0.185	0.011	4.233
1974	0.036	1.371	0.012	0.193	0.027	0.002	1.641
1975	0.021	1.327	0.673	0.004	0.078	0.000	2.102
1976	0.066	1.869	1.049	0.016	0.300	0.039	3.339

QUANTITY--MILLION BF

YEAR	N AMERIC	BRAZIL	W GERMNY	SWITZLND	OTH EURO	OTHERS	TOTAL
1967	0.004	0.000	1.261	0.041	0.019	0.077	1.403
1968	0.005	0.000	2.263	0.248	0.114	0.005	2.634
1969	0.001	0.000	1.191	0.168	0.528	0.042	1.930
1970	0.002	1.839	1.195	0.117	0.193	0.000	3.346
1971	0.015	1.485	0.460	0.000	0.165	0.000	2.126
1972	0.038	3.291	0.500	0.062	0.229	0.002	4.122
1973	0.057	4.782	0.379	0.098	0.408	0.015	5.739
1974	0.044	2.546	0.028	0.246	0.064	0.001	2.928
1975	0.017	1.196	0.673	0.010	0.080	0.000	1.976
1976	0.044	2.551	0.640	0.016	0.208	0.021	3.480

EXPORTS
SOFTWOOD LUMBER (2432)
SOUTH ATLANTIC

VALUE--MILLION CURRENT DOLLARS

YEAR	BAHAMAS	OTH N AM	S AMERIC	W GEMNY0	TH EURO	OTHERS	TOTAL
1967	3.303	0.821	0.005	1.283	0.608	0.258	6.278
1968	3.609	0.906	0.024	1.387	0.550	0.203	6.678
1969	4.435	1.162	0.019	1.221	0.553	0.156	7.546
1970	3.180	1.397	0.077	1.069	0.861	0.433	7.017
1971	2.312	1.335	0.002	1.429	0.393	0.445	5.916
1972	2.067	1.394	0.003	1.532	0.806	0.199	6.002
1973	2.397	1.149	0.002	2.699	1.472	0.494	8.213
1974	2.243	2.317	0.113	0.732	0.994	3.042	9.442
1975	1.864	1.660	0.097	1.963	2.058	1.625	9.308
1976	2.189	5.313	0.219	5.218	2.886	2.216	18.041

QUANTITY--MILLION BF

YEAR	BAHAMAS	OTH N AM	S AMERIC	W GEMNY0	TH EURO	OTHERS	TOTAL
1967	21.022	5.324	0.025	9.999	4.565	2.468	43.404
1968	21.625	5.351	0.163	10.579	3.899	1.769	43.725
1969	24.342	5.715	0.118	8.961	3.409	1.540	44.083
1970	18.318	8.376	0.227	6.991	5.690	4.038	43.639
1971	12.094	6.325	0.010	10.230	2.313	2.413	33.385
1972	9.632	6.219	0.007	9.568	5.030	1.009	31.466
1973	9.943	4.392	0.003	11.107	5.563	2.056	33.063
1974	7.163	8.720	0.369	2.055	3.119	10.710	32.125
1975	6.286	6.146	0.245	6.034	3.432	5.421	32.563
1976	5.942	16.555	1.231	13.968	8.616	9.619	55.911

EXPORTS
HARDWOOD LUMBER (2433)
SOUTH ATLANTIC

VALUE--MILLION CURRENT DOLLARS

YEAR	N AMERIC	U KINGDM	W GERMNY	OTH EURO	ASIA	OTHERS	TOTAL
1967	0.086	0.077	0.062	0.365	0.076	0.126	0.792
1968	0.053	0.040	0.116	0.265	0.047	0.072	0.593
1969	0.054	0.074	0.059	0.270	0.029	0.031	0.517
1970	0.077	0.042	0.060	0.216	0.004	0.069	0.469
1971	0.077	0.337	0.038	0.279	0.012	0.096	0.839
1972	0.242	0.786	0.070	0.253	0.125	0.040	1.515
1973	0.176	1.056	0.030	0.312	0.144	0.030	1.748
1974	0.260	0.471	0.122	0.452	0.000	0.019	1.325
1975	0.364	0.402	0.564	0.909	0.081	0.075	2.396
1976	1.570	0.307	0.942	1.664	0.517	0.010	5.030

QUANTITY--MILLION BF

YEAR	N AMERIC	U KINGDM	W GERMNY	OTH EURO	ASIA	OTHERS	TOTAL
1967	0.245	0.334	0.199	1.139	0.121	0.394	2.433
1968	0.146	0.271	0.402	0.874	0.121	0.237	2.051
1969	0.141	0.294	0.159	0.953	0.057	0.079	1.684
1970	0.234	0.181	0.207	0.564	0.007	0.200	1.393
1971	0.313	0.919	0.125	0.651	0.032	0.366	2.405
1972	0.610	2.361	0.192	0.703	0.149	0.148	4.163
1973	0.424	2.924	0.055	0.628	0.362	0.120	4.513
1974	0.639	0.960	0.178	0.762	0.000	0.042	2.581
1975	0.753	0.729	1.105	1.385	0.144	0.084	4.201
1976	2.917	0.540	1.614	2.376	0.728	0.006	8.180

EXPORTS
WOOD VENEER (6311)
SOUTH ATLANTIC

VALUE--MILLION CURRENT DOLLARS

YEAR	SWEDEN	NORWAY	W GERMNY	POLAND	OTH EURO	OTHERS	TOTAL
1967	0.000	0.000	0.008	0.000	0.013	0.010	0.031
1968	0.000	0.000	0.826	0.000	0.312	0.013	1.152
1969	0.025	0.000	1.053	0.000	0.050	0.050	1.178
1970	0.000	0.000	0.335	0.000	0.017	0.050	0.402
1971	0.010	0.000	0.643	0.000	0.000	0.013	0.666
1972	0.000	0.000	2.474	0.000	0.015	0.021	2.510
1973	0.275	0.071	4.240	0.000	0.334	0.049	4.969
1974	0.084	0.045	2.769	0.096	0.374	0.068	3.435
1975	0.224	0.156	3.209	0.218	0.301	0.009	4.117
1976	0.219	0.191	2.749	0.000	0.419	0.014	3.591

QUANTITY--MIL SQ FT

YEAR	SWEDEN	NORWAY	W GERMNY	POLAND	OTH EURO	OTHERS	TOTAL
1967	0.000	0.000	0.016	0.000	0.018	0.030	0.064
1968	0.000	0.000	0.666	0.000	0.250	0.015	0.932
1969	0.032	0.000	0.892	0.000	0.038	0.090	1.052
1970	0.000	0.000	0.605	0.000	0.036	0.032	0.673
1971	0.012	0.000	0.477	0.000	0.000	0.012	0.502
1972	0.000	0.000	1.938	0.000	0.016	0.009	1.963
1973	0.173	0.034	3.254	0.000	0.416	0.136	4.013
1974	0.054	0.021	2.239	0.085	0.429	0.100	2.929
1975	0.139	0.084	2.478	0.166	0.329	0.020	3.217
1976	0.154	0.118	2.584	0.000	0.408	0.018	3.283

EXPORTS
PLYWOOD (6312)
SOUTH ATLANTIC

VALUE--MILLION CURRENT DOLLARS

YEAR	BAHAMAS	OTH N AM	U KINGDM	W GERMNY	OTH EURO	OTHERS	TOTAL
1967	1.247	0.155	0.089	0.004	0.098	0.008	1.601
1968	1.646	0.320	0.095	0.000	0.234	0.005	2.299
1969	1.909	0.193	0.100	0.012	0.116	0.006	2.335
1970	1.222	0.369	0.056	0.000	0.096	0.005	1.746
1971	0.796	0.531	0.021	0.000	0.030	0.010	1.388
1972	0.921	0.452	0.042	0.000	0.021	0.007	1.443
1973	1.622	0.979	1.096	0.000	0.025	0.062	3.783
1974	1.272	1.700	1.151	0.109	0.318	0.055	4.605
1975	0.872	1.796	0.075	0.378	0.541	0.012	3.674
1976	1.168	1.851	0.953	1.733	0.469	0.114	6.288

QUANTITY--MIL SQ FT

YEAR	BAHAMAS	OTH N AM	U KINGDM	W GERMNY	OTH EURO	OTHERS	TOTAL
1967	10.983	1.076	0.127	0.033	0.192	0.075	12.485
1968	12.773	1.810	0.118	0.000	0.444	0.024	15.170
1969	13.297	1.036	0.166	0.023	0.224	0.008	14.753
1970	9.487	2.417	0.049	0.000	0.162	0.010	12.124
1971	5.690	3.851	0.020	0.000	0.037	0.037	9.635
1972	5.579	2.392	0.035	0.000	0.021	0.035	8.064
1973	5.952	4.196	8.219	0.000	0.168	0.143	18.678
1974	5.683	8.955	10.174	0.877	2.009	0.170	27.868
1975	5.008	8.747	0.182	0.436	4.226	0.046	18.644
1976	6.289	7.030	4.789	2.143	1.138	0.289	21.678

EXPORTS
RECONSTITUTED WOOD (6314)
SOUTH ATLANTIC

VALUE--MILLION CURRENT DOLLARS

YEAR	BAHAMAS	NETH ANT	OTH N AM	S AMERIC	EUROPE	OTHERS	TOTAL
1967	0.023	0.000	0.000	0.000	0.000	0.000	0.023
1968	0.016	0.000	0.000	0.000	0.000	0.000	0.016
1969	0.017	0.000	0.001	0.000	0.000	0.000	0.018
1970	0.004	0.000	0.002	0.000	0.000	0.000	0.006
1971	0.009	0.000	0.001	0.000	0.000	0.000	0.010
1972	0.016	0.044	0.016	0.000	0.000	0.000	0.076
1973	0.008	0.065	0.021	0.000	0.000	0.000	0.094
1974	0.022	0.024	0.046	0.000	0.001	0.000	0.094
1975	0.011	0.014	0.035	0.015	0.000	0.000	0.074
1976	0.005	0.054	0.068	0.070	0.000	0.022	0.220

QUANTITY--MIL SQ FT

YEAR	BAHAMAS	NETH ANT	OTH N AM	S AMERIC	EUROPE	OTHERS	TOTAL
1967	0.065	0.000	0.000	0.000	0.000	0.000	0.065
1968	0.109	0.000	0.000	0.000	0.000	0.000	0.109
1969	0.076	0.000	0.001	0.000	0.000	0.000	0.076
1970	0.015	0.000	0.012	0.000	0.000	0.000	0.027
1971	0.042	0.000	0.003	0.000	0.000	0.000	0.045
1972	0.056	0.200	0.048	0.000	0.000	0.000	0.305
1973	0.031	0.090	0.055	0.000	0.000	0.000	0.176
1974	0.064	0.060	0.117	0.000	0.015	0.000	0.256
1975	0.061	0.101	0.141	0.123	0.000	0.000	0.426
1976	0.022	0.197	0.274	0.478	0.000	0.001	0.971

EXPORTS
PULPWOOD CHIPS (631.8320)
SOUTH ATLANTIC

VALUE--MILLION CURRENT DOLLARS

YEAR	SWEDEN	FRANCE	ITALY	OTH EURO	JAPAN	OTHERS	TOTAL
1967	0.000	0.000	0.000	0.000	0.000	0.000	0.000
1968	0.000	0.000	0.000	0.000	0.000	0.000	0.000
1969	0.000	0.000	0.000	0.000	0.000	0.000	0.000
1970	0.000	0.000	0.000	0.000	0.000	0.000	0.000
1971	0.000	0.000	0.000	0.000	0.000	0.000	0.000
1972	0.000	0.000	0.000	0.000	0.000	0.000	0.000
1973	0.000	0.000	0.000	0.000	0.000	0.000	0.000
1974	0.000	0.000	0.000	0.000	0.663	0.000	0.663
1975	3.841	0.785	0.000	0.000	0.000	0.000	4.626
1976	0.000	0.000	1.600	0.000	0.000	0.000	1.600

QUANTITY--1000 STN

YEAR	SWEDEN	FRANCE	ITALY	OTH EURO	JAPAN	OTHERS	TOTAL
1967	0.000	0.000	0.000	0.000	0.000	0.000	0.000
1968	0.000	0.000	0.000	0.000	0.000	0.000	0.000
1969	0.000	0.000	0.000	0.000	0.000	0.000	0.000
1970	0.000	0.000	0.000	0.000	0.000	0.000	0.000
1971	0.000	0.000	0.000	0.000	0.000	0.000	0.000
1972	0.000	0.000	0.000	0.000	0.000	0.000	0.000
1973	0.000	0.000	0.000	0.000	0.000	0.000	0.000
1974	0.000	0.000	0.000	0.000	16.212	0.000	16.212
1975	77.283	16.026	0.000	0.000	0.000	0.000	93.309
1976	0.000	0.000	30.863	0.000	0.000	0.000	30.863

EXPORTS
WOOD PULP (2510)
SOUTH ATLANTIC

VALUE--MILLION CURRENT DOLLARS

YEAR	U KINGDM	FRANCE	W GERMNY	OTH EURO	ASIA	OTHERS	TOTAL
1967	13.827	5.921	10.828	17.376	8.565	7.569	64.090
1968	12.286	7.836	15.117	17.549	7.143	10.493	70.424
1969	7.435	4.789	16.792	19.553	6.132	9.816	64.516
1970	13.876	10.295	21.211	42.752	9.726	15.286	113.148
1971	9.318	11.055	20.621	27.298	7.266	13.261	88.819
1972	9.231	10.481	19.414	25.736	5.503	11.891	82.255
1973	6.318	11.820	24.425	31.352	8.853	17.013	99.781
1974	12.784	18.305	42.807	57.501	23.909	39.938	195.244
1975	16.489	22.087	42.168	67.825	21.039	46.726	216.355
1976	17.713	28.755	55.903	87.355	14.981	37.518	242.225

QUANTITY--1000 STN

YEAR	U KINGDM	FRANCE	W GERMNY	OTH EURO	ASIA	OTHERS	TOTAL
1967	102.887	38.559	73.782	106.799	62.246	66.203	450.476
1968	92.902	52.035	96.955	102.307	61.447	94.231	499.776
1969	54.781	27.606	99.430	113.730	48.361	81.918	425.825
1970	101.403	60.902	123.419	258.805	69.234	112.919	726.661
1971	62.751	62.681	107.616	147.952	55.731	94.534	531.266
1972	59.893	55.913	101.861	140.714	37.878	81.220	477.479
1973	34.363	55.963	120.668	162.083	42.710	93.180	508.966
1974	64.254	53.025	136.724	188.343	77.223	133.061	652.630
1975	69.783	61.907	112.067	204.079	58.265	161.381	667.461
1976	54.607	82.773	142.374	253.947	41.268	133.256	708.225

EXPORTS
NEWSPRINT (6411)
SOUTH ATLANTIC

VALUS--MILLION CURRENT DOLLARS

YEAR	N AMERIC	S AMERIC	INDONESA	OTH ASIA	NIGERIA	OTHERS	TOTAL
1967	0.013	0.226	0.000	0.022	0.000	0.029	0.290
1968	0.048	0.034	0.422	1.053	0.000	0.000	1.557
1969	0.020	0.027	0.996	0.044	0.000	0.000	1.086
1970	0.037	0.000	0.000	0.250	0.000	0.000	0.288
1971	0.085	0.009	0.000	0.011	0.000	0.000	0.105
1972	0.109	0.000	0.000	0.000	0.000	0.000	0.109
1973	0.215	0.000	0.000	0.000	0.000	0.000	0.215
1974	0.500	0.002	0.000	0.019	0.000	0.000	0.521
1975	0.658	1.087	0.000	0.026	1.944	0.000	3.714
1976	0.241	0.101	0.000	0.000	0.000	0.000	0.342

QUANTITY--1000 STN

YEAR	N AMERIC	S AMERIC	INDONESA	OTH ASIA	NIGERIA	OTHERS	TOTAL
1967	0.056	1.821	0.000	0.215	0.000	0.209	2.300
1968	0.255	0.304	3.916	9.379	0.000	0.000	13.854
1969	0.130	0.272	9.566	0.443	0.000	0.000	10.412
1970	0.232	0.000	0.000	2.032	0.000	0.001	2.266
1971	0.441	0.029	0.000	0.112	0.000	0.000	0.583
1972	0.901	0.000	0.000	0.000	0.000	0.000	0.901
1973	1.053	0.000	0.000	0.000	0.000	0.000	1.053
1974	1.803	0.007	0.000	0.055	0.000	0.000	1.866
1975	1.893	3.095	0.000	0.086	3.968	0.000	9.042
1976	0.634	0.362	0.000	0.000	0.000	0.000	0.996

EXPORTS
BUILDING BOARD (6416)
SOUTH ATLANTIC

VALUE--MILLION CURRENT DOLLARS

YEAR	BAHAMAS	OTH N AM	FRANCE	OTH EURO	ASIA	OTHERS	TOTAL
1967	0.225	0.088	0.061	0.075	0.109	0.019	0.577
1968	0.224	0.184	0.000	0.065	0.046	0.025	0.545
1969	0.201	0.144	0.000	0.048	0.078	0.420	0.891
1970	0.221	0.206	0.000	0.044	0.003	0.007	0.482
1971	0.112	0.262	0.233	0.061	0.004	0.107	0.779
1972	0.099	0.304	0.187	0.102	0.093	0.070	0.856
1973	0.187	0.482	0.193	0.135	0.042	0.018	1.057
1974	0.240	0.858	0.589	0.207	0.133	0.078	2.104
1975	0.169	0.796	0.224	0.137	0.167	0.061	1.555
1976	0.139	1.022	0.217	0.178	0.035	0.356	1.947

QUANTITY--1000 STN

YEAR	BAHAMAS	OTH N AM	FRANCE	OTH EURO	ASIA	OTHERS	TOTAL
1967	1.428	0.256	0.558	0.335	0.461	0.067	3.105
1968	1.107	0.529	0.000	0.259	0.159	0.107	2.162
1969	0.719	0.571	0.000	0.180	0.264	1.152	2.987
1970	0.869	1.158	0.000	0.176	0.008	0.022	2.232
1971	0.449	0.910	1.259	0.242	0.013	0.349	3.223
1972	0.433	1.127	0.938	0.414	0.276	0.204	3.391
1973	1.249	1.669	0.760	0.446	0.112	0.052	4.308
1974	1.385	2.419	1.625	0.685	0.253	0.100	6.467
1975	0.870	1.942	0.717	0.295	0.211	0.106	4.041
1976	0.700	2.264	0.653	0.210	0.065	0.413	4.306

EXPORTS
PAPER & PAPERBOARD (6410)
SOUTH ATLANTIC

VALUE--MILLION CURRENT DOLLARS

YEAR	N AMERIC	S AMERIC	EUROPE	ASIA	AFRICA	OTHERS	TOTAL
1967	12.989	10.173	33.140	6.297	5.169	0.993	68.760
1968	13.129	9.022	37.630	9.935	5.581	0.498	75.794
1969	6.509	9.329	43.474	7.292	6.069	0.511	73.183
1970	7.333	11.282	55.746	9.887	8.849	0.443	93.540
1971	9.964	13.154	55.866	10.212	6.922	0.635	96.753
1972	13.033	13.025	49.511	9.528	6.249	0.431	91.777
1973	14.788	17.652	62.912	10.040	10.303	0.986	116.682
1974	28.551	26.923	99.794	18.224	20.560	0.916	194.967
1975	26.833	31.924	87.629	23.154	17.384	1.254	188.177
1976	41.607	30.845	84.588	24.698	22.250	2.372	206.360

QUANTITY--1000 STN

YEAR	N AMERIC	S AMERIC	EUROPE	ASIA	AFRICA	OTHERS	TOTAL
1967	111.033	88.904	285.957	53.783	38.242	4.824	582.742
1968	115.812	75.800	344.483	92.586	42.753	2.972	674.405
1969	48.097	80.095	380.376	63.984	49.228	2.681	624.461
1970	49.153	81.494	441.287	78.981	69.147	2.098	722.159
1971	71.387	104.556	464.003	82.766	55.841	2.085	780.638
1972	93.747	99.945	358.337	71.054	47.851	1.714	672.649
1973	82.637	120.389	341.624	63.642	60.799	3.185	672.275
1974	98.062	127.531	385.725	66.707	86.151	3.133	767.329
1975	92.528	123.424	295.230	83.349	64.763	3.154	662.448
1976	140.622	97.394	310.872	99.121	64.335	5.772	738.118

IMPORTS
HARDWOOD LOGS (2423)
SOUTH ATLANTIC

VALUE--MILLION CURRENT DOLLARS

YEAR	GUATMALA	BRAZIL	OTH S AM	INDIA	IVRY CST	OTHERS	TOTAL
1967	0.000	0.047	0.008	0.000	0.093	0.086	0.234
1968	0.000	0.119	0.077	0.000	0.000	0.030	0.226
1969	0.000	0.003	0.017	0.000	0.000	0.000	0.020
1970	0.000	0.002	0.006	0.000	0.000	0.006	0.014
1971	0.002	0.000	0.039	0.000	0.000	0.002	0.043
1972	0.018	0.000	0.008	0.040	0.000	0.000	0.066
1973	0.144	0.000	0.002	0.087	0.000	0.019	0.252
1974	0.013	0.000	0.000	0.005	0.085	0.024	0.128
1975	0.103	0.007	0.009	0.105	0.000	0.000	0.223
1976	0.031	0.000	0.002	0.012	0.000	0.001	0.046

QUANTITY--MILLION BF

YEAR	GUATMALA	BRAZIL	OTH S AM	INDIA	IVRY CST	OTHERS	TOTAL
1967	0.000	0.229	0.055	0.000	0.523	0.441	1.248
1968	0.000	0.479	0.371	0.000	0.000	0.020	0.870
1969	0.000	0.031	0.090	0.000	0.000	0.000	0.121
1970	0.000	0.021	0.037	0.000	0.000	0.009	0.067
1971	0.012	0.000	0.053	0.000	0.000	0.004	0.068
1972	0.080	0.000	0.005	0.165	0.000	0.000	0.251
1973	0.341	0.000	0.001	0.429	0.000	0.026	0.796
1974	0.024	0.000	0.000	0.004	0.189	0.076	0.293
1975	0.267	0.015	0.011	0.021	0.000	0.000	0.314
1976	0.033	0.000	0.004	0.011	0.000	0.001	0.049

IMPORTS
SOFTWOOD LUMBER (2432)
SOUTH ATLANTIC

VALUE--MILLION CURRENT DOLLARS

YEAR	CANADA	OTH N AM	BRAZIL	TAIWAN	OTH ASIA	OTHERS	TOTAL
1967	0.407	0.012	0.126	0.000	0.000	0.000	0.544
1968	0.725	0.006	0.669	0.000	0.000	0.000	1.399
1969	0.254	0.008	0.876	0.000	0.000	0.003	1.140
1970	0.973	0.003	0.333	0.061	0.027	0.005	1.402
1971	3.393	0.010	0.104	0.424	0.105	0.000	4.036
1972	4.577	0.212	0.342	0.414	0.112	0.011	5.667
1973	6.118	1.400	1.220	0.652	0.129	0.019	9.538
1974	1.729	0.395	0.074	0.250	0.159	0.000	2.606
1975	0.554	0.324	0.020	0.245	0.157	0.000	1.300
1976	0.000	0.545	0.101	0.215	0.038	0.259	1.159

QUANTITY--MILLION BF

YEAR	CANADA	OTH N AM	BRAZIL	TAIWAN	OTH ASIA	OTHERS	TOTAL
1967	10.837	0.078	1.523	0.000	0.000	0.000	12.438
1968	13.777	0.121	5.738	0.000	0.000	0.000	19.636
1969	5.703	0.157	6.361	0.000	0.000	0.012	12.233
1970	25.710	0.014	2.233	0.251	0.138	0.138	28.464
1971	61.031	0.150	0.752	1.656	0.505	0.000	64.094
1972	56.875	2.302	3.694	1.532	0.496	0.043	64.943
1973	55.468	0.599	5.270	2.187	0.537	0.102	73.163
1974	22.326	2.686	0.270	0.650	0.511	0.000	26.443
1975	6.538	1.697	0.056	0.628	0.540	0.000	9.459
1976	0.000	2.364	0.593	0.415	0.130	1.163	4.665

IMPORTS
HARDWOOD LUMBER (2433)
SOUTH ATLANTIC

VALUE--MILLION CURRENT DOLLARS

YEAR	COLOMBIA	ECUADOR	BRAZIL	ASIA	AFRICA	OTHERS	TOTAL
1967	0.960	0.000	1.047	0.434	0.161	0.225	2.826
1968	0.483	0.001	1.518	0.373	0.095	0.099	2.570
1969	1.219	0.002	2.166	0.283	0.015	0.659	4.362
1970	0.577	0.054	2.494	0.225	0.000	0.422	3.772
1971	0.677	0.086	1.933	0.486	0.002	0.414	3.598
1972	1.191	0.314	3.270	1.053	0.000	0.541	6.369
1973	1.719	0.604	8.231	1.229	0.013	0.675	12.471
1974	0.785	0.210	7.212	1.325	0.031	0.301	9.864
1975	0.226	1.269	2.440	0.907	0.000	0.175	5.037
1976	0.120	3.991	4.529	0.812	0.107	0.437	9.996

QUANTITY--MILLION BF

YEAR	COLOMBIA	ECUADOR	BRAZIL	ASIA	AFRICA	OTHERS	TOTAL
1967	10.109	0.000	9.423	2.784	0.824	0.767	23.907
1968	5.354	0.002	13.799	2.268	0.495	0.456	22.375
1969	12.248	0.004	15.483	1.281	0.040	3.953	33.009
1970	6.577	0.611	19.525	0.839	0.000	2.535	30.086
1971	6.671	0.618	17.817	1.370	0.012	2.860	29.350
1972	11.828	2.019	24.798	3.097	0.000	2.738	44.480
1973	13.577	3.594	53.601	3.127	0.033	2.746	76.678
1974	6.543	0.888	30.221	2.830	0.079	1.418	41.979
1975	1.421	1.078	9.450	1.708	0.000	0.524	14.181
1976	0.745	2.779	18.367	1.422	0.364	1.457	25.134

IMPORTS
WOOD VENEER (6311)
SOUTH ATLANTIC

VALUE--MILLION CURRENT DOLLARS

YEAR	BRAZIL	OTH S AM	PHIL REP	OTH ASIA	AFRICA	OTHERS	TOTAL
1967	1.263	0.441	2.101	0.933	0.091	0.120	4.948
1968	1.617	0.849	4.073	1.552	0.528	0.042	8.660
1969	1.620	0.781	3.437	0.738	1.017	0.033	7.625
1970	2.472	0.925	2.953	0.414	0.263	0.006	7.033
1971	2.596	0.710	2.626	1.038	0.670	0.045	7.685
1972	3.663	1.000	4.909	2.015	1.457	0.138	13.182
1973	3.568	0.771	9.180	1.497	2.054	0.464	17.534
1974	2.790	0.761	10.049	2.625	1.912	0.862	18.998
1975	2.644	0.561	2.811	0.123	0.240	0.748	7.128
1976	5.056	1.001	6.721	0.116	0.143	0.876	13.913

QUANTITY--MIL SQ FT

YEAR	BRAZIL	OTH S AM	PHIL REP	OTH ASIA	AFRICA	OTHERS	TOTAL
1967	18.206	4.556	24.507	9.255	1.031	0.392	57.947
1968	27.424	7.001	45.818	18.279	4.022	0.172	102.717
1969	27.109	7.286	35.722	6.797	9.235	0.184	86.333
1970	25.852	7.480	37.001	6.371	1.834	0.055	78.594
1971	28.896	4.844	32.873	16.936	5.794	0.394	89.737
1972	38.476	9.813	60.945	25.092	14.774	1.852	150.954
1973	37.551	7.821	85.102	12.811	12.764	2.809	158.858
1974	29.432	6.355	60.093	16.771	9.120	3.493	125.264
1975	27.108	1.996	25.865	0.213	0.792	1.987	57.961
1976	40.148	6.642	50.274	0.686	0.534	1.546	99.829

IMPORTS
PLYWOOD (6312)
SOUTH ATLANTIC

VALUE--MILLION CURRENT DOLLARS

YEAR	EUROPE	KOR REP	TAIWAN	JAPAN	OTH ASIA	OTHERS	TOTAL
1967	0.391	2.548	3.298	3.696	2.676	0.044	12.652
1968	0.421	5.746	4.559	5.528	5.365	0.068	21.686
1969	0.713	12.745	3.729	4.711	5.535	0.094	27.526
1970	0.424	14.834	6.234	4.093	2.683	0.039	28.307
1971	0.170	20.671	11.931	4.589	3.511	0.029	40.901
1972	0.263	34.048	15.674	3.854	7.773	0.048	61.660
1973	0.348	34.524	17.271	3.765	12.937	0.227	69.073
1974	0.054	32.321	18.976	3.589	8.577	0.021	63.538
1975	0.043	28.335	16.112	2.913	3.173	0.011	50.587
1976	0.157	43.729	20.324	3.595	7.804	0.101	75.711

QUANTITY--MIL SQ FT

YEAR	EUROPE	KOR REP	TAIWAN	JAPAN	OTH ASIA	OTHERS	TOTAL
1967	3.044	22.880	34.188	23.762	27.390	0.394	111.658
1968	1.871	45.896	41.865	35.552	50.015	0.491	175.690
1969	4.348	103.924	35.061	26.424	50.802	0.523	221.081
1970	2.654	129.280	58.826	23.772	28.619	0.325	243.476
1971	1.210	171.922	106.214	27.211	35.601	0.144	342.302
1972	1.004	279.251	138.252	16.747	76.787	0.150	512.192
1973	1.224	183.423	101.902	10.168	91.620	0.795	389.133
1974	0.144	157.163	106.929	9.009	48.309	0.060	321.613
1975	0.125	168.852	107.614	10.328	20.140	0.061	307.120
1976	0.340	231.618	104.541	14.774	47.411	0.642	399.326

IMPORTS
RECONSTITUTED WOOD (6314)
SOUTH ATLANTIC

VALUE--MILLION CURRENT DOLLARS

YEAR	FINLAND	W GERMNY	SPAIN	OTH EURO	ASIA	OTHERS	TOTAL
1967	0.000	0.094	0.000	0.012	0.000	0.006	0.112
1968	0.000	0.138	0.000	0.007	0.000	0.003	0.149
1969	0.149	0.073	0.000	0.104	0.125	0.053	0.575
1970	0.010	0.096	0.000	0.035	0.000	0.031	0.173
1971	0.074	0.014	0.000	0.000	0.015	0.002	0.104
1972	0.256	0.062	0.004	0.010	0.025	0.003	0.360
1973	0.000	0.089	0.059	0.008	0.000	0.000	0.156
1974	0.000	0.167	0.043	0.000	0.001	0.000	0.211
1975	0.000	0.074	0.033	0.018	0.007	0.000	0.133
1976	0.000	0.099	0.023	0.032	0.002	0.000	0.156

QUANTITY--MIL SQ FT

YEAR	FINLAND	W GERMNY	SPAIN	OTH EURO	ASIA	OTHERS	TOTAL
1967	0.000	0.197	0.000	0.021	0.000	0.071	0.289
1968	0.000	0.285	0.000	0.014	0.000	0.024	0.322
1969	1.391	0.133	0.000	1.623	1.109	0.417	4.673
1970	0.090	0.196	0.000	0.068	0.000	0.234	0.588
1971	0.735	0.021	0.000	0.000	0.025	0.011	0.792
1972	3.903	0.165	0.009	0.020	0.060	0.004	4.101
1973	0.000	0.100	0.129	0.016	0.000	0.000	0.245
1974	0.000	0.147	0.065	0.000	0.001	0.000	0.212
1975	0.000	0.061	0.047	0.018	0.006	0.000	0.131
1976	0.000	0.100	0.031	0.035	0.004	0.000	0.169

IMPORTS
PULPWOOD (EXC CHIPS)(2421)
SOUTH ATLANTIC

VALUE--MILLION CURRENT DOLLARS

YEAR	BERMUDA	BAHAMAS	S AMERIC	EUROPE	ASIA	OTHERS	TOTAL
1967	0.041	3.957	0.000	0.000	0.000	0.000	3.999
1968	0.000	5.526	0.000	0.000	0.000	0.002	5.528
1969	0.000	5.510	0.000	0.000	0.000	0.000	5.510
1970	0.000	6.044	0.000	0.000	0.000	0.000	6.044
1971	0.000	6.177	0.000	0.000	0.000	0.000	6.177
1972	0.000	6.346	0.000	0.000	0.000	0.000	6.346
1973	0.000	4.257	0.000	0.000	0.000	0.000	4.257
1974	0.000	2.478	0.000	0.000	0.000	0.000	2.478
1975	0.000	1.160	0.000	0.000	0.000	0.000	1.160
1976	0.000	1.621	0.000	0.000	0.000	0.000	1.621

QUANTITY--1000 CORDS

YEAR	BERMUDA	BAHAMAS	S AMERIC	EUROPE	ASIA	OTHERS	TOTAL
1967	2.953	282.547	0.000	0.000	0.000	0.000	285.500
1968	0.000	312.148	0.000	0.000	0.000	0.074	312.222
1969	0.000	301.496	0.000	0.000	0.000	0.000	301.496
1970	0.000	316.187	0.000	0.000	0.000	0.000	316.187
1971	0.000	267.497	0.000	0.000	0.000	0.000	267.497
1972	0.000	276.422	0.000	0.000	0.000	0.000	276.422
1973	0.000	206.588	0.000	0.000	0.000	0.000	206.588
1974	0.000	134.777	0.000	0.000	0.000	0.000	134.777
1975	0.000	72.761	0.000	0.000	0.000	0.000	72.761
1976	0.000	101.344	0.000	0.000	0.000	0.000	101.344

IMPORTS
WOOD PULP (2510)
SOUTH ATLANTIC

VALUE--MILLION CURRENT DOLLARS

YEAR	CANADA	SWEDEN	OTH EURO	ASIA	AFRICA	OTHERS	TOTAL
1967	0.000	0.135	0.013	0.000	0.000	0.000	0.148
1968	0.007	0.076	0.014	0.000	0.000	0.000	0.097
1969	0.006	0.000	0.000	0.000	0.001	0.000	0.007
1970	0.000	0.042	0.000	0.005	0.000	0.000	0.047
1971	0.000	0.000	0.000	0.000	0.000	0.000	0.000
1972	0.000	0.000	0.002	0.000	0.000	0.000	0.003
1973	0.000	0.803	0.000	0.001	0.000	0.000	0.805
1974	0.000	0.913	0.000	0.000	0.000	0.000	0.913
1975	3.033	0.262	0.001	0.000	0.000	0.000	3.296
1976	8.366	2.692	0.001	0.000	0.000	0.002	11.061

QUANTITY--1000 STN

YEAR	CANADA	SWEDEN	OTH EURO	ASIA	AFRICA	OTHERS	TOTAL
1967	0.000	1.008	0.074	0.000	0.000	0.000	1.082
1968	0.057	0.639	0.074	0.000	0.000	0.000	0.770
1969	0.057	0.000	0.000	0.000	0.006	0.000	0.063
1970	0.000	0.221	0.000	0.018	0.000	0.000	0.239
1971	0.000	0.000	0.000	0.000	0.000	0.000	0.000
1972	0.000	0.002	0.017	0.000	0.000	0.000	0.020
1973	0.000	3.698	0.000	0.007	0.000	0.000	3.705
1974	0.000	3.030	0.000	0.000	0.000	0.000	3.030
1975	13.237	0.563	0.003	0.000	0.000	0.000	13.803
1976	33.003	8.471	0.002	0.000	0.000	0.007	41.483

IMPORTS
NEWSPRINT (6411)
SOUTH ATLANTIC

VALUE--MILLION CURRENT DOLLARS

YEAR	CANADA	S AMERIC	FINLAND	OTH EURO	ASIA	OTHERS	TOTAL
1967	5.785	0.000	1.234	0.000	0.000	0.000	7.019
1968	5.532	0.000	1.689	0.000	0.000	0.000	7.221
1969	5.542	0.000	2.657	0.000	0.000	0.000	8.199
1970	4.843	0.000	4.162	0.000	0.000	0.000	9.006
1971	6.597	0.000	4.554	0.000	0.000	0.000	11.151
1972	5.904	0.000	4.397	0.000	0.000	0.000	10.301
1973	5.357	0.000	8.047	0.000	0.000	0.000	13.404
1974	1.562	0.000	5.195	0.000	0.000	0.000	6.756
1975	2.604	0.000	0.212	0.000	0.000	0.000	2.816
1976	2.704	0.000	0.000	0.000	0.000	0.000	2.704

QUANTITY--1000 STN

YEAR	CANADA	S AMERIC	FINLAND	OTH EURO	ASIA	OTHERS	TOTAL
1967	45.192	0.000	9.307	0.000	0.000	0.000	54.499
1968	42.229	0.000	13.118	0.000	0.000	0.000	55.347
1969	51.452	0.000	19.151	0.000	0.000	0.000	70.603
1970	38.900	0.000	31.084	0.000	0.000	0.000	69.984
1971	62.174	0.000	34.150	0.000	0.000	0.000	96.324
1972	41.309	0.000	55.083	0.000	0.000	0.000	96.392
1973	35.693	0.000	58.311	0.000	0.000	0.000	94.004
1974	7.822	0.000	33.732	0.000	0.000	0.000	41.554
1975	11.508	0.000	1.210	0.000	0.000	0.000	12.718
1976	11.028	0.000	0.000	0.000	0.000	0.000	11.028

IMPORTS
BUILDING BOARD (6416)
SOUTH ATLANTIC

VALUE--MILLION CURRENT DOLLARS

YEAR	N AMERIC	S AMERIC	SWEDEN	POLAND	OTH EURO	OTHERS	TOTAL
1967	0.000	0.059	0.326	0.006	0.302	0.000	0.693
1968	0.000	0.008	0.791	0.019	0.402	0.003	1.223
1969	0.000	0.081	0.593	0.012	0.557	0.000	1.243
1970	0.000	0.113	0.462	0.047	0.343	0.000	0.964
1971	0.000	0.138	0.617	0.024	0.486	0.092	1.356
1972	0.004	0.174	1.821	0.200	1.025	0.334	3.557
1973	0.000	0.200	1.850	0.612	0.832	0.668	4.162
1974	0.057	0.731	0.985	0.902	1.044	0.517	4.236
1975	0.000	0.094	0.175	0.058	0.085	0.050	0.462
1976	0.000	0.453	1.587	0.143	0.425	0.089	2.697

QUANTITY--1000 STN

YEAR	N AMERIC	S AMERIC	SWEDEN	POLAND	OTH EURO	OTHERS	TOTAL
1967	0.000	0.510	4.919	0.090	4.784	0.002	10.306
1968	0.000	0.085	12.432	0.330	6.378	0.026	19.251
1969	0.000	1.503	9.018	0.224	8.480	0.000	19.225
1970	0.000	1.784	5.459	0.557	4.791	0.000	12.592
1971	0.000	1.995	10.161	0.338	6.854	1.587	20.936
1972	0.039	2.321	22.734	2.630	14.033	5.082	46.838
1973	0.000	1.944	16.328	7.022	8.406	7.560	41.261
1974	0.208	5.980	7.026	6.573	7.743	5.523	33.053
1975	0.000	0.762	1.202	0.652	0.813	0.500	3.930
1976	0.000	3.185	13.679	1.685	4.524	0.661	23.734

IMPORTS
PAPER & PAPERBOARD (6410)
SOUTH ATLANTIC

VALUE--MILLION CURRENT DOLLARS

YEAR	S AMERIC	SWEDEN	FINLAND	OTH EURO	ASIA	OTHERS	TOTAL
1967	0.000	0.598	0.401	0.074	0.026	0.000	1.100
1968	0.006	0.445	0.928	0.186	0.052	0.014	1.631
1969	0.000	0.410	0.259	0.381	0.057	0.000	1.106
1970	0.000	0.497	0.193	0.231	0.096	0.000	1.017
1971	0.000	0.116	0.071	0.122	0.137	0.001	0.448
1972	0.000	0.021	0.235	0.352	0.079	0.001	0.689
1973	0.000	0.199	0.072	0.233	0.122	0.202	0.829
1974	0.000	0.361	0.207	0.592	0.484	0.040	1.684
1975	0.025	0.319	0.110	0.355	0.208	0.034	1.052
1976	0.116	1.135	1.799	0.850	0.299	0.004	4.202

QUANTITY--1000 STN

YEAR	S AMERIC	SWEDEN	FINLAND	OTH EURO	ASIA	OTHERS	TOTAL
1967	0.000	4.087	1.641	0.367	0.066	0.000	6.161
1968	0.050	2.806	7.516	0.792	0.124	0.081	11.369
1969	0.000	2.602	1.251	1.936	0.095	0.000	5.884
1970	0.000	2.490	0.672	0.840	0.182	0.000	4.183
1971	0.000	0.586	0.309	0.263	0.279	0.007	1.443
1972	0.000	0.064	1.325	0.354	0.109	0.000	1.852
1973	0.000	0.469	0.403	0.499	0.126	1.874	3.371
1974	0.000	0.602	0.681	1.160	0.123	0.055	2.621
1975	0.081	0.504	0.267	0.145	0.127	0.052	1.176
1976	0.363	1.468	6.230	0.382	0.076	0.001	8.519

Appendix F

THE GULF REGION

Exports

Regional Overview

Not surprisingly, the Gulf region is very similar to the neighboring South Atlantic region in both the amount and mix of forest product exports. The Gulf region is a slightly larger exporter; in 1976 the Gulf's exports reached $567.7 million, whereas exports from the South Atlantic totaled $492.9 million. In 1976 the Gulf provided 14 percent of the value of U.S. total forest products exports and the South Atlantic provided 12 percent.

The Gulf's export growth rate during 1967-76 was exactly the same as that of the South Atlantic--each region increased the value of its exports by an average of 15 percent each year.

In 1976 the commodity mixes exported by the two regions were also very similar, with about 90 percent of the exports being pulp and paper products and the remaining 10 percent in solid wood commodities. In neither region did the commodity composition of exports change appreciably from 1967 to 1976.

Paper and paperboard was the most important Gulf export throughout 1967-76. In an average year, more than half the value of the region's forest product exports was composed of paper and paperboard.

Wood pulp in an average year contributed 38 percent of the total value of the Gulf's exports of forest products. The $229.5 million worth of pulp exported in 1976 represented 41 percent of the region's total exports.

The Gulf and South Atlantic region's most important solid wood product is softwood lumber. Despite being the Gulf region's third most important forest product export, softwood lumber in 1976 constituted only 5 percent of the value of the region's total forest product exports.

Paper and Paperboard

Paper and paperboard exports from the Gulf region during 1967-76 represented an average of 26 percent of the value of U.S. exports of paper and paperboard, making the region the second largest U.S. exporting region of that commodity.

Exports of paper and paperboard from the Gulf region increased steadily during 1967-76; the average yearly gain in value was 14 percent and in volume was 6 percent. The region's 1976 exports of paper and paperboard reached 872,000 tons (worth $225.9 million).

North American and European countries provided the largest foreign markets for Gulf paper and paperboard. In 1976 North American countries imported $75.1 million worth of these products and European countries $89.5 million, which were, respectively, 29 percent and 35 percent of the region's total paper and paperboard exports. Africa's share of Gulf paper and paperboard exports was 13 percent, and South America's share was 10 percent. Mexico was the leading importer in the latter half of the 1967-76 period. Its 1976 imports of 51,000 tons (valued at $28.9 million) represented 11 percent of the Gulf's paper and paperboard exports. Mexico's growth as an export market was rapid during the 1967-76 period. The average annual increase in volume was 18 percent and in value was 20 percent.

The United Kingdom was the major importer of Gulf paper and paperboard in the early years of the period. In 1967 its share of the region's exports was 28 percent (by value), but by 1976 that share had dwindled to 10 percent. During the period the volume of United Kingdom imports of paper and paperboard declined by an average of 5 percent per year.

Wood Pulp

Through most of the 1967-76 period, the Gulf was the largest U.S. wood pulp exporting region. In an average year, the region supplied 27 percent (by value) of all U.S. pulp exports.

The 707,000 tons of wood pulp exported by the region in 1976 was valued at $229.5 million, more than three and a half times the value of 1967 exports. The average export growth rates during the period were annually 16 percent per year in value and 6 percent in volume.

European countries provided markets for an average of 62 percent (by value) of the region's pulp exports during 1967-76. The United Kingdom, West Germany, and Italy were the largest European importers of the Gulf's wood pulp.

The United Kingdom was the largest single importer of pulp, taking, in 1976, 19 percent of the volume and 17 percent of the value of the region's wood pulp exports.

West Germany showed tremendous growth as an export market during 1967-76. Its 1976 imports of pulp--89,000 tons (worth $30.9 million)--represented increases over the 1967 volume and value of 378 percent and 1,152 percent, respectively. The equivalent annual increases are 19 percent in volume and 32 percent in nominal value.

In 1976, West Germany and Italy each imported thirteen percent (by value) of the Gulf region's pulp exports. Italian imports from the region, of $30.1 million and 95,000 tons, represented increases over 1967 of 307 percent in value and 166 percent in volume.

From 1967 to 1976, European countries other than United Kingdom, West Germany, and Italy collectively declined in importance as markets for Gulf wood pulp. In 1967 the group of other European countries imported 28 percent of the value of the region's pulp exports, but by 1976 that share had slipped to 17 percent. Gulf exports to other European countries did, however, increase modestly, by an annual average of 2 percent in volume and by 10 percent in value.

In 1976 Asian countries as a group imported 75,000 tons of Gulf wood pulp (worth $24.9 million) which represented 11 percent of the Gulf region's pulp exports.

Softwood Lumber

The Gulf region, in 1976, exported 78.8 million board-feet of softwood lumber (valued at $25.9 million), which represented 6 percent of total U.S. exports of softwood lumber. For the 1967-76 period, the region's share of total softwood lumber exports averaged 6 percent.

The region's softwood lumber exports from 1967 to 1975 were gradually declining in volume, but in 1976 there was an increase of 75 percent over the 1975 volume of 45.1 million board-feet. Contrary to the decline in volumes, the nominal value of softwood lumber exports was generally increasing during 1967-76, and the 1976 value was more than twice the 1975 value.

The major foreign markets for Gulf softwood lumber during 1967-76 were European countries, principally West Germany, Spain, and Italy.

West Germany's average share of the value of the region's softwood lumber exports during the study period was 29 percent. West German imports of Gulf lumber generally rose during the period, and 1976 was a peak year in both volume and value; the region's exports to West Germany were 25.6 million board-feet (worth $9.4 million).

Italy was the second largest European importer; in 1976 Italy imported from the Gulf region 8.2 million board-feet (valued at $3 million). Softwood lumber exports to Italy generally declined during 1967-76. The volume shipped to Italy in 1976 was less than one-half that shipped in 1967, and the nominal value of exports increased by an average of only 3 percent annually.

Spain also declined in imports of softwood lumber from the Gulf region; in 1976 Spain was the destination for 10 percent (by value) of the region's total softwood lumber exports, but in 1967 that share had been 16 percent. The 1976 volume of 7.5 million board-feet represented a drop of 25 percent from the 1967 volume, or an average annual decline of 3 percent.

North American countries imported an increasingly large share of the Gulf region's softwood lumber exports during 1974-76. In 1976 these countries imported 23.2 million board-feet (worth $6 million), or 24 percent of the Gulf's total softwood lumber exports. In that year, Jamaica, which had not imported Gulf softwood lumber in earlier years, accounted for two-thirds of the softwood lumber imported by North American countries.

Newsprint

The Gulf region is the largest U.S. newsprint-exporting region. The 1975 exports of 59,000 tons (valued at $17.3 million) represented 35 percent of the value of all U.S. exports of newsprint. For the entire 1967-76 period, the region averaged 45 percent of U.S. exports.

The region's newsprint exports during 1967-76 grew at an average annual rate of 7 percent in volume and 18 percent in value.

Mexico was the largest market for Gulf newsprint throughout 1967-76, and it was also one of the fastest growing. The $14.2 million worth of newsprint exported to Mexico in 1976 represented 82 percent of the region's total forest product exports.

Venezuela was also a major destination for newsprint from the Gulf region. During 1967-76, an average of 15 percent of the value of the region's newsprint exports were shipped to Venezuela.

In 1976 countries outside of Central and South America purchased only 2 percent of the Gulf region's newsprint exports.

Building Board

Over the course of the 1967-76 period the Gulf region declined substantially as an exporter of building board, relative to the performance of other regions. In 1967 the Gulf was responsible for nearly 30 percent of the U.S. exports of building board, but by 1976 that share was down to only 8 percent.

In 1976 the 7,054 tons of building board made uo two-thirds of the 1967 volume, and the 1976 nominal value of $2.5 million was only 6 percent higher than the 1967 value.

European countries, which generally imported the largest share of the region's building board exports, experienced a decline from 1967 to 1976 of 65 percent in the volume imported from the Gulf. The principal European markets are the United Kingdom and the Netherlands. The United Kingdom in 1976 imported $271 thousand and the Netherlands $552 thousand worth of Gulf building board, which represented 11 percent and 22 percent, respectively, of the region's total building board exports.

The Middle Eastern countries of Saudi Arabia and Bahrain were relatively large importers of Gulf building board. In 1976 the $551,000 worth imported by Saudi Arabia was 22 percent of the Gulf region's total building board exports, and the $320,000 imported by tiny Bahrain represented 13 percent.

Together, all non-European nations more than doubled the volume of their imports of building board from the Gulf during the period. The non-European countries accounted for 58 percent (by value) of the region's total 1976 building board exports.

Hardwood Lumber

The Gulf's exports of hardwood lumber declined substantially during 1967-76, going from 20.4 million board-feet in 1967 to 11.4 million board-feet in 1976. The nominal value increased from $4.6 million to $6 million, or an average annual growth rate of only 3 percent.

The Gulf's share of the value of U.S. hardwood lumber exports went from 17 percent in 1967 to 7 percent in 1976.

European countries were the destinations of nearly 80 percent (by value) of the region's 1976 hardwood lumber exports. Nordic countries were the largest European importers, and took 31 percent of the region's exports. The percentage shares imported by the United Kingdom and Italy were 15 and 8, respectively, and together all other European countries imported 25 percent of the Gulf's hardwood lumber exports.

All the European countries together imported 8.3 billion board-feet of Gulf hardwood lumber (valued at $4.7 million) in 1976.

Japan was the major non-European importer of hardwood lumber from the Gulf region. Its 1976 imports totaled 673,000 board-feet (worth $270,000), which was 5 percent of the region's total hardwood lumber exports.

Hardwood Logs

The Gulf region was the fastest growing hardwood log-exporting region in the United States from 1967 to 1976. The volume of log exports from the region grew at an average annual rate of 29 percent during the period, and the nominal value grew at an average of 39 percent per year. The 1976 volume and value of hardwood log exports was 11.1 million board-feet (worth $11.1 million), which represented 17 percent of total U.S. hardwood log exports.

The largest importer of hardwood logs from the Gulf region in 1976 was West Germany, which took 70 percent of the value of the region's log exports. West Germany exhibited spectacular growth in its imports of Gulf logs during the 1967-76 period, when it averaged growth in volume of 61 percent and in value of 65 percent annually.

Italy was the largest hardwood log importer in 1967, when it consumed 43 percent (by value) of the region's hardwood log exports, but by 1976 Italy's share had dwindled to only 4 percent. All European countries combined imported 9.4 million board-feet (worth $9.3 million), which represented 84 percent of the Gulf's total hardwood log exports.

Other Forest Products

The remaining forest products (pulpwood except chips, softwood logs, wood veneer, plywood, reconstituted wood, pulpwood chips, and other industrial roundwood) together constituted only 2 percent of the value of the region's total forest product exports in 1976.

Among this group of commodities, the veneer and panel products (wood veneer, plywood, reconstituted wood) are notable for their rapid growth as export items from 1967 to 1976. In 1976 exports of these products from the Gulf region were valued at $9.3 million, which represented an increase of 1,810 percent over their 1967 value. The equivalent annual growth rate for exports of veneer and panel products was 38 percent.

Europe was the destination of $7.5 million worth (or 80 percent) of the Gulf's 1976 exports of veneer and panel products.

Imports

Regional Overview

The 1976 imports of forest products to the Gulf region totaled $228.8 million, which represented only 4 percent of the U.S. forest product imports.

The Gulf experienced a modest rate of growth in imports during 1967-76. The 1976 nominal value was 136 percent higher than the 1967 value which is equivalent to an annual increase of 10 percent.

The mix of commodities imported by the region is dissimilar to the import mix of the United States as a whole, in that the Gulf region imports more solid wood products than it does pulp and paper goods.

In 1976, $130.3 million (or 58 percent) of the region's forest product imports was in solid wood products, principally plywood ($91 million), hardwood lumber ($24.7 million), and softwood lumber ($13 million). The remaining 42 percent ($98.5 million) was in pulp and paper products, principally newsprint ($66.7 million) and wood pulp ($22.6 million).

Plywood

Nearly one-fourth of the value of total U.S. imports of plywood in 1976 was imported by the Gulf region. The 614.3 million square feet (valued at $91 million) was an increase of 69 percent over the 1967 volume and of 204 percent over the 1967 nominal value. The equivalent annual growth rates are 6 percent in volume and 13 percent in nominal value.

In every year of the 1967-76 period, four Asian countries--the Philippines, Korea, Taiwan, and Japan--were the sources of no less than 95 percent of the region's total plywood imports.

Korea was the leading exporter to the region in terms of volume throughout the period. In 1976 Korea exported 313.2 million square feet (valued at $40.3 million), which represented 51 percent of the volume, and 44 percent of the value of the Gulf's total plywood imports. During 1967-76, the volume of Korean plywood exports grew at an average annual rate of 9 percent, and the nominal value at an average rate of 15 percent per year.

Japan's plywood exports to the region were erratic during the period in terms of absolute amounts, but were fairly stable in terms of Japan's relative share of the region's imports. In 1976 the Gulf imported 124.1 million square feet of Japanese plywood valued at $28.3 million, which was 31 percent of the Gulf's total plywood imports.

Taiwan recorded the fastest growth rate of the major plywood exporters. The 1976 exports of 151.2 million square feet (worth $19.4 million) represented increases over the 1967 value and volume of 565 percent and 269 percent, respectively. In annual terms, value increased at a 21 percent rate and volume at 16 percent.

Newsprint

Although the Gulf region was not a large importer of newsprint relative to other U.S. regions, newsprint did constitute 29 percent of the Gulf's total imports of forest products in 1976.

From 1967 to 1976, the volume of newsprint imports declined by 13 percent, or by about 2 percent per year. The nominal value rose at an average of 5 percent annually. The 1976 imports of newsprint totaled 301,000 tons (valued at $66.7 million).

Canada and Finland were the only significant foreign suppliers of newsprint to the region during 1967-76. Canada, on average, provided 85 percent (by value), and Finland supplied the remainder of the Gulf's imports of newsprint.

In 1976 Canada shipped 299,000 tons (worth $66 million), or 99 percent of the region's total newsprint imports. Finland did not export any newsprint to the Gulf in that year, but Sweden exported 2,544 tons (worth $700,000).

Hardwood Lumber

The Gulf region was the largest U.S. importer of hardwood lumber in 1976, both in quantity and in value. The 76.2 million board-feet (valued at $24.7 million) represented 26 percent of the U.S. imports of hardwood lumber.

While the United States as a whole declined in hardwood lumber imports during 1967-76, the Gulf region registered average growth rates of 6 percent in volume and 13 percent in nominal value.

Brazil showed spectacular growth in exports to the region in 1967-76, and, as a result, was the leading foreign supplier of hardwood lumber to the Gulf in 1976. The country was responsible for 30 percent ($7.5 million) of the value of hardwood lumber imported by the Gulf region, and 47 percent (35.4 million board-feet) of the volume. The average annual growth rates for imports from Brazil during 1967-76 were 32 percent in volume and 28 percent in value.

South American countries other than Brazil exported 9.9 million board-feet (worth $2.8 million) to the Gulf region in 1976.

Malaysia was a steady source of supply for the Gulf region throughout 1967-76. In an average year, Malaysia provided nearly one-fourth of the value of the region's total hardwood lumber imports. The quantity of lumber imports in 1967 was about the same as in 1976, when the Gulf imported 13 million board-feet of hardwood lumber from Malaysia. The nominal value of that lumber, however, had increased by more than 200 percent, from $1.7 million in 1967 to $5.2 million in 1976.

Asian countries other than Malaysia supplied the Gulf region with about the same dollar-amount of hardwood lumber as did Malaysia. In 1976 these other Asian countries exported 9.3 million board-feet (worth $5.1 million).

African countries, in 1976, exported $2 million worth of hardwood lumber to the Gulf region. The 5.5 million board-feet of African exports was only 1 percent higher than the 1967 export volume.

Wood Pulp

Relative to other U.S. regions, the Gulf is an insignificant wood pulp importer. In 1976 the region's imports were only 2 percent (by value) of the U.S. total wood pulp imports. In that same year, however, the $22.6 million of wood pulp imported by the Gulf was 10 percent of the region's total forest product imports.

Imports of wood pulp to the region increased considerably over the 1967-76 period, by an average of 5 percent per year in quantity and 18 percent annually in nominal value.

The Republic of South Africa was the principal source of the region's wood pulp imports throughout 1967-76. In an average year, South Africa provided 93 percent of the value of the region's total wood pulp imports. The 78,000 tons imported in 1976 were worth $21.1 million.

In that same year, Canada, Sweden, and Finland together shipped 3,238 tons (worth $995,000) which represented only 5 percent of the Gulf's total pulp imports.

Softwood Lumber

In 1976 the Gulf's softwood lumber imports of 79.2 million board-feet (worth $13 million) were 6 percent of the region's total forest product imports, but only 1 percent of U.S. total softwood lumber imports.

The region's imports fluctuated quite a bit during the 1967-76 period, but were generally on an upward trend. The average annual growth rate for softwood lumber imports during the period was 4 percent in quantity and 14 percent in value.

Canada was the principal foreign supplier of softwood lumber to the region throughout 1967-76. In 1976 the imports from Canada of 74.8 million board-feet (worth $11.3 million) represented 87 percent (by value) of the region's total softwood lumber imports.

Brazil was also a steady exporter to the Gulf region. In an average year during the study period, Brazil shipped 12 percent of the value of the Gulf's softwood lumber imports.

Other Forest Products

The nine commodities (pulpwood except chips, softwood logs, hardwood logs, wood veneer, reconstituted wood, pulpwood chips, other industrial roundwood, building board, and paper and paperboard) which have not been discussed above constituted only 4 percent (by value) of the region's total forest products imports in 1976.

Of the region's 1976 imports, paper and paperboard constituted 25 percent of this group of commodities. The 6,724 tons of 1976 paper and paperboard imports were valued at $3.4 million. Those figures represent gains over 1967 of 42 percent in quantity and 228 percent in value. European countries, led by the Nordic group, were the sources of 97 percent of the Gulf's paper and paperboard imports in 1976. Finland's exports of $1.8 million represented 54 percent of the region's imports, and Sweden shipped $790,000 worth of paper and paperboard, giving it 23 percent of the Gulf import market. All other European countries were responsible for 19 percent of the region's paper and paperboard imports.

The Gulf was one of the largest U.S. building board-importing regions. The 1976 imports of $5.8 million were 22 percent of the U.S. total building board imports. From 1967 to 1976, the region increased its imports of building board at an average annual rate of 6 percent in volume and 13 percent in value.

Brazil was the source of the largest share of Gulf building board imports in 1976. In that year, Brazil's $2.8 million worth of exports represented nearly half the region's total building board imports.

The Nordic countries were the source of 37 percent (by value) of the Gulf's building board imports in 1967, but by 1976 that share had dropped to 8 percent. The volume of imports from Scandinavia declined by an average of 11 percent per year during 1967-76. Other European countries, as well as Argentina and the Republic of South Africa, considerably increased their exports during the period to fill the gap created by declining Nordic exports.

The Gulf region's import of hardwood logs during 1967-76 is an interesting case because of the precipitous decline in both volume and value. The 1976 value of the region's hardwood log imports was only 2 percent of the 1967 value; volume declined by 99 percent. In 1967 the region imported 36 percent of all U.S. hardwood log imports, but by 1976 that share was only 2 percent.

The decline in hardwood log imports can be traced directly to two South American countries--Colombia and Brazil. In 1967 they provided 80 percent of the value of the region's hardwood log imports, but by 1976, neither Brazil nor Colombia was exporting any hardwood logs to the Gulf. In 1976 the region's imported hardwood logs were supplied largely by Belize in Central America and Ghana in West Africa.

EXPORTS
SOFTWOOD LOGS (2422)
GULF

VALUE--MILLION CURRENT DOLLARS

YEAR	N AMERIC	W GERMNY	OTH EURO	JAPAN	OTH ASIA	OTHERS	TOTAL
1967	0.004	0.000	0.006	0.448	0.000	0.001	0.459
1968	0.122	0.001	0.005	0.118	0.168	0.000	0.414
1969	0.000	0.000	0.019	0.000	0.000	0.001	0.020
1970	0.010	0.114	0.027	0.092	0.003	0.002	0.248
1971	0.089	0.000	0.000	0.095	0.000	0.027	0.211
1972	0.029	0.011	0.003	0.000	0.005	0.000	0.048
1973	0.041	0.008	0.046	0.000	0.011	0.000	0.106
1974	0.018	0.005	0.010	0.120	0.000	0.014	0.167
1975	0.046	0.231	0.048	0.000	0.290	0.019	0.634
1976	0.152	0.259	0.258	1.142	0.051	0.046	1.908

QUANTITY--MILLION BF

YEAR	N AMERIC	W GERMNY	OTH EURO	JAPAN	OTH ASIA	OTHERS	TOTAL
1967	0.018	0.000	0.017	4.738	0.000	0.004	4.777
1968	1.786	0.007	0.017	1.457	1.854	0.000	5.121
1969	0.000	0.000	0.065	0.000	0.000	0.003	0.067
1970	0.031	1.149	0.086	0.019	0.041	0.005	1.332
1971	0.681	0.000	0.000	0.907	0.000	0.011	1.599
1972	0.184	0.020	0.013	0.000	0.009	0.000	0.225
1973	0.225	0.038	0.057	0.000	0.022	0.000	0.342
1974	0.095	0.022	0.030	0.469	0.000	0.012	0.627
1975	0.095	0.368	0.076	0.000	0.379	0.030	0.946
1976	0.277	0.275	0.407	4.664	0.081	0.080	5.782

EXPORTS
HARDWOOD LOGS (2423)
GULF

VALUE--MILLION CURRENT DOLLARS

YEAR	W GERMNY	SWITZLND	ITALY	OTH EURO	JAPAN	OTHERS	TOTAL
1967	0.088	0.000	0.244	0.097	0.068	0.065	0.563
1968	0.465	0.002	0.910	0.227	0.141	0.025	1.770
1969	0.548	0.022	0.207	0.163	0.195	0.032	1.166
1970	0.270	0.071	0.224	0.128	0.102	0.311	1.106
1971	0.172	0.041	0.078	0.165	0.538	0.418	1.412
1972	0.174	0.070	0.215	0.278	0.799	0.428	1.964
1973	1.888	0.086	0.660	0.344	0.367	0.658	4.004
1974	1.408	0.036	0.500	0.286	0.153	0.585	2.967
1975	2.877	0.015	0.128	0.175	0.000	0.874	4.070
1976	7.774	0.789	0.407	0.329	0.000	1.835	11.134

QUANTITY--MILLION BF

YEAR	W GERMNY	SWITZLND	ITALY	OTH EURO	JAPAN	OTHERS	TOTAL
1967	0.102	0.000	0.425	0.214	0.245	0.163	1.150
1968	0.417	0.006	1.244	0.400	0.242	0.053	2.363
1969	0.442	0.026	0.367	0.249	0.251	0.269	1.605
1970	0.307	0.045	0.490	0.244	0.117	1.782	2.986
1971	0.179	0.078	0.188	0.388	0.465	2.612	3.909
1972	0.203	0.096	0.380	0.285	0.738	0.353	2.055
1973	2.327	0.109	0.955	0.351	0.227	0.671	4.640
1974	1.841	0.042	0.603	0.466	0.107	0.559	3.618
1975	2.916	0.066	0.187	0.256	0.000	0.544	3.969
1976	7.506	0.884	0.594	0.393	0.000	1.760	11.138

EXPORTS
SOFTWOOD LUMBER (2432)
GULF

VALUE--MILLION CURRENT DOLLARS

YEAR	W GERMNY	SPAIN	ITALY	OTH EURO	JAPAN	OTHERS	TOTAL
1967	2.499	1.329	2.208	0.913	0.367	1.144	8.460
1968	2.343	1.242	2.259	1.047	0.546	1.208	8.646
1969	1.406	1.330	2.174	0.878	0.870	1.190	7.848
1970	2.580	1.619	2.331	0.928	1.753	1.087	10.299
1971	2.324	0.829	0.866	0.406	3.567	1.255	9.246
1972	2.872	0.842	1.042	0.248	9.800	0.397	15.201
1973	6.906	1.314	4.841	1.271	0.522	1.578	16.433
1974	2.512	1.893	4.384	0.716	0.093	3.002	12.599
1975	6.079	0.619	1.665	0.497	0.000	4.073	12.933
1976	9.422	2.659	2.969	1.707	0.021	9.153	25.931

QUANTITY--MILLION BF

YEAR	W GERMNY	SPAIN	ITALY	OTH EURO	JAPAN	OTHERS	TOTAL
1967	17.447	9.953	16.529	5.577	1.555	8.083	59.144
1968	15.504	9.176	16.539	9.740	2.443	12.322	65.724
1969	8.038	9.869	14.572	5.600	3.375	7.037	48.490
1970	13.502	10.861	14.223	4.586	7.513	7.240	57.924
1971	11.702	5.456	5.576	2.121	12.566	8.467	45.888
1972	14.182	5.041	5.678	0.973	27.910	2.387	56.171
1973	23.126	5.833	16.873	4.221	0.627	6.639	57.319
1974	9.159	6.444	13.948	2.502	0.122	11.154	43.328
1975	17.420	2.359	5.745	1.379	0.000	18.205	45.109
1976	25.553	7.471	8.157	4.522	0.023	33.125	78.852

EXPORTS
HARDWOOD LUMBER (2433)
GULF

VALUE--MILLION CURRENT DOLLARS

YEAR	SCANDINA	U KINGDM	ITALY	OTH EURO	JAPAN	OTHERS	TOTAL
1967	0.609	1.447	0.327	0.888	0.250	1.127	4.648
1968	0.914	1.089	0.232	0.560	0.225	0.836	3.856
1969	0.794	0.941	0.243	0.499	0.248	0.740	3.465
1970	1.163	1.244	0.284	0.691	0.183	0.941	4.507
1971	0.867	0.733	0.250	0.430	0.205	0.636	3.120
1972	0.855	0.732	0.343	0.577	1.523	0.764	4.793
1973	1.412	0.889	0.372	0.712	0.464	0.540	4.390
1974	2.105	0.775	0.465	0.771	0.437	0.707	5.260
1975	1.433	0.855	0.272	0.775	0.244	1.782	5.361
1976	1.826	0.908	0.476	1.493	0.270	0.985	5.958

QUANTITY--MILLION BF

YEAR	SCANDINA	U KINGDM	ITALY	OTH EURO	JAPAN	OTHERS	TOTAL
1967	2.763	6.002	1.407	5.258	0.568	4.441	20.440
1968	3.203	4.164	0.915	2.334	0.598	3.592	14.805
1969	2.632	3.574	1.156	1.797	0.514	3.393	13.067
1970	3.564	4.160	1.035	2.584	0.515	3.338	15.195
1971	2.675	2.410	0.967	1.384	0.719	1.947	10.103
1972	2.490	2.404	1.052	1.713	2.413	2.236	12.308
1973	3.489	2.235	1.263	1.613	1.886	1.677	12.163
1974	3.908	1.565	0.952	1.275	1.113	1.452	10.266
1975	2.814	1.776	1.049	1.410	0.579	4.192	11.820
1976	3.157	1.815	0.995	2.330	0.673	2.416	11.385

EXPORTS
WOOD VENEER (6311)
GULF

VALUE--MILLION CURRENT DOLLARS

YEAR	SCANDINA	W GERMNY	SWITZLND	ITALY	OTH EURO	OTHERS	TOTAL
1967	0.037	0.016	0.002	0.000	0.005	0.058	0.117
1968	0.020	0.044	0.003	0.000	0.002	0.132	0.201
1969	0.045	0.000	0.000	0.000	0.008	0.068	0.121
1970	0.028	0.013	0.010	0.000	0.028	0.098	0.176
1971	0.015	0.012	0.000	0.000	0.029	0.039	0.094
1972	0.000	0.006	0.000	0.000	0.071	0.088	0.166
1973	0.000	0.006	0.000	0.000	0.055	0.248	0.309
1974	0.000	0.001	0.000	0.000	0.003	0.048	0.052
1975	0.000	0.598	0.000	0.000	0.009	0.191	0.799
1976	0.186	1.987	0.159	0.135	0.042	0.034	2.541

QUANTITY--MIL SQ FT

YEAR	SCANDINA	W GERMNY	SWITZLND	ITALY	OTH EURO	OTHERS	TOTAL
1967	0.024	0.023	0.002	0.000	0.006	0.219	0.274
1968	0.019	0.042	0.003	0.000	0.003	0.820	0.886
1969	0.026	0.000	0.000	0.000	0.010	0.773	0.809
1970	0.017	0.014	0.009	0.000	0.036	1.546	1.621
1971	0.014	0.012	0.000	0.000	0.038	0.608	0.671
1972	0.000	0.005	0.000	0.000	0.098	1.210	1.312
1973	0.000	0.004	0.000	0.000	0.083	5.008	5.094
1974	0.000	0.001	0.000	0.000	0.003	0.509	0.513
1975	0.000	0.454	0.000	0.000	0.008	0.625	1.088
1976	0.089	2.226	0.131	0.073	0.359	0.282	3.161

EXPORTS
PLYWOOD (6312)
GULF

VALUE--MILLION CURRENT DOLLARS

YEAR	N AMERIC	DENMARK	U KINGDM	NETHLNDS	OTH EURO	OTHERS	TOTAL
1967	0.199	0.055	0.021	0.000	0.081	0.034	0.389
1968	0.166	0.000	0.004	0.000	0.002	0.034	0.206
1969	0.186	0.000	0.025	0.000	0.065	0.038	0.314
1970	0.218	0.000	0.040	0.000	0.021	0.060	0.339
1971	0.187	0.000	0.009	0.000	0.002	0.010	0.207
1972	0.402	0.000	0.157	0.004	0.043	0.120	0.727
1973	0.663	0.045	0.743	0.036	0.991	0.061	2.539
1974	0.831	0.338	1.245	0.196	0.905	0.370	3.885
1975	0.689	0.597	0.101	1.573	0.442	0.874	4.277
1976	1.012	0.741	0.764	1.948	1.507	0.502	6.473

QUANTITY--MIL SQ FT

YEAR	N AMERIC	DENMARK	U KINGDM	NETHLNDS	OTH EURO	OTHERS	TOTAL
1967	1.815	0.442	0.277	0.000	0.406	0.209	3.148
1968	1.133	0.000	0.048	0.000	0.006	0.189	1.375
1969	0.958	0.000	0.371	0.000	0.603	0.246	2.178
1970	1.846	0.000	0.595	0.000	0.105	0.543	3.090
1971	1.595	0.000	0.164	0.000	0.020	0.053	1.832
1972	2.598	0.000	1.728	0.032	0.420	0.494	5.272
1973	3.764	0.153	5.216	0.395	6.186	0.242	15.955
1974	5.201	2.002	9.740	1.620	7.172	1.535	27.271
1975	3.980	3.418	0.658	7.972	3.321	5.616	24.965
1976	5.741	3.737	4.348	10.393	7.467	2.328	34.014

EXPORTS
RECONSTITUTED WOOD (6314)
GULF

VALUE--MILLION CURRENT DOLLARS

YEAR	N AMERIC	S AMERIC	EUROPE	INDONESA	OTH ASIA	OTHERS	TOTAL
1967	0.005	0.000	0.000	0.000	0.000	0.003	0.008
1968	0.005	0.000	0.000	0.000	0.019	0.000	0.024
1969	0.008	0.000	0.000	0.000	0.000	0.000	0.008
1970	0.003	0.000	0.000	0.000	0.000	0.002	0.006
1971	0.005	0.000	0.000	0.019	0.000	0.000	0.024
1972	0.016	0.000	0.000	0.001	0.000	0.000	0.017
1973	0.054	0.000	0.000	0.000	0.000	0.000	0.054
1974	0.057	0.000	0.003	0.003	0.000	0.000	0.062
1975	0.068	0.013	0.036	0.280	0.017	0.086	0.500
1976	0.044	0.165	0.000	0.000	0.081	0.000	0.291

QUANTITY--MIL SQ FT

YEAR	N AMERIC	S AMERIC	EUROPE	INDONESA	OTH ASIA	OTHERS	TOTAL
1967	0.007	0.000	0.000	0.000	0.000	0.015	0.022
1968	0.014	0.000	0.000	0.000	0.040	0.000	0.054
1969	0.012	0.000	0.000	0.000	0.000	0.000	0.012
1970	0.009	0.000	0.000	0.000	0.000	0.007	0.016
1971	0.023	0.000	0.000	0.069	0.000	0.000	0.091
1972	0.067	0.000	0.000	0.002	0.000	0.000	0.069
1973	0.044	0.000	0.000	0.000	0.000	0.000	0.044
1974	0.209	0.000	0.006	0.007	0.000	0.000	0.222
1975	0.167	0.043	0.208	0.074	0.039	0.362	0.892
1976	0.309	1.058	0.000	0.000	0.309	0.000	1.677

EXPORTS
WOOD PULP (2510)
GULF

VALUE--MILLION CURRENT DOLLARS

YEAR	U KINGDM	W GERMNY	ITALY	EUROPE	ASIA	OTHERS	TOTAL
1967	13.851	2.470	7.403	16.936	7.764	12.994	61.418
1968	16.845	2.805	8.680	21.612	8.593	14.679	73.214
1969	17.661	4.504	11.897	20.661	7.911	16.434	79.068
1970	27.750	11.258	18.185	37.981	13.821	37.184	146.179
1971	19.378	6.373	12.014	27.118	9.536	25.530	99.950
1972	19.009	5.367	8.412	24.980	6.550	29.030	93.348
1973	18.817	5.281	8.558	22.259	9.172	40.830	104.918
1974	31.327	11.634	24.976	39.399	24.460	77.663	209.459
1975	44.130	28.263	23.953	61.218	38.715	64.583	260.862
1976	37.994	30.918	30.150	39.822	24.921	65.716	229.522

QUANTITY--1000 STN

YEAR	U KINGDM	W GERMNY	ITALY	EUROPE	ASIA	OTHERS	TOTAL
1967	91.049	18.616	57.057	101.451	51.069	85.123	404.366
1968	115.093	25.456	67.844	130.761	66.982	99.619	505.756
1969	143.502	41.145	95.714	156.424	66.182	118.854	621.821
1970	185.357	80.252	135.540	244.177	93.889	241.806	981.021
1971	122.479	42.823	75.239	161.329	56.255	154.530	612.659
1972	117.038	36.586	58.142	154.765	41.184	186.034	593.750
1973	120.387	35.946	57.706	135.900	48.492	221.572	620.003
1974	139.428	43.583	97.132	157.225	86.036	262.623	786.026
1975	141.793	76.951	71.045	183.267	115.680	181.478	770.214
1976	137.593	88.967	94.741	122.985	75.352	186.984	706.622

EXPORTS
NEWSPRINT (6411)
GULF

VALUE--MILLION CURRENT DOLLARS

YEAR	MEXICO	VENEZUEL	PERU	ARGENTIN	OTH S AM	OTHERS	TOTAL
1967	2.173	0.467	0.673	0.000	0.298	0.429	4.040
1968	4.622	1.689	0.547	0.000	0.512	1.618	8.988
1969	3.005	1.697	0.891	0.007	0.033	3.475	9.108
1970	5.723	1.733	0.099	0.000	0.162	2.645	10.361
1971	4.588	0.792	0.749	1.842	0.055	2.144	10.170
1972	5.421	0.855	1.507	0.814	0.291	0.762	9.650
1973	5.219	1.099	0.356	0.889	0.178	0.209	7.950
1974	14.882	2.885	0.000	0.961	0.320	1.411	20.458
1975	13.667	3.804	0.299	0.482	0.158	1.059	19.469
1976	14.209	2.026	0.012	0.098	0.515	0.396	17.256

QUANTITY-- 1000 STN

YEAR	MEXICO	VENEZUEL	PERU	ARGENTIN	OTH S AM	OTHERS	TOTAL
1967	15.847	3.902	5.424	0.000	2.764	3.557	31.495
1968	27.269	13.735	4.416	0.000	4.696	15.089	65.205
1969	21.585	13.788	7.713	0.071	0.308	29.303	72.768
1970	39.811	12.943	0.817	0.000	1.363	24.074	79.007
1971	29.528	5.754	5.960	16.951	0.402	19.431	78.026
1972	34.412	5.608	11.064	7.612	2.632	6.442	67.770
1973	30.288	6.654	2.183	8.002	1.621	1.739	50.489
1974	55.492	13.427	0.000	4.267	1.472	4.250	78.909
1975	52.132	13.375	1.095	1.491	0.500	3.852	72.445
1976	49.079	6.767	0.056	0.329	1.580	1.370	59.180

EXPORTS
BUILDING BOARD (6416)
GULF

VALUE--MILLION CURRENT DOLLARS

YEAR	N AMERIC	S AMERIC	EUROPE	ASIA	AFRICA	OTHERS	TOTAL
1967	0.170	0.069	1.893	0.237	0.021	0.003	2.394
1968	0.409	0.070	1.609	0.415	0.038	0.013	2.554
1969	0.220	0.014	1.089	0.298	0.022	0.008	1.651
1970	0.211	0.048	0.433	0.190	0.010	0.005	0.896
1971	0.166	0.032	0.727	0.439	0.013	0.030	1.407
1972	0.145	0.061	0.615	0.421	0.053	0.127	1.422
1973	0.356	0.058	0.645	0.664	0.022	0.045	1.789
1974	0.331	0.035	0.837	1.127	0.103	0.031	2.464
1975	0.278	0.076	0.731	1.174	0.066	0.001	2.327
1976	0.340	0.070	1.067	1.040	0.016	0.007	2.540

QUANTITY--1000 STN

YEAR	N AMERIC	S AMERIC	EUROPE	ASIA	AFRICA	OTHERS	TOTAL
1967	0.766	0.314	8.476	0.870	0.096	0.009	10.532
1968	1.388	0.416	6.862	1.414	0.207	0.072	10.359
1969	1.104	0.050	4.906	1.119	0.088	0.048	7.314
1970	0.931	0.107	1.855	0.758	0.042	0.026	3.719
1971	0.704	0.079	2.809	1.270	0.057	0.127	5.046
1972	0.517	0.255	2.426	1.437	0.182	0.438	5.255
1973	1.725	0.203	2.701	1.857	0.073	0.124	6.684
1974	1.530	0.044	2.390	2.864	0.281	0.080	7.189
1975	0.963	0.094	2.357	6.833	0.146	0.004	10.395
1976	1.199	0.554	2.933	2.330	0.030	0.008	7.054

EXPORTS
PAPER & PAPERBOARD (6410)
GULF

VALUE--MILLION CURRENT DOLLARS

YEAR	N AMERIC	S AMERIC	EUROPE	ASIA	AFRICA	OTHERS	TOTAL
1967	16.958	9.527	44.878	3.323	5.229	1.531	81.446
1968	25.227	11.931	62.553	4.588	5.311	1.881	111.490
1969	29.536	9.546	70.528	4.350	6.076	2.120	122.156
1970	34.567	11.232	68.177	7.560	8.329	2.977	132.841
1971	33.956	10.348	90.084	11.666	10.388	3.481	159.923
1972	38.528	9.601	86.152	11.311	11.720	2.548	159.861
1973	41.378	10.472	71.283	15.403	17.902	3.822	160.259
1974	66.435	31.183	116.229	27.824	37.413	14.803	293.887
1975	62.378	23.146	84.417	34.854	36.316	9.213	250.325
1976	75.071	25.839	89.474	17.386	34.491	13.597	255.858

QUANTITY--1000 STN

YEAR	N AMERIC	S AMERIC	EUROPE	ASIA	AFRICA	OTHERS	TOTAL
1967	90.053	66.402	322.266	20.407	30.554	6.784	536.467
1968	156.258	86.537	479.102	29.795	32.890	8.416	792.998
1969	197.344	67.105	558.546	31.203	39.845	8.862	902.906
1970	241.868	76.871	513.122	53.881	60.175	12.587	958.504
1971	242.536	63.037	675.628	90.896	70.228	13.193	1155.518
1972	240.731	50.553	651.967	74.225	76.257	10.665	1104.398
1973	226.208	48.216	481.451	70.626	88.479	14.883	929.863
1974	243.285	89.403	540.241	100.232	131.310	41.910	1146.381
1975	193.180	62.443	310.028	119.862	116.606	22.684	824.803
1976	248.729	76.063	340.764	59.691	115.115	31.721	872.084

IMPORTS
HARDWOOD LOGS (2423)
GULF

VALUE--MILLION CURRENT DOLLARS

YEAR	COLOMBIA	BRAZIL	OTH S AM	ASIA	AFRICA	OTHERS	TOTAL
1967	1.112	0.517	0.004	0.000	0.153	0.251	2.037
1968	0.997	1.408	0.031	0.001	0.010	0.357	2.804
1969	0.848	0.442	0.126	0.006	0.158	0.130	1.709
1970	0.774	0.413	0.042	0.005	0.083	0.274	1.590
1971	0.493	0.484	0.001	0.000	0.003	0.175	1.155
1972	0.236	0.764	0.000	0.379	0.021	0.003	1.404
1973	0.020	0.062	0.107	0.001	0.012	0.119	0.320
1974	0.000	0.000	0.117	0.129	0.157	0.005	0.408
1975	0.000	0.005	0.034	0.000	0.000	0.010	0.049
1976	0.000	0.000	0.000	0.000	0.020	0.029	0.050

QUANTITY--MILLION BF

YEAR	COLOMBIA	BRAZIL	OTH S AM	ASIA	AFRICA	OTHERS	TOTAL
1967	8.702	3.034	0.025	0.000	0.755	1.870	14.386
1968	8.367	11.769	0.239	0.001	0.080	3.092	23.548
1969	6.577	4.619	1.287	0.009	0.756	0.958	14.205
1970	6.431	4.815	0.637	0.007	0.334	1.785	14.009
1971	3.580	7.053	0.006	0.000	0.009	1.157	11.806
1972	1.674	7.933	0.000	0.166	0.083	0.012	9.868
1973	0.086	0.586	1.367	0.002	0.032	0.197	2.270
1974	0.000	0.000	1.320	0.278	0.866	0.070	2.534
1975	0.000	0.013	0.055	0.000	0.000	0.012	0.080
1976	0.000	0.000	0.000	0.000	0.036	0.052	0.088

IMPORTS
SOFTWOOD LUMBER (2432)
GULF

VALUE--MILLION CURRENT DOLLARS

YEAR	CANADA	HONDURAS	OTH N AM	BRAZIL	ASIA	OTHERS	TOTAL
1967	3.205	0.025	0.001	0.609	0.008	0.009	3.858
1968	5.782	0.018	0.099	1.079	0.037	0.012	7.025
1969	7.941	0.017	0.065	1.251	0.036	0.046	9.356
1970	5.132	0.020	0.022	1.115	0.042	0.014	6.344
1971	12.519	0.065	0.066	1.302	0.157	0.013	14.122
1972	23.594	0.426	0.105	2.195	0.367	0.009	26.697
1973	28.233	2.830	0.365	4.168	0.518	0.046	36.160
1974	16.167	0.309	0.109	1.900	0.176	0.103	18.765
1975	7.909	0.256	0.065	0.920	0.084	0.018	9.253
1976	11.306	0.111	0.051	1.377	0.092	0.047	12.983

QUANTITY--MILLION BF

YEAR	CANADA	HONDURAS	OTH N AM	BRAZIL	ASIA	OTHERS	TOTAL
1967	47.571	0.377	0.011	6.249	0.048	0.101	54.357
1968	73.191	0.272	0.299	11.325	0.262	0.027	85.376
1969	89.825	0.282	0.290	10.681	0.235	0.189	101.501
1970	86.554	0.270	0.170	7.549	0.364	0.032	94.939
1971	173.938	1.028	0.308	8.456	0.752	0.158	184.640
1972	221.757	3.437	0.730	12.052	1.491	0.028	239.496
1973	175.517	31.395	3.129	16.133	1.622	0.198	227.994
1974	109.680	2.642	2.469	4.553	0.448	0.903	120.694
1975	63.665	2.629	0.445	2.623	0.261	0.024	69.647
1976	74.798	0.462	0.185	3.232	0.305	0.174	79.155

IMPORTS
HARDWOOD LUMBER (2433)
GULF

VALUE--MILLION CURRENT DOLLARS

YEAR	BRAZIL	OTH S AM	MALAYSIA	OTH ASIA	AFRICA	OTHERS	TOTAL
1967	0.830	2.082	1.733	1.446	1.109	1.268	8.468
1968	1.195	1.526	3.065	1.920	1.014	0.948	9.667
1969	2.057	1.583	3.731	1.935	1.950	0.361	11.616
1970	2.113	1.309	2.432	1.393	1.401	1.176	9.824
1971	2.012	1.363	3.556	1.962	1.906	0.839	11.638
1972	4.146	1.520	4.225	2.876	2.034	1.353	16.154
1973	8.197	4.725	5.040	5.764	1.809	1.495	27.029
1974	7.178	5.519	7.053	6.414	2.312	2.069	30.546
1975	5.803	3.217	2.660	2.840	1.073	1.964	17.556
1976	7.513	2.804	5.218	5.146	2.009	2.053	24.743

QUANTITY--MILLION BF

YEAR	BRAZIL	OTH S AM	MALAYSIA	OTH ASIA	AFRICA	OTHERS	TOTAL
1967	2.926	11.701	13.387	6.814	5.386	4.803	45.018
1968	5.186	10.410	20.344	11.049	5.279	3.406	55.673
1969	9.323	11.953	26.615	9.376	9.004	1.558	67.829
1970	15.286	8.000	15.788	5.423	6.423	4.259	55.179
1971	14.566	8.342	21.109	8.305	8.665	3.106	64.093
1972	33.118	9.842	24.923	12.469	9.010	4.743	94.106
1973	59.000	25.533	23.793	16.614	6.447	4.980	136.366
1974	37.828	23.299	27.023	14.025	7.758	5.308	115.242
1975	26.400	11.795	9.926	5.811	3.217	4.457	61.607
1976	35.438	9.882	13.014	9.269	5.454	3.144	76.202

IMPORTS
WOOD VENEER (6311)
GULF

VALUE--MILLION CURRENT DOLLARS

YEAR	COLOMBIA	EUROPE	PHIL REP	OTH ASIA	AFRICA	OTHERS	TOTAL
1967	0.094	0.052	0.624	0.369	0.591	0.209	1.939
1968	0.144	0.058	0.671	0.270	0.384	0.475	2.001
1969	0.144	0.072	0.814	0.310	0.119	0.406	1.866
1970	0.165	0.046	0.216	0.282	0.133	0.172	1.014
1971	0.488	0.055	0.368	0.597	0.075	0.421	2.003
1972	0.573	0.128	2.197	0.936	0.121	0.246	4.201
1973	0.652	0.189	2.895	1.397	0.694	0.549	6.375
1974	0.467	0.126	1.011	0.661	0.763	0.695	3.723
1975	0.087	0.068	0.055	0.129	0.007	0.215	0.561
1976	0.268	0.432	0.085	0.148	0.271	0.238	1.442

QUANTITY--MIL SQ FT

YEAR	COLOMBIA	EUROPE	PHIL REP	OTH ASIA	AFRICA	OTHERS	TOTAL
1967	0.442	0.062	7.187	3.942	5.443	2.584	19.661
1968	0.705	0.073	7.531	2.865	2.913	6.182	20.268
1969	0.732	0.047	8.106	1.912	0.800	4.449	16.047
1970	2.402	0.046	2.241	2.993	1.402	1.965	11.048
1971	13.427	0.043	4.458	6.320	0.744	4.291	29.284
1972	10.200	0.092	28.436	9.820	1.296	2.694	52.538
1973	10.372	0.070	27.810	10.878	5.322	5.996	60.449
1974	6.784	0.317	5.810	3.558	2.449	6.109	25.026
1975	0.257	0.035	0.494	0.751	0.045	2.724	4.308
1976	0.723	0.781	0.848	0.723	1.158	2.380	6.612

IMPORTS
PLYWOOD (6312)
GULF

VALUE--MILLION CURRENT DOLLARS

YEAR	PHIL REP	KOR REP	TAIWAN	JAPAN	OTH ASIA	OTHERS	TOTAL
1967	4.289	11.015	2.910	10.463	0.411	0.866	29.954
1968	4.317	18.560	5.218	19.562	0.733	0.926	49.316
1969	5.140	19.522	6.506	18.052	0.957	1.410	51.587
1970	2.753	17.747	11.155	13.644	0.541	1.031	46.871
1971	4.683	23.585	17.389	19.928	0.529	1.222	67.338
1972	2.855	33.187	23.922	23.720	0.245	1.543	85.471
1973	4.593	39.462	19.197	20.124	0.685	1.992	86.053
1974	0.969	29.070	14.194	19.590	0.773	1.459	66.055
1975	0.657	24.799	11.129	11.767	0.388	0.433	49.173
1976	1.358	40.259	19.356	28.279	0.905	0.864	91.019

QUANTITY--MIL SQ FT

YEAR	PHIL REP	KOR REP	TAIWAN	JAPAN	OTH ASIA	OTHERS	TOTAL
1967	41.726	96.901	27.341	66.215	3.607	6.697	242.487
1968	36.269	153.632	45.826	100.835	6.537	5.442	348.541
1969	43.639	152.380	62.017	94.633	8.093	10.292	371.054
1970	25.716	147.849	99.674	80.375	4.666	6.852	365.131
1971	42.995	193.798	146.280	107.954	5.134	7.661	503.822
1972	26.207	274.399	198.238	114.183	2.840	7.230	623.097
1973	39.854	230.289	104.240	68.714	7.096	6.606	456.800
1974	5.748	147.335	66.589	54.202	4.075	4.589	282.537
1975	4.831	143.429	70.371	44.730	2.954	1.393	267.709
1976	9.930	208.959	100.891	82.772	5.653	1.657	409.863

IMPORTS
WOOD PULP (2510)
GULF

VALUE--MILLION CURRENT DOLLARS

YEAR	CANADA	SWEDEN	FINLAND	OTH EURO	REP S AF	OTHERS	TOTAL
1967	0.581	0.000	0.021	0.011	4.386	0.002	5.001
1968	0.482	0.002	0.065	0.022	5.897	0.000	6.469
1969	0.159	0.000	0.000	0.000	5.982	0.000	6.142
1970	0.328	0.000	0.000	0.131	7.938	0.000	8.398
1971	0.232	0.000	0.000	0.138	7.261	0.276	7.907
1972	0.151	0.014	0.000	0.471	7.711	0.004	8.352
1973	0.012	0.000	0.000	0.253	11.626	0.000	11.892
1974	0.503	2.211	0.567	0.000	21.032	0.000	24.313
1975	0.098	0.008	0.000	0.000	13.774	0.000	13.880
1976	0.134	0.121	0.740	0.041	21.215	0.354	22.605

QUANTITY--1000 STN

YEAR	CANADA	SWEDEN	FINLAND	OTH EURO	REP S AF	OTHERS	TOTAL
1967	11.301	0.000	0.209	0.066	37.212	0.010	48.799
1968	8.627	0.055	0.689	0.124	51.064	0.000	60.559
1969	2.793	0.000	0.000	0.000	50.722	0.000	53.515
1970	5.829	0.000	0.000	1.125	67.291	0.000	74.245
1971	3.891	0.000	0.000	1.180	62.316	1.646	69.033
1972	2.607	0.160	0.000	4.029	62.290	0.059	69.146
1973	0.084	0.000	0.000	1.501	64.731	0.000	66.316
1974	3.085	5.363	1.243	0.000	82.501	0.000	92.193
1975	0.313	0.019	0.000	0.000	49.776	0.000	50.108
1976	0.413	0.400	2.425	0.290	78.120	1.763	83.411

IMPORTS
NEWSPRINT (6411)
GULF

VALUE--MILLION CURRENT DOLLARS

YEAR	CANADA	OTH NAM	SWEDEN	FINLANDO	TH EURO	OTHERS	TOTAL
1967	37.741	0.000	0.000	4.914	0.000	0.000	42.655
1968	36.991	0.000	0.000	6.512	0.000	0.000	43.503
1969	29.079	0.000	0.000	6.926	0.000	0.000	36.005
1970	32.581	0.000	0.000	8.521	0.000	0.000	41.102
1971	29.427	0.000	0.000	8.581	0.000	0.000	38.009
1972	37.916	0.000	0.000	11.496	0.000	0.000	49.412
1973	46.000	0.000	0.000	14.916	0.000	0.000	60.916
1974	58.921	0.000	0.000	9.828	0.000	0.000	68.749
1975	62.859	0.000	0.000	1.102	0.000	0.000	63.961
1976	66.009	0.000	0.700	0.000	0.000	0.000	66.709

QUANTITY--1000 STN

YEAR	CANADA	OTH NAM	SWEDEN	FINLANDO	TH EURO	OTHERS	TOTAL
1967	303.937	0.000	0.000	41.918	0.000	0.000	345.855
1968	294.044	0.000	0.000	54.066	0.000	0.000	348.109
1969	230.003	0.000	0.000	56.635	0.000	0.000	286.638
1970	235.685	0.000	0.000	68.393	0.000	0.000	304.078
1971	224.570	0.000	0.000	98.135	0.000	0.000	322.705
1972	258.253	0.000	0.000	84.339	0.000	0.000	342.592
1973	293.853	0.000	0.000	108.084	0.000	0.000	401.937
1974	309.772	0.000	0.000	68.347	0.000	0.000	378.119
1975	271.286	0.000	0.000	6.164	0.000	0.000	277.450
1976	298.675	0.001	2.544	0.000	0.000	0.000	301.220

IMPORTS
BUILDING BOARD (6416)
GULF

VALUE--MILLION CURRENT DOLLARS

YEAR	BRAZIL	ARGENTIN	SCANDINA	OTH EURO	REP S AF	OTHERS	TOTAL
1967	0.879	0.003	0.725	0.209	0.008	0.152	1.975
1968	1.142	0.000	1.288	0.220	0.179	0.353	3.181
1969	0.745	0.000	1.921	0.456	1.797	0.590	5.509
1970	0.614	0.000	1.067	0.203	0.876	0.207	2.987
1971	1.424	0.027	2.281	0.216	0.069	0.422	4.439
1972	2.415	0.105	1.689	0.577	0.771	0.292	5.849
1973	3.249	0.811	2.399	0.916	1.204	0.477	9.057
1974	4.000	1.050	1.866	0.599	0.598	0.601	8.714
1975	1.556	0.339	0.236	0.162	0.000	0.295	2.588
1976	2.849	0.373	0.484	0.889	0.458	0.786	5.837

QUANTITY--1000 STN

YEAR	BRAZIL	ARGENTIN	SCANDINA	OTH EURO	REP S AF	OTHERS	TOTAL
1967	10.398	0.017	10.350	3.879	0.110	1.761	26.516
1968	14.251	0.000	18.734	3.997	2.384	3.887	43.253
1969	8.981	0.000	26.833	7.700	23.526	6.491	73.531
1970	8.272	0.000	13.432	3.248	11.562	23.501	60.014
1971	17.851	0.578	31.704	3.591	1.231	3.918	58.872
1972	28.459	1.905	20.843	8.440	10.222	3.831	73.700
1973	30.071	8.600	19.610	9.989	14.839	4.064	87.174
1974	28.347	9.536	12.049	4.749	6.933	4.758	66.371
1975	11.308	3.554	1.506	1.919	0.000	2.300	20.586
1976	18.991	3.356	3.540	9.375	4.111	5.683	45.057

IMPORTS
PAPER & PAPERBOARD (6410)
GULF

VALUE--MILLION CURRENT DOLLARS

YEAR	SWEDEN	NORWAY	FINLAND	OTH EURO	ASIA	OTHERS	TOTAL
1967	0.255	0.043	0.494	0.170	0.034	0.030	1.026
1968	0.141	0.045	1.433	0.211	0.009	0.043	1.882
1969	0.121	0.077	1.942	0.207	0.018	0.088	2.453
1970	0.104	0.039	2.925	0.193	0.362	0.015	3.638
1971	0.071	0.039	1.939	0.278	0.048	0.015	2.390
1972	0.089	0.094	0.989	0.187	0.060	0.006	1.425
1973	0.449	0.178	1.053	0.232	0.026	0.026	1.962
1974	1.034	0.194	1.918	0.616	0.080	0.433	4.275
1975	0.216	0.011	1.365	0.503	0.012	0.072	2.179
1976	0.790	0.013	1.822	0.641	0.032	0.071	3.368

QUANTITY--1000 STN

YEAR	SWEDEN	NORWAY	FINLAND	OTH EURO	ASIA	OTHERS	TOTAL
1967	1.430	0.149	2.270	0.580	0.165	0.145	4.738
1968	0.690	0.186	10.167	0.531	0.022	0.228	11.825
1969	0.635	0.287	13.554	0.491	0.039	0.441	15.448
1970	0.416	0.132	20.429	0.472	2.557	0.066	24.072
1971	0.271	0.121	13.680	0.761	0.078	0.074	14.986
1972	0.273	0.266	5.958	0.407	0.083	0.027	7.014
1973	1.021	0.430	4.184	0.500	0.038	0.179	6.352
1974	1.798	0.345	4.558	0.784	0.129	2.799	10.414
1975	0.349	0.020	2.747	0.365	0.002	0.262	3.745
1976	1.270	0.020	4.853	0.358	0.005	0.218	6.724

Appendix G

THE NORTH ATLANTIC REGION

Exports

Regional Overview

Although the North Atlantic region is a large importer of forest products, it is also a considerable exporter and, in fact, was the third largest forest-product-exporting region in the United States for 1967-76. In 1976 this region exported forest products valued at $499.3 million, which was 12 percent of the value of U.S. forest product exports. The region's share of U.S. exports was fairly stable throughout the period, averaging 14 percent of the U.S. export value. From 1967 to 1976, the nominal value of the North Atlantic forest exports grew at an average annual rate of 13 percent.

The ratio of solid wood product exports to pulp and paper exports for the region was much lower than that for the United States as a whole. Only 25 percent of the region's 1976 exports were of solid wood commodities, while nearly 50 percent of the U.S. exports in that year fell into the solid-wood category.

The most important regional export in 1976 was paper and paperboard, which accounted for 61 percent ($307 million) of the value of the region's total forest product exports. Wood pulp exports in 1976 totaled $47 million, which represented 9 percent of the region's forest product exports. Softwood logs (6 percent of the region's forest product exports), hardwood logs (7 percent), softwood lumber (3 percent), hardwood lumber (4 percent), and wood veneer (5 percent) were the largest components of the region's $126 million worth of solid wood exports.

Paper and Paperboard

The North Atlantic was the largest paper and paperboard exporting region in the United States during 1967-76, accounting for an average of 28 percent of the value of U.S. exports of paper and paperboard. The 1976 volume of regional exports was, however, only 14 percent of the volume of U.S. paper and paperboard exports, indicating that the North Atlantic is exporting a higher-valued, perhaps more highly processed product than are other U.S. regions. The volume of the region's paper and paperboard exports grew at an average annual rate of 4 percent during the study period.

Paper and paperboard were exported from the region to every continent. European countries as a group were the largest importers, taking a combined share of 40 percent of the value of the region's 1976 exports. Canada was the single largest country of destination for paper and paperboard, importing 107,000 short tons (worth $46.9 million) in 1976. Australia in that year imported 41,000 tons (worth $31.7 million), which was 10 percent of the region's exports. European countries of importance in 1976 included the United Kingdom (with 9 percent of the value of the region's paper and paperboard exports), France (6 percent), West Germany (5 percent), and the Netherlands (5 percent).

The country structure of the region's export trade was very similar in 1967 to the situation in 1976. The breakdown by continent for the 1967-76 period, in terms of average percentage (by value) of the region's paper and paperboard exports is as follows: North America, 17 percent (of which Canada represented 64 percent); South America, 12 percent; Europe, 45 percent; Asia, 9 percent; Africa, 6 percent; and Oceania, 10 percent (of which Australia represented 91 percent).

Wood Pulp

Although the North Atlantic is not a major U.S. wood pulp-exporting region, its exports of wood pulp, in 1976, represented 9 percent (by value) of the region's total forest product exports. In that year, the region exported 122,000 tons (valued at $48 million).

The volume of the region's wood pulp exports grew during the 1967-76 period at an average rate of 6 percent per year, while the nominal value grew at an annual average rate of 13 percent.

European countries were the destinations of 82 percent (by value) of the region's wood pulp exports in 1976. About one-third of the wood pulp destined for Europe went to the United Kingdom, which imported 29,000 tons of wood pulp valued at $12.5 million. Italy, Spain, and West Germany were also major destinations for North Atlantic wood pulp; together these three countries in 1976 imported 33,000 tons (valued at $13.3 million), which was 28 percent of the region's wood pulp exports.

The share of North Atlantic exports that was shipped to Europe changed very little between 1967 and 1976, but the shares taken by the various European countries did so dramatically. The United Kingdom, in particular, increased its share of the region's exports from 8 percent in 1967 to 27 percent in 1976, whereas West Germany and Italy declined from 16 percent to 7 percent and from 21 percent to 15 percent, respectively. Spain's share increased slightly, but that of the other European countries dropped from 36 percent in 1967 to 27 percent in 1976.

Hardwood Logs

Throughout 1967-76, the North Atlantic was the major U.S. hardwood log-exporting region. The region exported, on average, 52 percent of the value and 72 percent of the volume of U.S. hardwood logs. Hardwood logs constituted only 2 percent (by value) of the U.S. forest products export trade in 1976, and 7 percent of North Atlantic forest product exports.

Over this period, the volume of hardwood log exports from the region dropped by 40 percent, while the nominal value increased from $19.7 million to $33 million.

Nearly three-fourths of the region's hardwood log exports were shipped to just four countries in 1976. West Germany was the largest importer, accounting for 40 percent of the value of the region's 1976 exports. About 10 percent of the value of the region's exports was shipped to each of Canada, Switzerland, and Italy.

Canada, in 1976, imported nearly half the volume of the region's hardwood log exports, but only 12 percent of the value, indicating that Canada is receiving a lower-grade log than are other, more distant countries.

The percentage market shares captured by the region's different trading partners changed substantially from 1967 to 1976. Canada and Italy both declined in importance as importers of hardwood logs, each losing about 14 percent of the region's total hardwood log export market. Switzerland's share fluctuated during the period, but by 1976 had returned to its 1967 level of 11 percent of the value of the region's exports. West Germany increased its share of the region's hardwood log exports from 29 percent in 1967 to 40 percent in 1976. During the same period, the group of European countries other than West Germany, Switzerland, and Italy increased its share from 8 percent to 20 percent.

Softwood Logs

The North Atlantic region, in 1976, exported 258 million board-feet of softwood logs (valued at $28.6 million). That value was only 3 percent of the total U.S. softwood log exports, and 6 percent of the region's total forest product exports.

The region's softwood log exports increased modestly during the period 1967-76. Volumes increased at an average of 3 percent per year, whereas nominal values increased at an average annual rate of 11 percent.

Canada has been the only major importer of North Atlantic softwood logs, taking an average of 98 percent of the value of the region's exports during the study period. In 1975-76, Italy imported most of the residual not shipped to Canada.

Wood Veneer

The North Atlantic region was the origin of more than half the U.S. exports of wood veneer in 1976, shipping out 103.7 million square feet (worth $24.8 million). That year was a peak for the region; since 1967 veneer exports had been steadily increasing, by an annual average of 43 percent in value and 36 percent in volume.

Nearly all the region's veneer exports in 1976 were shipped to European countries. West Germany was the single most important country, taking nearly half (in value terms) of the region's exports in 1976. Switzerland and the United Kingdom also accounted for substantial shares (14 percent and 6 percent, respectively) of the North Atlantic wood veneer. All other European countries together imported 28 percent of the region's 1976 exports.

Canada was the only significant destination outside of Europe for North Atlantic veneer but imported only 2 percent of the region's exports in 1976.

Since 1967 the country structure of the North Atlantic's veneer export trade has shifted considerably. Canadian and West German shares dropped substantially from 10 percent to 2 percent and from 56 percent to 48 percent, respectively. Exports to European countries other than the United Kingdom and West Germany grew the fastest, and those countries' combined share increased from 24 percent to 42 percent of the value of the North Atlantic's veneer exports.

Other Forest Products

Together, the remaining forest products (newsprint, building board, pulpwood, softwood lumber, hardwood lumber, plywood, reconstituted wood, pulpwood chips, and other industrial roundwood) represented 12 percent of value of the North Atlantic region's total forest product exports in 1976.

The most important of these products--hardwood lumber--accounted for 4 percent of the region's 1976 forest product exports. North Atlantic hardwood lumber exports of 56 million board-feet (worth $22 million) represetned one-fourth (by value) of the total U.S. hardwood lumber exports in 1976. The region's principal trading partners include Canada (33 percent of the region's hardwood lumber exports) and various European countries (together totaling 58 percent), principally France (10 percent), Spain (10 percent), and the Netherlands (8 percent). Countries outside Canada and Europe imported 10 percent of the region's 1976 exports.

Softwood lumber exports (53 million board-feet worth $14.4 million in 1976) were also shipped principally to Canada and Europe. Canada, in 1976, imported 26.8 million board-feet (valued at $5.7 million), which represented 40 percent of the region's softwood lumber exports. Exports to Italy showed the fastest growth rates during the 1967-76 period, and, by 1976, Italy was the destination of 35 percent of the value of the region's exports. Other European countries took 12 percent (by value) of the North Atlantic's 1976 softwood lumber exports.

The North Atlantic region was the second largest U.S. newsprint-exporting region throughout 1967-76. On average, the region shipped 32 percent of the U.S. total exports of newsprint. The 1976 volume was 45,000 tons (valued at $15 million). European and South American countries were the destinations for 45 percent and 35 percent, respectively, of the region's 1976 newsprint exports. Venezuela was the largest importer of newsprint. Exports of $4.1 million to Venezuela in 1976 represented 20 percent of the North Atlantic's total newsprint exports.

Imports

Regional Overview

The North Atlantic region was the second largest forest product importing region in the United States throughout 1967-76. On average, the value of the region's imports represented 22 percent of the value of U.S. forest product imports. North Atlantic imports grew (in nominal value terms) at an average rate of 9 percent per year during the study period, going from $476.5 million in 1967 to $1 billion in 1976.

The region's import commodity mix, like that of the United States as a whole, is heavily skewed toward pulp and paper products. Only about 35 percent of the value of the region's 1976 imports was in the solid wood products, the rest being wood pulp, newsprint, building board, and paper and paperboard.

Of the $655.3 million worth of pulp and paper products imported by the region in 1976, $308.5 million (47 percent) was in newsprint, $269.7 million (41 percent) was in wood pulp, $71.4 million (11 percent) was in paper and paperboard, and $5.6 million (1 percent) was in building board.

Of the $350.1 million in solid wood product imports, $195.6 million (56 percent) was in softwood lumber, $108.7 million (31 percent) was in wood veneer, $20.6 million (6 percent) was in hardwood lumber, and $10.8 million (3 percent) was in plywood. Only $13 million (4 percent) was in the relatively unprocessed wood commodities (logs, pulpwood, and pulpwood chips).

The commodity mix between solid wood and pulp and paper products was essentially the same in 1976 as it was in 1967, when pulp and paper products constituted 67 percent of the value of the region's imports. However, in 1967 newsprint made up a larger share (56 percent) of the value of the non-solid wood imports, and wood pulp was not as important (35 percent of the non-solid wood imports) as in 1976.

Paper and Paperboard

In 1976 paper and paperboard constituted only 7 percent of the value of the North Atlantic's total forest product imports. However, the $71.4 million imported by the region represented one-third of all the paper and paperboard imported by the United States in 1976.

The region's imports of paper and paperboard grew modestly during 1967-76, gaining on average 5 percent per year in volume and 14 percent per year in nominal value. The United States as a whole had import growth rates of 7 percent and 19 percent for volume and value, respectively.

European nations are the largest exporters of paper and paperboard to the North Atlantic region. During the study period, European countries supplied 71 percent of the value and 63 percent of the volume of the region's imports in an average year. The Nordic countries of Sweden, Norway, Finland, and Denmark usually provided more than half the value of the region's imports from Europe. In 1976 the region imported 58,000 tons (valued at $19.7 million) from this area and 14,000 tons (valued at $26.9 million) from the rest of Europe. Obviously the unit values differ greatly between the two European areas, indicating that the Nordic countries are lower-valued, perhaps less-processed product mix than is the rest of Europe. This relationship was true throughout 1967-76.

Canada has been the other major supplier of paper and paperboard to the North Atlantic region, and substantially increased its share of the region's import market over the study period. In 1967 Canada provided only 6 percent of the value of the region's imports, but by 1976 was responsible for 29 percent of the value and 46 percent of the volume of North Atlantic imports. Over the period the nominal value of Canadian imports grew at an average rate of 129 percent per year, while the volume grew at 29 percent annually.

Newsprint

It should be no surprise that the North Atlantic region, with its high population and large urban centers, is a major importer of newsprint. In fact, newsprint is the region's most valuable forest product import, and the North Atlantic is the second largest regional importer of newsprint in the United States. Regional imports of newsprint in 1976 were valued at $308.5 million, which represented 18 percent of total U.S. newsprint imports.

The quantity of newsprint imported to the North Atlantic declined by about 15 percent during 1967-76, going from 1.36 million tons in 1967 to 1.13 million tons in 1976. Import quantities peaked at 1.43 million tons in 1968.

Canada is the leading exporter of newsprint to the United States as a whole, and the North Atlantic region is no exception. Canada was the origin of over 89 percent of the value of regional newsprint imports for 1967-76 and, in 1974-76, Canada's share averaged over 97 percent. The residual was imported from European countries, principally the Nordic nations of Sweden and Finland. Although this residual is small relative to Canadian imports, the 1976 value of newsprint imports from Europe was over $14 million. Other than Sweden and Finland, Norway ($2.11 million) and France ($1.2 million) were important European suppliers of newsprint in 1976.

Wood Pulp

The North Atlantic region is the second largest wood pulp-importing region in the United States--during 1967-76 the region imported, on average, about one-fourth the value of U.S. annual wood pulp imports. In 1976, 865,000 tons (valued at $269.7 million) were imported to the region.

While the United States as a whole increased the quantity of wood pulp imports by 25 percent over the decade, the North Atlantic region experienced a decline of about 5 percent. However, the nominal value of wood pulp imports to the region increased by 239 percent, which is the equivalent to an increase of approximately 10 percent per year.

During 1967-76, the principal source of wood pulp flowing into the North Atlantic region--Canada--increased its share of the import market considerably. Canada's average percentage share of the value of regional wood pulp imports during 1967-76 was 89 percent, and during 1972-76 was 94 percent. Volumes of wood pulp imports from Canada increased at an average annual rate of only 1 percent, but imports from the rest of the world declined by 13 percent per year.

Sweden and Finland were responsible for most of the region's wood pulp imports from countries other than Canada. In 1976 these two countries supplied only 6 percent of the value of the region's imports.

Both volumes and values of wood pulp imports from Sweden and Finland fluctuated considerably throughout 1967-76, but generally showed a downtrend. The combined volume of wood pulp imported from the two countries in 1976 was less than one-third the combined volume in 1967. In the rest of the world, there was a similar trend in countries that usually contributed less than 1 percent of the value of the region's wood pulp imports.

Softwood Lumber

Throughout 1967-76, nearly one-fifth of the value of the North Atlantic region's total forest product imports was comprised of softwood lumber. However, the region moved during this period from being the second largest to the fourth largest (in value terms) softwood-lumber-importing region in the United States. In 1967 the $90.5 million worth of softwood lumber imported by the region was 28 percent of the value of U.S. softwood lumber imports, but in 1976, the $195.6 million worth of regional imports were only 14 percent of the value of U.S. imports.

The volume of North Atlantic softwood lumber imports dropped sharply at the beginning of the period, going from 5.5 billion board-feet in 1967 to 1.7 billion board-feet in 1968. During 1968-76 volumes fluctuated, but generally declined, to 1.3 billion board-feet in 1976.

Virtually all of the region's softwood lumber imports are of Canadian origin. In only one year did imports from Canada comprise less than 99 percent of the value of the region's total softwood lumber imports. Much of the remainder of the softwood lumber imported to the North Atlantic, which totaled only $539,000 in 1976, was supplied by various European and South American countries.

Plywood

In 1976 the North Atlantic region imported 830.6 million square feet of plywood (valued at $108.7 million), which represented 28 percent of the entire country's plywood imports in that year. The region's plywood imports constituted 11 percent (by value) of the region's total forest product imports.

The volume of the region's plywood imports grew substantially over the 1967-76 period, peaking in 1973 at 967 million square feet. This was 16 percent greater than the 1976 volume and more than twice the 1967 volume. In nominal value terms, the region's 1976 plywood imports represented a nearly 300 percent increase over 1967 imports.

At the beginning of the study period, Finland and Japan were the major suppliers of plywood to the region, but their combined share has dwindled from 49 percent to 9 percent of the value of the region's imports. Korea has filled the void by increasing the quantity of its plywood shipments to the region by an average of 19 percent annually, so that in 1976 imports from Korea represented about 60 percent of both quantity and value of regional plywood imports. Taiwan was a stable source of plywood throughout the period, supplying, on average, 16 percent (by value) of the North Atlantic's plywood imports.

Other Forest Products

The remaining forest products (plywood, softwood logs, hardwood logs, hardwood lumber, wood veneer, reconstituted wood, pulpwood chips, and other industrial roundwood) accounted for only about 5 percent of the value of the region's imports in 1976. However, for several of these commodities, the region's imports constituted a relatively large share of the U.S. imports of the particular commodity.

Hardwood logs are the most outstanding example. The 9.5 million board-feet imported by the region in 1976 were valued at $2.2 million, which is only two-tenths of 1 percent of the value of the region's total forest products imports for that year, but it is 78 percent of the value of the U.S. 1976 imports of hardwood logs. Canada has been the leading supplier of hardwood logs to the region, accounting for 53 percent of the value of the region's imports in 1976. Substantial quantities are also imported from the West African nations of Ghana and the Ivory Coast, which together supply 23 percent (in value terms) of the region's 1976 imports of hardwood logs. Over the study period, Canada has substantially increased its import market share, while the share captured by the two African countries steadily declined.

The North Atlantic region imported 14 percent (by value) of all the wood veneer imported by the United States in 1976. However, the $10.8 million of regional veneer imports represented only 1 percent of the region's total forest products imports.

Canada was the source of one-third of the region's veneer imports in 1976, accounting for 119.5 million square feet (valued at $3.5 million). The Philippine Republic provided 23 percent of the 1976 value of the region's imports, and European countries--notably France (12 percent) and West Germany (6 percent)--provided another 30 percent.

The North Atlantic was the largest U.S. pulpwood-importing region for 1974-76. Virtually all of the pulpwood imported to the region originated in Canada. In 1976 imports to the region from Canada amounted to 188,000 cords (valued at $7.8 million), which represented two-thirds (by value) of all pulpwood imported by the United States in that year.

EXPORTS
SOFTWOOD LOGS (2422)
NORTH ATLANTIC

VALUE--MILLION CURRENT DOLLARS

YEAR	CANADA	S AMERIC	ITALY	OTH EURO	ASIA	OTHERS	TOTAL
1967	10.851	0.000	0.032	0.011	0.069	0.005	10.968
1968	11.417	0.000	0.002	0.016	0.000	0.000	11.435
1969	12.718	0.009	0.023	0.027	0.004	0.004	12.786
1970	12.148	0.000	0.020	0.104	0.377	0.000	12.648
1971	13.539	0.002	0.034	0.028	0.000	0.003	13.606
1972	15.552	0.000	0.042	0.064	0.113	0.002	15.772
1973	18.300	0.000	0.022	0.078	0.008	0.000	18.408
1974	16.996	0.020	0.111	0.172	0.022	0.032	17.354
1975	20.189	0.000	1.008	0.210	0.164	0.019	21.589
1976	26.959	0.000	0.726	0.364	0.590	0.005	28.645

QUANTITY--MILLION BF

YEAR	CANADA	S AMERIC	ITALY	OTH EURO	ASIA	OTHERS	TOTAL
1967	204.250	0.000	0.125	0.037	0.949	0.018	205.380
1968	216.061	0.000	0.005	0.024	0.000	0.000	216.090
1969	226.432	0.011	0.091	0.188	0.011	0.005	226.739
1970	215.688	0.000	0.159	0.540	5.116	0.000	221.503
1971	241.511	0.011	0.271	0.281	0.000	0.004	242.079
1972	280.221	0.000	0.103	0.203	1.161	0.003	281.690
1973	277.364	0.000	0.031	0.192	0.007	0.000	277.594
1974	199.297	0.025	0.166	0.571	0.015	0.031	200.105
1975	175.673	0.000	1.827	0.452	0.094	0.024	178.070
1976	255.837	0.000	0.977	0.687	0.418	0.005	257.924

EXPORTS
HARDWOOD LOGS (2423)
NORTH ATLANTIC

VALUE--MILLION CURRENT DOLLARS

YEAR	CANADA	W GERMNY	SWITZRLD	ITALY	OTH EURO	OTHERS	TOTAL
1967	5.046	5.742	2.252	4.621	1.565	0.450	19.676
1968	4.129	8.011	2.738	5.533	2.588	0.317	23.317
1969	3.566	6.524	1.434	2.536	3.357	1.407	18.824
1970	3.372	6.553	2.100	2.512	2.634	0.754	17.926
1971	2.818	5.805	0.692	1.659	2.541	0.512	14.028
1972	4.139	6.989	1.539	2.081	2.905	2.994	20.647
1973	5.983	10.801	1.054	3.693	4.856	1.546	27.933
1974	5.500	10.400	1.101	2.132	7.668	0.794	27.595
1975	3.146	8.549	2.132	3.593	4.800	0.833	23.054
1976	3.992	13.267	3.484	3.216	6.749	2.271	32.977

QUANTITY--MILLION BF

YEAR	CANADA	W GERMNY	SWITZRLD	ITALY	OTH EURO	OTHERS	TOTAL
1967	70.621	6.053	2.252	5.403	2.263	0.402	86.995
1968	58.791	7.544	2.677	5.955	2.661	0.276	77.905
1969	44.273	6.865	1.647	2.802	3.455	1.072	60.113
1970	38.257	7.162	2.479	2.839	3.142	0.666	54.545
1971	31.495	6.037	0.802	1.850	2.793	0.473	43.451
1972	44.779	6.860	1.512	1.974	2.706	2.169	59.999
1973	52.378	12.395	1.331	3.719	5.523	1.226	76.572
1974	41.805	12.280	1.335	2.536	8.486	0.517	66.959
1975	23.110	7.572	3.451	3.015	4.796	0.681	42.625
1976	25.271	12.139	4.885	3.783	7.287	0.943	54.308

EXPORTS
SOFTWOOD LUMBER (2432)
NORTH ATLANTIC

VALUE--MILLION CURRENT DOLLARS

YEAR	CANADA	W GERMNY	ITALY	OTH EURO	ASIA	OTHERS	TOTAL
1967	1.858	1.614	0.303	0.404	0.054	0.156	4.389
1968	1.997	2.042	0.424	0.492	0.200	0.182	5.336
1969	2.256	1.571	0.667	0.344	0.068	0.160	5.066
1970	2.051	1.517	1.060	0.504	0.182	0.165	5.480
1971	2.344	1.161	0.875	0.304	1.370	0.309	6.362
1972	2.723	1.045	0.943	0.244	5.030	0.157	10.142
1973	3.891	1.592	2.603	1.887	0.203	0.288	10.464
1974	6.326	0.844	3.667	5.897	0.093	0.173	17.000
1975	3.987	0.921	1.634	0.832	0.295	0.161	7.830
1976	5.690	0.803	5.100	0.804	0.559	1.436	14.393

QUANTITY--MILLION BF

YEAR	CANADA	W GERMNY	ITALY	OTH EURO	ASIA	OTHERS	TOTAL
1967	20.955	10.703	1.769	2.498	0.295	0.876	37.096
1968	22.979	13.018	2.795	3.234	0.977	0.847	43.849
1969	24.419	10.206	4.240	2.593	0.313	0.700	42.471
1970	22.872	9.761	6.810	3.406	0.616	0.895	44.359
1971	21.060	6.344	5.701	2.010	3.865	1.561	40.541
1972	21.932	5.652	5.999	1.261	12.518	0.816	48.178
1973	25.188	6.415	11.160	9.226	1.050	1.077	54.117
1974	31.975	2.496	12.754	27.514	0.383	0.363	75.484
1975	19.613	2.551	5.701	2.636	1.373	0.467	32.341
1976	26.790	2.019	15.251	2.141	1.250	5.508	52.959

EXPORTS
HARDWOOD LUMBER (2433)
NORTH ATLANTIC

VALUE--MILLION CURRENT DOLLARS

YEAR	CANADA	NETHRLND	FRANCE	SPAIN	OTH EURO	OTHERS	TOTAL
1967	1.906	0.031	0.010	0.194	1.681	1.245	5.067
1968	1.021	0.062	0.022	0.152	1.412	1.339	4.008
1969	1.458	0.079	0.032	0.082	0.925	1.630	4.207
1970	0.811	0.079	0.088	0.163	1.857	3.516	6.514
1971	1.336	0.036	0.026	0.617	2.081	5.382	9.478
1972	2.906	0.136	0.249	1.093	2.350	21.814	29.549
1973	3.077	0.169	0.437	1.282	4.012	1.393	10.370
1974	3.895	3.281	0.938	1.516	6.608	1.266	17.505
1975	3.367	2.340	1.579	2.330	4.234	0.790	14.639
1976	7.316	1.644	2.141	2.210	6.507	2.145	21.962

QUANTITY--MILLION BF

YEAR	CANADA	NETHRLND	FRANCE	SPAIN	OTH EURO	OTHERS	TOTAL
1967	17.222	0.123	0.023	0.845	5.325	3.416	26.955
1968	9.859	0.237	0.041	0.706	4.455	3.276	18.574
1969	16.089	0.253	0.096	0.364	2.686	4.106	23.595
1970	5.620	0.229	0.158	0.659	5.533	9.065	21.264
1971	10.094	0.124	0.058	1.910	6.136	14.508	32.830
1972	17.412	0.516	0.528	3.098	6.395	34.325	62.274
1973	18.321	0.444	0.771	3.114	9.157	2.334	34.140
1974	16.813	5.962	1.144	2.436	11.356	2.294	40.005
1975	16.722	3.591	0.930	4.441	6.546	1.371	33.602
1976	34.911	2.343	1.049	3.894	10.872	3.180	56.248

EXPORTS
WOOD VENEER (6311)
NORTH ATLANTIC

VALUE--MILLION CURRENT DOLLARS

YEAR	CANADA	U KINGDM	W GERMNY	SWITZLND	OTH EURO	OTHERS	TOTAL
1967	0.094	0.080	0.556	0.112	0.130	0.021	0.993
1968	0.204	0.114	1.003	0.166	0.151	0.183	1.821
1969	0.157	0.151	2.437	0.046	0.274	0.120	3.184
1970	0.101	0.058	4.122	0.016	0.235	0.041	4.574
1971	0.049	0.174	4.534	0.063	0.532	0.046	5.398
1972	0.040	0.180	3.977	0.277	1.240	0.034	5.747
1973	0.093	0.731	4.903	1.711	2.236	0.047	9.720
1974	0.034	0.960	9.367	1.214	3.278	0.505	15.358
1975	0.059	0.711	11.665	2.975	2.853	0.253	18.517
1976	0.440	1.416	11.843	3.545	6.997	0.576	24.817

QUANTITY--MIL SQ FT

YEAR	CANADA	U KINGDM	W GERMNY	SWITZLND	OTH EURO	OTHERS	TOTAL
1967	1.545	0.133	0.843	0.114	0.150	0.059	2.844
1968	8.631	0.175	2.229	0.234	0.459	3.004	14.731
1969	3.289	0.128	2.721	0.029	0.342	0.491	7.000
1970	3.785	0.131	3.835	0.019	0.420	0.030	8.220
1971	1.082	0.199	3.850	0.041	0.698	0.030	5.900
1972	0.948	0.228	2.994	0.293	1.067	0.678	6.208
1973	4.628	0.745	4.666	1.452	2.148	0.049	13.689
1974	1.570	0.977	9.482	1.222	3.804	0.426	17.482
1975	2.027	0.795	9.799	2.935	3.139	0.220	18.915
1976	11.409	1.316	12.088	3.793	6.708	0.686	36.000

EXPORTS
PLYWOOD (6312)
NORTH ATLANTIC

VALUE--MILLION CURRENT DOLLARS

YEAR	CANADA	U KINGDM	W GERMNY	OTH EURO	ASIA	OTHERS	TOTAL
1967	0.106	0.016	0.006	0.054	0.045	0.220	0.446
1968	0.122	0.028	0.008	0.132	0.017	0.140	0.448
1969	0.114	0.079	0.002	0.187	0.038	0.164	0.584
1970	0.007	0.165	0.000	0.183	0.033	0.079	0.467
1971	0.041	0.098	0.052	0.210	0.010	0.121	0.533
1972	0.027	0.220	0.000	0.170	0.006	0.105	0.528
1973	0.540	0.206	0.096	0.535	0.144	0.186	1.710
1974	0.638	0.055	0.224	0.353	0.063	0.121	1.453
1975	1.332	0.155	0.011	0.455	0.070	0.158	2.181
1976	0.713	0.053	0.040	0.361	0.614	0.251	2.031

QUANTITY--MIL SQ FT

YEAR	CANADA	U KINGDM	W GERMNY	OTH EURO	ASIA	OTHERS	TOTAL
1967	1.127	0.040	0.011	0.146	0.179	1.094	2.597
1968	1.538	0.085	0.020	0.504	0.042	0.692	2.880
1969	1.133	0.109	0.003	0.351	0.089	0.374	2.058
1970	0.026	0.330	0.000	0.223	0.089	0.362	1.029
1971	0.185	0.134	0.063	0.136	0.029	0.508	1.055
1972	0.137	0.322	0.000	0.575	0.043	0.372	1.449
1973	3.559	0.615	0.239	1.201	0.239	0.550	6.403
1974	3.589	0.235	0.378	0.779	0.114	0.496	5.592
1975	9.505	0.130	0.015	0.871	0.069	0.608	11.198
1976	4.494	0.092	0.103	1.043	1.736	0.919	8.386

EXPORTS
RECONSTITUTED WOOD (6314)
NORTH ATLANTIC

VALUE--MILLION CURRENT DOLLARS

YEAR	CANADA	SWEDEN	W GERMNY	OTH EURO	ASIA	OTHERS	TOTAL
1967	0.005	0.000	0.043	0.000	0.002	0.015	0.065
1968	0.020	0.006	0.024	0.001	0.000	0.000	0.052
1969	0.000	0.004	0.080	0.000	0.000	0.008	0.091
1970	0.000	0.000	0.017	0.000	0.000	0.004	0.021
1971	0.000	0.000	0.080	0.028	0.040	0.002	0.151
1972	0.000	0.000	0.000	0.063	0.078	0.000	0.141
1973	0.283	0.019	0.004	0.007	0.014	0.005	0.332
1974	0.297	0.053	0.042	0.119	0.024	0.015	0.550
1975	0.341	0.043	0.020	0.066	0.018	0.023	0.510
1976	0.054	0.000	0.070	0.082	0.015	0.039	0.260

QUANTITY--MIL SQ FT

YEAR	CANADA	SWEDEN	W GERMNY	OTH EURO	ASIA	OTHERS	TOTAL
1967	0.004	0.000	0.027	0.000	0.002	0.023	0.056
1968	0.094	0.003	0.011	0.001	0.000	0.000	0.110
1969	0.000	0.002	0.022	0.000	0.000	0.004	0.028
1970	0.000	0.000	0.008	0.002	0.000	0.012	0.022
1971	0.000	0.000	0.036	0.016	0.036	0.001	0.089
1972	0.000	0.000	0.000	0.048	0.063	0.000	0.111
1973	1.560	0.019	0.002	0.004	0.011	0.002	1.597
1974	1.412	0.020	0.013	0.113	0.060	0.009	1.628
1975	0.652	0.020	0.006	0.046	0.008	0.021	0.752
1976	0.504	0.000	0.018	0.068	0.075	0.092	0.757

EXPORTS
PULPWOOD CHIPS (631.8320)
NORTH ATLANTIC

VALUE--MILLION CURRENT DOLLARS

YEAR	CANADA	BERMUDA	VENEZUEL	EUROPE	AUSTRALA	OTHERS	TOTAL
1967	0.000	0.000	0.000	0.000	0.000	0.000	0.000
1968	0.005	0.000	0.000	0.000	0.000	0.000	0.005
1969	0.000	0.000	0.000	0.000	0.000	0.000	0.000
1970	0.000	0.000	0.000	0.000	0.000	0.000	0.000
1971	0.000	0.000	0.022	0.000	0.000	0.000	0.022
1972	0.000	0.000	0.000	0.000	0.000	0.000	0.000
1973	0.000	0.000	0.000	0.000	0.000	0.000	0.000
1974	0.048	0.000	0.000	0.000	0.000	0.000	0.048
1975	0.171	0.001	0.000	0.000	0.000	0.000	0.172
1976	0.714	0.000	0.000	0.000	0.001	0.000	0.715

QUANTITY--1000 STN

YEAR	CANADA	BERMUDA	VENEZUEL	EUROPE	AUSTRALA	OTHERS	TOTAL
1967	0.000	0.000	0.000	0.000	0.000	0.000	0.000
1968	0.411	0.000	0.000	0.000	0.000	0.000	0.411
1969	0.000	0.000	0.000	0.000	0.000	0.000	0.000
1970	0.000	0.000	0.000	0.000	0.000	0.000	0.000
1971	0.000	0.000	0.343	0.000	0.000	0.000	0.343
1972	0.000	0.000	0.000	0.000	0.000	0.000	0.000
1973	0.000	0.000	0.000	0.000	0.000	0.000	0.000
1974	1.972	0.000	0.000	0.000	0.000	0.000	1.972
1975	7.084	0.032	0.000	0.000	0.000	0.000	7.116
1976	24.267	0.000	0.000	0.000	0.036	0.000	24.303

EXPORTS
PULPWOOD(EXC CHIPS) (2421)
NORTH ATLANTIC

VALUE--MILLION CURRENT DOLLARS

YEAR	CANADA	S AMERIC	W GERMNY	OTH EURO	ASIA	OTHERS	TOTAL
1967	1.162	0.000	0.000	0.000	0.000	0.000	1.162
1968	0.537	0.000	0.000	0.000	0.000	0.000	0.537
1969	0.748	0.000	0.000	0.000	0.000	0.000	0.748
1970	0.348	0.000	0.002	0.000	0.000	0.000	0.350
1971	0.000	0.000	0.000	0.000	0.000	0.000	0.000
1972	0.220	0.000	0.000	0.000	0.000	0.000	0.220
1973	0.197	0.000	0.000	0.000	0.000	0.000	0.197
1974	1.523	0.000	0.000	0.000	0.000	0.000	1.523
1975	0.832	0.000	0.000	0.000	0.000	0.000	0.832
1976	0.087	0.000	0.000	0.000	0.005	0.000	0.092

QUANTITY--1000 CORDS

YEAR	CANADA	S AMERIC	W GERMNY	OTH EURO	ASIA	OTHERS	TOTAL
1967	60.356	0.000	0.000	0.000	0.000	0.000	60.356
1968	26.346	0.000	0.000	0.000	0.000	0.000	26.346
1969	37.960	0.000	0.000	0.000	0.000	0.000	37.960
1970	17.785	0.000	0.034	0.000	0.000	0.000	17.819
1971	0.000	0.000	0.000	0.000	0.000	0.000	0.000
1972	11.260	0.000	0.000	0.000	0.000	0.000	11.260
1973	7.205	0.000	0.000	0.000	0.000	0.000	7.205
1974	44.927	0.000	0.000	0.000	0.000	0.000	44.927
1975	24.181	0.000	0.000	0.000	0.000	0.000	24.181
1976	2.468	0.000	0.000	0.000	0.018	0.000	2.486

EXPORTS
WOOD PULP (2510)
NORTH ATLANTIC

VALUE--MILLION CURRENT DOLLARS

YEAR	U KINGDM	W GERMNY	SPAIN	ITALY	OTH EURO	OTHERS	TOTAL
1967	1.164	2.183	0.221	2.848	5.002	2.338	13.755
1968	2.351	2.348	0.272	1.527	6.629	1.810	14.937
1969	1.887	2.014	0.192	1.328	6.537	2.010	13.970
1970	3.180	2.627	0.408	2.831	10.446	3.343	22.835
1971	1.837	2.422	0.896	1.893	7.654	7.112	21.814
1972	1.411	1.905	1.267	2.093	5.875	10.111	22.662
1973	5.051	1.750	1.520	3.745	5.887	5.809	23.761
1974	12.457	3.743	3.637	5.657	12.003	11.178	48.676
1975	12.712	4.067	2.769	5.631	12.959	7.982	46.119
1976	12.497	3.407	2.756	7.139	12.543	8.702	47.044

QUANTITY--1000 STN

YEAR	U KINGDM	W GERMNY	SPAIN	ITALY	OTH EURO	OTHERS	TOTAL
1967	6.781	9.777	0.996	10.410	26.360	18.380	72.704
1968	16.513	12.212	0.919	5.836	39.954	14.807	90.240
1969	12.060	7.616	0.664	5.369	37.437	13.977	77.123
1970	19.733	12.457	1.543	12.599	61.311	22.584	130.228
1971	9.226	11.369	3.367	6.551	40.495	51.012	122.020
1972	10.306	9.534	4.680	8.824	32.600	71.024	136.967
1973	23.717	5.826	5.839	14.014	22.036	30.923	102.354
1974	22.602	8.451	11.294	14.389	31.804	39.872	127.413
1975	27.483	9.708	6.065	13.002	31.479	23.354	111.090
1976	28.843	8.696	6.387	17.945	34.168	26.311	122.349

EXPORTS
NEWSPRINT (6411)
NORTH ATLANTIC

VALUE--MILLION CURRENT DOLLARS

YEAR	N AMERIC	VENEZUEL	OTH S AM	EUROPE	ASIA	OTHERS	TOTAL
1967	1.163	0.491	0.013	0.712	1.229	0.005	3.613
1968	1.262	0.053	0.064	0.306	1.439	0.004	3.127
1969	0.691	0.161	0.001	0.593	0.956	0.008	2.310
1970	0.454	0.105	0.024	1.196	0.750	0.007	2.535
1971	0.481	0.233	0.104	0.922	1.967	0.026	3.734
1972	0.250	2.109	0.597	2.284	2.982	0.202	8.424
1973	0.656	2.512	1.608	1.841	0.921	0.179	7.718
1974	5.649	1.745	2.024	6.723	7.228	5.791	29.160
1975	3.277	3.289	4.279	5.309	4.288	3.804	24.245
1976	0.887	3.931	2.745	5.499	0.610	1.323	14.995

QUANTITY--1000 STN

YEAR	N AMERIC	VENEZUEL	OTH S AM	EUROPE	ASIA	OTHERS	TOTAL
1967	9.276	3.818	0.063	6.085	11.088	0.023	30.352
1968	9.504	0.299	0.427	2.487	11.482	0.017	24.215
1969	4.862	0.860	0.010	4.837	6.305	0.053	16.927
1970	3.314	0.445	0.158	10.315	5.409	0.051	19.692
1971	3.245	1.272	0.759	6.877	17.741	0.174	30.068
1972	1.529	12.015	4.509	17.175	25.799	1.062	62.088
1973	2.893	12.606	9.314	9.327	4.758	1.198	40.098
1974	20.033	5.952	7.355	22.299	20.246	16.991	92.877
1975	9.133	11.339	12.849	14.990	14.667	10.090	73.069
1976	2.645	11.728	7.649	16.135	2.659	4.125	44.941

EXPORTS
BUILDING BOARD (6416)
NORTH ATLANTIC

VALUE--MILLION CURRENT DOLLARS

YEAR	CANADA	S AMERIC	EUROPE	ASIA	AFRICA	OTHERS	TOTAL
1967	0.204	0.084	0.511	0.179	0.032	0.253	1.263
1968	0.038	0.071	0.582	0.224	0.021	0.242	1.177
1969	0.083	0.071	1.356	0.170	0.090	0.189	1.960
1970	0.081	0.065	1.654	0.128	0.097	0.144	2.169
1971	0.092	0.016	1.179	0.132	0.131	0.179	1.730
1972	0.247	0.015	0.849	0.149	0.025	0.169	1.455
1973	0.763	0.044	1.219	0.237	0.039	0.234	2.535
1974	0.739	0.144	1.551	0.655	0.029	0.386	3.503
1975	0.486	0.120	1.561	0.600	0.105	0.444	3.316
1976	0.624	0.320	1.464	1.496	0.086	0.364	4.354

QUANTITY--1000 STN

YEAR	CANADA	S AMERIC	EUROPE	ASIA	AFRICA	OTHERS	TOTAL
1967	0.866	0.253	1.479	0.661	0.080	0.659	3.998
1968	0.276	0.190	2.022	0.631	0.077	0.625	3.820
1969	0.296	0.170	4.429	0.562	0.225	0.538	6.221
1970	0.534	0.201	4.893	0.399	0.244	0.421	6.691
1971	0.504	0.037	4.627	0.391	0.405	0.549	6.513
1972	0.927	0.042	2.861	0.339	0.093	0.474	4.737
1973	2.507	0.114	3.436	0.446	0.100	0.473	7.077
1974	2.677	0.209	4.964	1.156	0.025	0.614	9.646
1975	1.575	0.252	3.612	0.972	0.148	0.540	7.099
1976	2.192	0.196	4.088	2.238	0.113	0.560	9.387

EXPORTS
PAPER & PAPERBOARD (6410)
NORTH ATLANTIC

VALUE--MILLION CURRENT DOLLARS

YEAR	N AMERIC	S AMERIC	EUROPE	ASIA	AFRICA	OTHERS	TOTAL
1967	19.924	11.662	47.282	10.174	6.785	9.428	105.255
1968	21.828	15.604	59.803	9.272	7.149	11.444	125.100
1969	21.891	15.034	60.272	10.314	8.583	10.332	126.426
1970	21.009	17.063	69.398	11.717	9.031	13.477	141.694
1971	21.050	15.977	69.134	10.871	8.726	15.256	141.014
1972	24.781	17.283	67.856	12.614	8.373	14.138	145.045
1973	33.246	19.774	90.534	21.495	12.953	22.150	200.152
1974	45.802	41.549	122.258	48.312	27.024	34.545	319.490
1975	53.739	31.867	120.838	23.057	15.877	23.927	269.305
1976	66.010	41.527	122.698	21.686	21.082	33.952	306.955

QUANTITY--1000 STN

YEAR	N AMERIC	S AMERIC	EUROPE	ASIA	AFRICA	OTHERS	TOTAL
1967	91.338	23.121	117.333	21.231	16.376	21.506	290.906
1968	88.707	28.482	130.730	21.596	17.769	27.600	314.884
1969	75.363	26.559	133.869	22.352	21.961	24.178	304.280
1970	85.534	27.612	138.932	23.343	20.067	27.435	322.922
1971	79.410	25.682	134.391	19.756	20.527	33.312	313.078
1972	93.265	27.746	124.388	21.040	18.267	28.739	313.444
1973	77.143	31.387	136.620	39.940	23.324	42.701	351.114
1974	96.733	57.452	175.008	82.105	41.768	51.164	504.230
1975	114.738	40.210	139.070	33.848	23.871	34.120	385.856
1976	145.624	46.353	122.388	27.813	27.186	43.469	412.834

IMPORTS
SOFTWOOD LOGS (2422)
NORTH ATLANTIC

VALUE--MILLION CURRENT DOLLARS

YEAR	CANADA	MEXICO	GUATMALA	BRAZIL	U KINGDM	OTHERS	TOTAL
1967	0.025	0.000	0.000	0.000	0.000	0.000	0.025
1968	0.003	0.000	0.000	0.000	0.000	0.000	0.003
1969	0.004	0.000	0.004	0.000	0.000	0.000	0.008
1970	0.003	0.004	0.000	0.005	0.000	0.000	0.013
1971	0.019	0.000	0.000	0.000	0.000	0.000	0.019
1972	0.003	0.000	0.000	0.000	0.001	0.000	0.005
1973	0.203	0.000	0.000	0.000	0.000	0.000	0.203
1974	0.626	0.000	0.000	0.000	0.000	0.000	0.626
1975	0.915	0.000	0.000	0.000	0.000	0.000	0.915
1976	1.896	0.000	0.000	0.000	0.000	0.000	1.896

QUANTITY--MILLION BF

YEAR	CANADA	MEXICO	GUATMALA	BRAZIL	U KINGDM	OTHERS	TOTAL
1967	0.697	0.000	0.000	0.000	0.000	0.000	0.697
1968	0.032	0.000	0.000	0.000	0.000	0.000	0.032
1969	0.041	0.000	0.017	0.000	0.000	0.000	0.058
1970	0.039	0.005	0.000	0.019	0.000	0.000	0.063
1971	0.367	0.000	0.000	0.000	0.000	0.000	0.367
1972	0.076	0.000	0.000	0.000	0.003	0.000	0.079
1973	3.659	0.000	0.000	0.000	0.000	0.000	3.658
1974	11.306	0.000	0.000	0.000	0.000	0.000	11.306
1975	12.322	0.000	0.000	0.000	0.000	0.000	12.322
1976	22.603	0.000	0.000	0.000	0.000	0.000	22.603

IMPORTS
HARDWOOD LOGS (2423)
NORTH ATLANTIC

VALUE--MILLION CURRENT DOLLARS

YEAR	CANADA	BRAZIL	INDIA	IVRY CST	GHANA	OTHERS	TOTAL
1967	0.154	0.494	0.005	0.435	0.248	0.597	1.933
1968	0.143	0.382	0.000	0.391	0.302	0.465	1.684
1969	0.148	0.027	0.088	0.419	0.478	0.362	1.523
1970	0.381	0.008	0.220	0.611	0.310	0.301	1.831
1971	0.619	0.023	0.054	0.200	0.198	0.207	1.301
1972	0.539	0.002	0.126	0.197	0.294	0.177	1.334
1973	0.707	0.001	0.108	0.411	0.251	0.227	1.705
1974	1.163	0.000	0.257	0.397	0.472	0.166	2.455
1975	0.883	0.000	0.104	0.307	0.127	0.215	1.636
1976	1.162	0.000	0.099	0.336	0.185	0.402	2.184

QUANTITY--MILLION BF

YEAR	CANADA	BRAZIL	INDIA	IVRY CST	GHANA	OTHERS	TOTAL
1967	1.245	1.060	0.008	2.214	1.046	3.404	8.977
1968	1.273	0.344	0.000	1.986	1.143	1.614	6.361
1969	1.490	0.078	0.051	1.963	1.631	1.375	6.589
1970	3.759	0.010	0.097	3.253	1.125	0.948	9.192
1971	4.970	0.036	0.035	1.056	0.725	0.707	7.529
1972	4.025	0.006	0.047	0.783	1.022	0.428	6.311
1973	6.420	0.001	0.038	1.023	0.543	0.443	8.468
1974	9.894	0.000	0.095	0.759	0.634	1.137	12.520
1975	6.736	0.000	0.031	0.558	0.281	0.835	8.441
1976	8.237	0.000	0.051	0.507	0.273	0.467	9.535

IMPORTS
SOFTWOOD LUMBER (2432)
NORTH ATLANTIC

VALUE--MILLION CURRENT DOLLARS

YEAR	CANADA	ECUADOR	OTH S AM	U KINGDM	OTH EURO	OTHERS	TOTAL
1967	90.024	0.040	0.239	0.000	0.019	0.179	90.501
1968	135.270	0.052	0.181	0.001	0.017	0.127	135.648
1969	142.390	0.092	0.018	0.009	0.026	0.069	142.604
1970	104.039	0.034	0.017	0.000	0.014	0.179	104.285
1971	152.928	0.052	0.014	0.005	0.008	0.063	153.071
1972	206.754	0.061	0.348	0.000	0.006	0.105	207.275
1973	266.823	0.074	0.317	0.382	0.986	0.343	268.925
1974	158.749	0.116	0.084	1.379	0.856	0.435	161.620
1975	113.665	0.256	0.010	0.000	0.035	0.130	114.096
1976	195.049	0.324	0.020	0.000	0.071	0.124	195.588

QUANTITY--MILLION BF

YEAR	CANADA	ECUADOR	OTH S AM	U KINGDM	OTH EURO	OTHERS	TOTAL
1967	5548.006	0.060	2.827	0.000	0.061	2.145	5553.099
1968	1741.439	0.112	2.157	0.007	0.043	1.537	1745.295
1969	1615.771	0.145	0.142	0.043	0.045	0.734	1616.880
1970	1493.565	0.069	0.066	0.001	0.026	2.245	1495.971
1971	1825.749	0.099	0.090	0.076	0.021	0.371	1826.406
1972	1969.025	0.101	3.681	0.002	0.024	0.642	1973.475
1973	1818.692	0.099	2.159	0.499	7.888	1.529	1830.867
1974	1160.521	0.098	0.442	1.485	1.436	1.334	1165.316
1975	892.302	0.084	0.013	0.000	0.472	0.202	893.073
1976	1260.542	0.073	0.063	0.000	0.214	0.226	1261.118

IMPORTS
HARDWOOD LUMBER (2433)
NORTH ATLANTIC

VALUE--MILLION CURRENT DOLLARS

YEAR	ECUADOR	BRAZIL	THAILAND	OTH ASIA	GHANA	OTHERS	TOTAL
1967	1.103	0.705	1.010	2.438	1.483	12.821	19.561
1968	1.408	0.520	0.587	3.411	0.896	12.433	19.254
1969	1.406	0.612	0.957	3.678	1.092	13.196	20.942
1970	1.020	0.910	0.932	3.398	0.996	13.162	20.418
1971	1.286	1.164	0.477	2.627	0.961	12.815	19.332
1972	1.800	0.982	0.838	2.777	1.381	14.812	22.590
1973	1.946	1.621	2.509	5.180	1.276	17.120	29.652
1974	3.043	1.760	2.752	6.806	1.560	16.694	32.615
1975	3.871	1.606	1.317	4.016	0.811	7.319	18.939
1976	1.169	1.436	2.883	3.241	0.921	10.925	20.575

QUANTITY--MILLION BF

YEAR	ECUADOR	BRAZIL	THAILAND	OTH ASIA	GHANA	OTHERS	TOTAL
1967	1.491	4.914	1.371	11.056	8.364	69.221	96.417
1968	1.763	2.106	0.780	16.497	5.214	64.975	91.335
1969	1.657	2.922	1.278	16.838	5.948	71.897	100.539
1970	1.239	5.466	1.419	13.316	5.267	69.407	96.114
1971	1.484	8.500	0.913	8.353	5.071	62.784	87.106
1972	2.074	5.439	1.490	9.992	6.994	59.912	85.902
1973	2.239	8.009	3.299	13.817	4.176	55.054	86.594
1974	3.090	5.741	2.965	12.937	3.345	47.622	75.701
1975	2.771	5.212	1.156	8.847	2.114	20.738	40.838
1976	0.632	4.152	2.016	6.756	1.874	30.118	45.548

IMPORTS
WOOD VENEER (6311)
NORTH ATLANTIC

VALUE--MILLION CURRENT DOLLARS

YEAR	CANADA	FRANCE	W GERMNY	OTH EURO	PHIL REP	OTHERS	TOTAL
1967	1.231	0.242	0.619	0.216	0.985	0.604	3.899
1968	1.951	0.296	0.826	0.543	1.674	0.952	6.242
1969	1.666	0.322	0.607	0.352	2.361	1.096	6.405
1970	1.412	0.413	0.739	0.546	0.812	1.114	5.035
1971	2.337	0.325	0.452	0.485	1.834	0.917	6.350
1972	2.749	0.395	0.521	0.883	2.011	0.873	7.432
1973	3.571	0.593	1.142	1.273	3.190	0.978	10.747
1974	3.081	1.119	0.991	1.406	3.708	0.923	11.227
1975	1.711	1.141	0.986	1.184	1.165	0.897	7.084
1976	3.536	1.340	0.636	1.631	2.483	1.165	10.791

QUANTITY--MIL SQ FT

YEAR	CANADA	FRANCE	W GERMNY	OTH EURO	PHIL REP	OTHERS	TOTAL
1967	53.294	0.643	1.355	0.339	12.359	3.850	71.840
1968	85.575	1.178	1.076	0.510	19.306	8.176	115.821
1969	71.487	0.660	1.101	0.327	24.729	7.091	105.394
1970	74.641	0.985	1.157	0.605	10.210	3.236	90.834
1971	114.623	1.245	0.779	0.524	23.495	6.915	147.582
1972	118.939	0.954	0.734	1.143	23.464	5.881	151.114
1973	116.142	0.987	0.667	1.486	25.661	3.583	148.526
1974	89.459	1.623	0.720	1.939	20.959	2.332	117.033
1975	63.821	1.550	0.913	0.886	11.739	2.297	81.207
1976	119.479	1.630	0.662	1.152	18.924	3.969	145.815

IMPORTS
PLYWOOD (6312)
NORTH ATLANTIC

VALUE--MILLION CURRENT DOLLARS

YEAR	FINLAND	PHIL REP	KOR REP	TAIWAN	JAPAN	OTHERS	TOTAL
1967	7.727	4.665	6.938	5.668	10.226	1.633	36.857
1968	9.869	5.703	17.210	6.906	13.527	2.748	55.962
1969	11.723	5.714	20.024	7.002	13.210	2.399	60.073
1970	8.982	5.659	23.876	4.943	9.774	1.320	54.555
1971	8.398	5.040	28.255	10.672	8.510	2.155	63.031
1972	10.914	4.760	38.649	17.430	9.007	4.428	85.188
1973	13.445	9.683	55.915	20.311	6.662	8.462	114.478
1974	7.750	5.226	35.652	13.539	4.992	8.396	75.555
1975	3.639	2.170	49.906	14.482	4.247	6.021	80.464
1976	4.333	8.138	65.074	19.319	5.577	6.235	108.676

QUANTITY--MIL SQ FT

YEAR	FINLAND	PHIL REP	KOR REP	TAIWAN	JAPAN	OTHERS	TOTAL
1967	69.164	44.512	71.713	65.326	64.711	13.228	328.655
1968	59.083	45.267	137.152	58.262	66.342	24.818	390.923
1969	87.650	43.596	165.369	59.737	70.194	19.843	446.389
1970	65.983	54.436	204.268	40.943	54.223	9.824	429.677
1971	54.354	50.603	224.063	88.695	45.214	20.015	482.943
1972	54.416	46.495	317.655	144.360	40.266	40.886	644.080
1973	46.992	70.878	328.810	119.806	22.447	63.331	652.264
1974	22.370	29.182	191.544	62.248	13.481	49.718	368.543
1975	11.496	15.513	305.609	87.874	13.915	29.504	463.912
1976	13.836	49.009	336.658	98.156	14.656	44.253	556.568

IMPORTS
RECONSTITUTED WOOD (6314)
NORTH ATLANTIC

VALUE--MILLION CURRENT DOLLARS

YEAR	U KINGDM	FRANCE	W GERMNY	OTH EURO	ASIA	OTHERS	TOTAL
1967	0.097	0.000	0.099	0.047	0.001	0.041	0.257
1968	0.070	0.000	0.082	0.079	0.000	0.004	0.235
1969	0.071	0.000	0.073	0.317	0.017	0.020	0.498
1970	0.064	0.002	0.135	0.020	0.245	0.017	0.482
1971	0.012	0.000	0.199	0.006	0.301	0.062	0.580
1972	0.187	0.000	0.133	0.022	0.014	0.112	0.469
1973	0.184	0.000	0.264	0.001	0.028	0.432	0.909
1974	0.213	0.203	0.260	0.033	0.017	0.635	1.361
1975	0.199	0.592	0.255	0.081	0.027	0.343	1.498
1976	0.135	0.373	0.192	0.301	0.051	0.458	1.510

QUANTITY--MIL SQ FT

YEAR	U KINGDM	FRANCE	W GERMNY	OTH EURO	ASIA	OTHERS	TOTAL
1967	0.063	0.000	0.142	0.037	0.000	0.315	0.558
1968	0.051	0.000	0.128	0.057	0.000	0.055	0.291
1969	0.043	0.000	0.102	2.500	0.105	0.167	2.918
1970	0.084	0.001	0.199	0.020	0.146	0.096	0.546
1971	0.030	0.000	0.320	0.000	0.204	0.314	0.868
1972	0.095	0.000	0.153	0.106	0.016	0.581	0.950
1973	0.083	0.000	0.268	0.003	0.062	3.245	3.661
1974	0.637	0.055	0.203	0.037	0.046	4.906	5.884
1975	0.058	0.172	0.184	0.031	0.024	2.278	2.747
1976	0.038	0.085	0.121	0.084	0.043	2.043	2.413

IMPORTS
PULPWOOD CHIPS (631.8320)
NORTH ATLANTIC

VALUE--MILLION CURRENT DOLLARS

YEAR	CANADA	ARGENTIN	FRANCE	W GERMNY	BR PAC I	OTHERS	TOTAL
1967	0.149	0.000	0.000	0.000	0.000	0.000	0.149
1968	0.202	0.000	0.000	0.000	0.000	0.000	0.202
1969	0.301	0.000	0.000	0.000	0.000	0.000	0.301
1970	0.027	0.000	0.000	0.000	0.000	0.000	0.027
1971	0.128	0.000	0.000	0.000	0.000	0.000	0.128
1972	0.120	0.000	0.000	0.000	0.000	0.000	0.120
1973	0.566	0.000	0.000	0.000	0.001	0.000	0.567
1974	0.879	0.000	0.000	0.000	0.000	0.000	0.880
1975	1.468	0.001	0.000	0.000	0.000	0.000	1.469
1976	0.996	0.000	0.000	0.001	0.000	0.000	0.996

QUANTITY--1000 STN

YEAR	CANADA	ARGENTIN	FRANCE	W GERMNY	BR PAC I	OTHERS	TOTAL
1967	11.359	0.000	0.000	0.000	0.000	0.000	11.359
1968	15.062	0.000	0.000	0.000	0.000	0.000	15.062
1969	16.412	0.000	0.000	0.000	0.000	0.000	16.412
1970	4.500	0.000	0.000	0.000	0.000	0.000	4.500
1971	8.570	0.000	0.000	0.000	0.000	0.000	8.570
1972	7.411	0.000	0.000	0.000	0.000	0.000	7.411
1973	39.830	0.000	0.000	0.000	0.055	0.000	39.885
1974	57.183	0.000	0.003	0.000	0.000	0.000	57.186
1975	66.971	0.004	0.000	0.000	0.000	0.000	66.975
1976	52.560	0.000	0.000	0.013	0.000	0.000	52.573

IMPORTS
PULPWOOD (EXC CHIPS)(2421)
NORTH ATLANTIC

VALUE--MILLION CURRENT DOLLARS

YEAR	CANADA	NICARAGA	S AMERIC	EUROPE	ASIA	OTHERS	TOTAL
1967	2.477	0.000	0.000	0.000	0.000	0.000	2.478
1968	1.896	0.000	0.000	0.000	0.000	0.000	1.896
1969	1.849	0.000	0.000	0.000	0.000	0.000	1.849
1970	1.976	0.000	0.000	0.000	0.000	0.000	1.976
1971	0.753	0.000	0.000	0.000	0.000	0.000	0.753
1972	0.639	0.000	0.000	0.000	0.000	0.000	0.639
1973	0.766	0.000	0.000	0.000	0.000	0.000	0.766
1974	3.677	0.000	0.000	0.000	0.000	0.000	3.677
1975	6.090	0.000	0.000	0.000	0.000	0.000	6.090
1976	7.811	0.000	0.000	0.000	0.000	0.000	7.811

QUANTITY--1000 CORDS

YEAR	CANADA	NICARAGA	S AMERIC	EUROPE	ASIA	OTHERS	TOTAL
1967	141.108	0.002	0.000	0.000	0.000	0.000	141.110
1968	109.044	0.000	0.000	0.000	0.000	0.000	109.044
1969	101.257	0.000	0.000	0.000	0.000	0.000	101.257
1970	107.679	0.000	0.000	0.000	0.000	0.000	107.679
1971	39.884	0.000	0.000	0.000	0.000	0.000	39.884
1972	35.363	0.000	0.000	0.000	0.000	0.000	35.363
1973	34.918	0.000	0.000	0.000	0.000	0.000	34.918
1974	132.884	0.000	0.000	0.000	0.000	0.000	132.884
1975	181.320	0.000	0.000	0.000	0.000	0.000	181.320
1976	187.552	0.000	0.000	0.000	0.000	0.000	187.552

IMPORTS
WOOD PULP (2510)
NORTH ATLANTIC

VALUE--MILLION CURRENT DOLLARS

YEAR	CANADA	SWEDEN	FINLAND	OTH EURO	TUNISIA	OTHERS	TOTAL
1967	92.028	14.694	5.219	0.611	0.389	0.047	112.987
1968	94.461	9.578	6.074	0.862	0.510	0.000	111.486
1969	106.595	5.588	4.151	0.409	0.190	0.002	116.934
1970	103.570	3.378	3.785	0.793	0.140	0.039	111.705
1971	109.035	2.316	2.849	0.872	0.370	0.000	115.442
1972	113.628	3.113	1.784	0.513	0.535	0.073	119.645
1973	159.466	10.765	1.265	0.798	1.040	0.073	173.406
1974	248.410	11.620	0.006	1.165	0.820	0.000	262.023
1975	217.034	5.280	1.025	0.285	0.515	0.000	224.140
1976	251.919	13.340	3.371	0.464	0.621	0.028	269.743

QUANTITY--1000 STN

YEAR	CANADA	SWEDEN	FINLAND	OTH EURO	TUNISIA	OTHERS	TOTAL
1967	718.539	129.961	59.235	2.077	1.746	0.621	911.178
1968	743.938	84.969	71.794	3.405	2.190	0.000	906.297
1969	840.822	45.662	45.235	1.739	0.877	0.017	934.351
1970	754.831	25.208	40.904	3.532	0.577	0.079	825.132
1971	766.895	16.154	28.250	2.961	1.651	0.000	815.911
1972	840.976	24.079	19.431	1.731	1.735	0.226	888.178
1973	973.326	54.809	10.715	4.577	3.592	0.244	1047.263
1974	988.470	38.876	0.013	3.132	19.752	0.001	1050.244
1975	690.775	14.965	2.493	0.686	0.987	0.000	709.906
1976	808.530	42.459	10.728	1.816	1.565	0.139	865.236

IMPORTS
NEWSPRINT (6411)
NORTH ATLANTIC

VALUE--MILLION CURRENT DOLLARS

YEAR	CANADA	S AMERIC	SWEDEN	FINLAND	OTH EURO	OTHERS	TOTAL
1967	160.283	0.000	0.002	19.174	0.014	0.000	179.473
1968	169.931	0.000	0.060	20.580	0.006	0.000	190.577
1969	171.225	0.000	0.120	19.455	0.006	0.000	190.805
1970	176.646	0.000	0.805	16.856	0.000	0.000	194.307
1971	170.521	0.000	1.635	11.882	0.000	0.000	184.038
1972	194.765	0.000	0.000	12.101	0.000	0.000	206.866
1973	202.321	0.000	0.000	16.449	0.000	0.000	218.769
1974	257.464	0.042	0.000	6.249	0.000	0.000	263.755
1975	231.085	0.000	0.000	1.290	0.000	0.000	232.376
1976	294.134	0.000	8.274	0.867	5.225	0.000	308.500

QUANTITY--1000 STN

YEAR	CANADA	S AMERIC	SWEDEN	FINLAND	OTH EURO	OTHERS	TOTAL
1967	1193.731	0.000	0.020	168.497	0.148	0.000	1362.396
1968	1247.148	0.000	0.523	176.616	0.058	0.000	1424.344
1969	1195.926	0.000	0.837	162.009	0.106	0.000	1358.879
1970	1205.785	0.000	6.906	138.612	0.000	0.000	1351.303
1971	1167.306	0.000	13.599	110.292	0.000	0.000	1291.197
1972	1258.864	0.000	0.000	94.312	0.000	0.000	1353.176
1973	1216.125	0.000	0.000	123.284	0.000	0.000	1339.409
1974	1259.706	0.017	0.000	40.774	0.000	0.000	1300.496
1975	948.319	0.000	0.000	5.449	0.000	0.000	953.768
1976	1106.666	0.000	28.268	3.861	17.839	0.000	1156.635

IMPORTS
BUILDING BOARD (6416)
NORTH ATLANTIC

VALUE--MILLION CURRENT DOLLARS

YEAR	CANADA	BRAZIL	SCANDINA	OTH EURO	ASIA	OTHERS	TOTAL
1967	0.883	0.508	5.232	0.489	0.085	0.415	7.612
1968	1.717	0.641	4.374	0.598	0.099	1.450	8.879
1969	1.449	0.535	3.413	0.470	0.073	0.874	6.812
1970	2.069	0.949	1.828	0.387	0.023	0.118	5.373
1971	2.003	0.731	2.382	0.490	0.062	0.991	6.659
1972	22.670	1.024	3.777	1.054	0.034	1.778	30.336
1973	3.547	1.366	3.131	1.360	0.075	1.597	11.076
1974	2.424	2.241	2.635	1.346	0.263	0.739	9.648
1975	1.136	1.816	0.200	0.386	0.039	0.103	3.680
1976	1.130	1.998	0.711	1.446	0.056	0.276	5.617

QUANTITY--1000 STN

YEAR	CANADA	BRAZIL	SCANDINA	OTH EURO	ASIA	OTHERS	TOTAL
1967	8.771	5.870	69.840	8.777	0.836	5.778	99.872
1968	17.943	7.259	58.290	10.437	0.885	19.694	114.507
1969	13.996	6.707	46.267	7.800	0.352	11.899	87.022
1970	21.392	11.692	23.378	6.135	0.117	1.852	64.566
1971	20.139	9.380	31.034	7.323	0.227	15.939	84.042
1972	20.161	11.422	45.992	16.011	0.210	26.270	120.066
1973	27.684	13.302	25.984	15.956	0.691	18.520	102.138
1974	14.815	16.279	18.009	9.753	2.237	7.789	68.882
1975	8.467	13.436	1.326	4.724	0.331	1.618	29.902
1976	7.042	13.140	5.665	15.930	0.367	2.610	44.754

IMPORTS
PAPER & PAPERBOARD (6410)
NORTH ATLANTIC

VALUE--MILLION CURRENT DOLLARS

YEAR	CANADA	S AMERIC	SCANDINA	OTH EURO	ASIA	OTHERS	TOTAL
1967	1.192	0.000	11.040	7.847	1.468	0.011	21.558
1968	2.012	0.131	10.498	7.541	1.783	0.016	21.982
1969	3.460	0.684	9.612	7.122	1.523	0.007	22.408
1970	8.278	0.000	8.921	8.164	2.009	0.012	27.384
1971	9.025	0.000	7.805	7.718	1.455	0.012	26.015
1972	8.914	0.003	10.193	11.815	1.784	0.012	32.721
1973	16.060	0.319	19.900	15.316	2.154	0.013	53.761
1974	18.614	0.192	33.074	19.808	4.527	0.036	76.250
1975	11.048	0.003	18.610	18.519	2.607	0.075	50.863
1976	20.837	0.261	19.670	26.940	3.635	0.057	71.400

QUANTITY--1000 STN

YEAR	CANADA	S AMERIC	SCANDINA	OTH EURO	ASIA	OTHERS	TOTAL
1967	6.251	0.000	65.705	9.639	3.996	0.003	85.594
1968	10.940	1.089	62.073	10.757	1.005	0.016	85.880
1969	16.624	5.674	55.959	8.276	0.884	0.003	87.420
1970	41.666	0.005	47.164	9.374	1.234	0.006	99.448
1971	52.594	0.000	39.134	7.959	0.852	0.026	100.565
1972	54.654	0.003	53.565	10.379	0.813	0.007	119.421
1973	95.543	1.228	90.642	14.033	0.729	0.052	202.226
1974	63.525	0.574	91.434	12.531	2.413	0.052	170.529
1975	37.310	0.013	50.726	9.445	0.854	0.069	98.417
1976	63.955	0.775	58.192	13.645	1.043	0.126	137.736

Appendix H

THE SOUTH PACIFIC REGION

Imports

Regional Overview

Compared with other U.S. regions, the South Pacific is not a large forest products exporter. The region's 1976 exports worth $210.2 million represented 5 percent of the total U.S. exports of forest products.

The region increased its exports substantially during the 1967-76 period. The nominal value of the region's forest product exports grew at an average rate of 18 percent per year.

The South Pacific's export commodity mix is very similar to that of the United States as a whole, in that the same four commodities--softwood logs, softwood lumber, wood pulp, and paper and paperboard--together make up a large percentage of each area's total forest product exports. In the South Pacific region those four commodities constituted 88 percent (by value) of the region's total forest product exports.

Wood pulp is the most important export. The 1976 shipments were worth $85.3 million, or 41 percent of the region's forest product exports. Softwood logs ($35.8 million), softwood lumber ($32.1 million), and paper and paperboard ($31.4 million) each constituted about 15 percent of the region's exports.

The region's export commodity structure changed somewhat from 1967 to 1976. Although the same four products again accounted for a large share (87 percent by value) of the region's forest product exports, the share accounted for by each of the commodities was quite different in 1967 than in 1976. In 1967 softwood lumber was the most important export, with 33 percent of the region's forest product exports, followed by wood pulp (23 percent), paper and paperboard (21 percent), and softwood logs (10 percent).

Wood Pulp

The South Pacific was the fastest-growing major U.S. exporter of wood pulp during 1967-76. The nominal value of the region's exports increased by an average of 26 percent per year and the volume grew at 12 percent average rate. The rapid growth enabled the region to provide 10 percent of the value of the U.S. total 1976 pulp exports, which was more than double the region's 1967 share.

Europe and Asia each accounted for about 45 percent of the $85.3 million worth of pulp exported from the South Pacific region in 1976. Japan was the most important single-country market, importing $31.6 million of pulp in 1976, or 37 percent of the South Pacific's total pulp exports.

France, Italy, and the Netherlands were the largest European markets, and together they were responsible for 69 percent of the Europe-bound South Pacific pulp exports and 31 percent of the region's total pulp exports in 1976. Exports to France were valued at $10.5 million; to Italy, $8.4 million; and to the Netherlands, $7.3 million.

Pulp exports to Europe grew tremendously from 1967 to 1976. The average export value growth was 37 percent per year, and volume growth took place at an annual rate of 21 percent.

Softwood Logs

The Pacific Northwest is by far the largest softwood log exporter in the United States, but all other regions combined exported more than $75 million worth of softwood logs in 1976. Of that residual value, the South Pacific region exported $35.8 million, or 47 percent.

The South Pacific's softwood log exports grew substantially during the 1967-76 period, averaging an increase of 26 percent per year in nominal value, and 12 percent annually in volume.

Japan and Korea were the principal foreign markets for South Pacific logs. Japan imported 129.6 million board-feet (worth $31.6 million), which represented 88 percent of the region's total softwood log exports in 1976.

Korea which did not import logs from the region in 1967, imported 19.9 million board-feet (valued at $4.1 million) in 1976. From 1968 to 1976, the nominal value of softwood log exports to Korea had increased at an average annual rate of 20 percent.

Softwood Lumber

From 1967-76, softwood lumber exports from the South Pacific declined by nearly 25 percent in volume, and their nominal value increased by 108 percent, equivalent to an annual gain of about 8 percent.

The decreasing absolute amount of exports led to a decline in the region's share of U.S. softwood lumber exports. The South Pacific's 1976 shipments of 107.5 million board-feet (worth $32.1 million) represented 7 percent of all U.S. softwood lumber exports. In 1967 the region's share of U.S. exports was 15 percent.

Mexico was the most important foreign market for the region's softwood lumber during 1967-76. In 1976 exports to Mexico of 59.9 million board-feet (worth $10.3 million) represented 32 percent of the South Pacific's total softwood lumber exports.

The Netherlands and West Germany were the leading European importers during 1967-76. In 1976 the Netherlands imported 13 percent, and West Germany 18 percent, of the value of the region's total softwood lumber exports. Other European countries, as a group, imported 11 percent of the region's 1976 exports.

Asian countries, in 1976, imported 13 percent of the region's softwood lumber, and Japan accounted for almost half the Asian imports.

Oceanic countries, particularly New Zealand, provided another important market for South Pacific lumber. In 1976 exports to these countries represented 11 percent of the region's total softwood lumber exports.

Paper and Paperboard

The South Pacific region was not a strong paper and paperboard exporter during 1967-76. In an average year, the region shipped out only 2 percent of the value of all U.S. exports of paper and paperboard.

The region exhibited modest growth in its export of paper and paperboard during 1967-76. The 1976 shipments of 55,000 tons (worth $31.4 million) represented increases over 1967 of 26 percent in volume and 218 percent in nominal value. The equivalent annual growth rates are 3 percent in volume and 14 percent in nominal value.

Mexico and Asian countries were the principal paper and paperboard export markets for the South Pacific during 1967-76. Together these countries imported 78 percent (by value) of the region's 1976 exports.

In an average year, Mexico was the destination of 27 percent of the value of the region's paper and paperboard exports. The 1976 exports, worth $6.4 million, represented 20 percent of the region's exports.

Japan, in 1976, imported 11,000 tons (worth $7.3 million), which represented 23 percent of the South Pacific's total paper and paperboard exports. Although the 1976 nominal export value was 194 percent greater than the 1967 value of paper and paperboard exports to Japan, the 1976 volume was 20 percent less than the 1967 volume. One possible explanation for this anomaly lies in the great diversity of commodities which constitute the paper and paperboard group. That is, it is quite possible that Japan imported a very different mix of paper and paperboard products in 1967 than it did in 1976. However, at the four-digit level or aggregation, there is no way to test this hypothesis.

The Philippines were another important Asian market in the 1967-76 period. In 1967 the Philippines imported 8 percent (by value) of the South Pacific's total paper and paperboard exports.

Asian countries other than Japan and the Philippines rapidly expanded their purchases of South Pacific paper and paperboard from 1967 to 1976. The volume of exports shipped to these countries grew at an average annual rate of 6 percent, while the nominal value increased by an average of 20 percent per year. In 1967 exports to these "other" Asian nations represented 16 percent of the region's paper and paperboard exports, but by 1976 that share had grown to 27 percent.

Pulpwood Chips

Pulpwood chips was a rapidly growing export commodity for the South Pacific region in 1967-76. Only $1,300 worth of chips were exported in 1967, but by 1976 the value of chip exports was $12.7 million.

Japan began importing pulpwood chips from the region in 1968 and has been the dominant chip importer ever since. In 1976 less than three-tenths of 1 percent of the value of the South Pacific's pulpwood chip exports was sent to countries other than Japan.

Plywood

Despite substantial growth in its absolute export of plywood, the South Pacific was not as important a plywood exporter relative to other regions, in 1976 as it was in 1967. In 1976 exports of 25.7 million square feet (valued at $4.8 million) represented only 3 percent of all U.S. plywood exports. In 1967 that share had been 13 percent. The decline came about even while the region was growing in export value at an average rate of 15 percent per year and in volume at 8 percent annually.

Mexico was the principal foreign market for the region's exports throughout 1967-76. In 1976 Mexico accounted for 37 percent of the value of the region's exports.

European countries, especially the Netherlands and West Germany, were also strong importers of South Pacific plywood. In 1976 the Netherlands imported 14 percent; West Germany, 12 percent; and other European countries, 9 percent of the value of the region's plywood exports.

Plywood exports to Asia were fairly erratic during 1967-76, and accounted for anywhere from 5 percent to 38 percent (by value) of the region's total plywood exports. In 1976 Asia's share of the region's exports was 10 percent.

Reconstituted Wood

In 1976 the South Pacific region exported 7.1 million square feet of reconstituted wood (valued at $1.3 million), which represented 7 percent of all U.S. reconstituted wood exports.

The region exhibited very substantial export growth, especially in the latter years of the period 1967-76. The average annual growth rate in nominal value was 36 percent, and in volume was 46 percent.

Mexico was the largest and most stable foreign market for the South Pacific region's reconstituted wood. During 1967-71, Mexico accounted for an average of 96 percent of the value of the region's reconstituted wood exports, but in 1972-76 that average share dropped to 74 percent.

In the latter part of the study period, Asia and Oceania imported most of what Mexico did not import. Oceania had two unusual years in 1973 and 1974, when it imported 29 percent and 58 percent, respectively, of the South Pacific's reconstituted wood exports. In the following two years, that share dropped to 3 percent and 1 percent, respectively.

Other Forest Products

Only 3 percent (by value) of the South Pacific region's 1976 total forest product exports was represented by the remaining seven commodities (pulpwood, except chips, hardwood logs, hardwood lumber, wood veneer, newsprint, building board, and other industrial roundwood).

The total value of these exports in 1976, $68 million, was 37 percent greater than the 1967 export value. The average annual growth rate in nominal values for the period was 4 percent.

The region's building board exports represented 5 percent (by value) of U.S. total building board exports in 1976. Mexico, the United Kingdom, and Japan were major importers of the region's building board.

Newsprint exports from the region grew rapidly during 1967-76, but contributed only 3 percent of the value of total U.S. newsprint exports in 1976. The average annual growth rates for the region's newsprint exports were 23 percent in nominal value, and 11 percent in volume. Mexico was the largest importer of newsprint from the South Pacific, and in an average year accounted for 83 percent of the region's newsprint exports.

The region was an important U.S. exporter of hardwood lumber in 1970-72, but by 1976 only 1 percent of U.S. hardwood lumber exports were shipped from there. From 1967 to 1972 Japan was the largest importer of the region's hardwood lumber, but from 1973 to 1976 Mexico imported an average of 71 percent (by value) of the South Pacific's total hardwood lumber exports.

Imports

Regional Overview

The South Pacific showed little growth in forest product imports over the 1967-76 period. The average annual rate of growth in the nominal value of imports was 7 percent.

The 1976 imports were valued at $241 million, which represented 5 percent of U.S. total forest product imports.

More than three-fourths of the value of the region's 1976 imports was comprised of just commodities--newsprint (54 percent) and plywood (23 percent). The remainder of the South Pacific imports was split evenly between solid wood products and pulp and paper products.

In 1967 newsprint and plywood had been even more prominent in the region's forest product import mix. In that year newsprint represented 61 percent and plywood 23 percent of the value of the region's forest product imports.

Newsprint

Newsprint imports into the South Pacific region declined in volume from 1967 to 1976 by 18 percent and increased in nominal value by 65 percent (equivalent to an annual increase of just 6 percent).

The decline in imports did not significantly alter the South Pacific's position vis-à-vis other newsprint importing regions. In 1967 the South Pacific was responsible for 9 percent (by value) of the total U.S. imports of newsprint, and in 1976 that share was 8 percent. The 1976 imports totaled 561,000 tons (valued at $131.1 million).

Canada and Finland were the major sources of newsprint during 1967-76. Canada, in an average year, provided 91 percent (by value) of the region's total newsprint imports. In 1976, however, Canada was the only exporter of any significance; Finland did not export newsprint to the region in that year.

Finland, in 1967-74, supplied about 11 percent of the value of the South Pacific region's imports, but in 1975-76 imports from Finland dropped sharply. The average import volume for those two years was only 16 percent of the previous eight years' average.

Plywood

The South Pacific region's imports of plywood grew very little from 1967 to 1976. As a result, the region's share of the value of all U.S. plywood imports dropped from 21 percent in 1967 to 14 percent in 1976. The 1976 imports of 382.7 million square feet were worth $55.4 million, an increase over the 1967 imports of 3 percent in volume and 87 percent in nominal value.

Asian countries were responsible for nearly all of the region's plywood imports throughout 1967-76. In an average year, four countries--the Philippines, Korea, Taiwan, and Japan--together provided 94 percent of the value of the region's imports.

Korea was the fastest-growing exporter during 1967-76, and the largest source of plywood for the region in 1976. The region's 1976 imports of 175.6 million square feet (valued at $22.2 million) were the result of nine years of growth, averaging 6 percent per year in volume and 14 percent annually in nominal value. In 1976 Korea was the source of 40 percent of the value of the region's total plywood imports.

Japan was the leading exporter of plywood to the South Pacific for much of the 1967-76 period, but the volume of its exports has steadily declined. The 1976 volume of 62.7 million square feet was only 55 percent of the 1967 volume. The nominal value increased by only 19 percent over the period. In 1976 Japan captured 27 percent (by value) of the South Pacific market for imported plywood.

Taiwan was a steady source of plywood throughout the 1967-76 period. Its 1976 export volume to the South Pacific was just 2 percent lower than the 1967 volume. In an average year, 18 percent of the South Pacific's plywood imports originated in Taiwan.

The Philippines declined as an exporter of plywood to the region during the study period. By 1976, volume had declined by 11 percent from its 1967 level, and nominal value had increased by 53 percent (equivalent to an annual increase of 5 percent). The Philippines supplied an average of 12 percent of the value of the South Pacific's imports.

Paper and Paperboard

In 1976 the South Pacific region imported 20,000 tons of paper and paperboard (valued at $14.3 million). Those figures represented a decrease over the 1967 imports of 17 percent in volume and an increase of 204 percent in nominal value. The 1976 imports were 7 percent of the value of all U.S. paper and paperboard imports.

The roles of the region's various trading partners changed quite a bit from 1967 to 1976. Canada was the source of half the value of the region's paper and paperboard imports in 1967, but, by 1976, Canada's share was only 7 percent. During the decade, imports from Canada declined by 60 percent in nominal value, and 80 percent in volume.

Japan, despite decreasing the volume of its paper and paperboard exports to the South Pacific, increased the nominal value by 358 percent, equivalent to an average annual increase of 18 percent. Japan supplied one-fourth of the value of the region's paper and paperboard imports in 1967, and 39 percent in 1976.

Korea displayed the fastest export growth during 1967-76, increasing its exports of paper and paperboard to the South Pacific by 36 percent per year in nominal value and 26 percent annually in volume. The 572 tons, valued at $2.6 million, represented 18 percent of the South Pacific's paper and paperboard imports in 1976, up from only 4 percent in 1967.

Finland also considerably expanded its paper and paperboard exports to the region during the study period. While increasing its exports at the average rates of 30 percent per year in nominal value and 21 percent per year in volume, Finland became the second largest source of paper and paperboard for the South Pacific, providing 19 percent of the value of the region's imports in 1976.

Softwood Lumber

Softwood lumber imports to the South Pacific increased at a faster rate than to any other U.S. region during 1967-76. Volume increased at an average rate of 13 percent, and nominal value at 25 percent. The 1976 imports totaled 82.4 million board-feet (worth $15.1 million). The region imported only 1 percent of U.S. total softwood lumber imports.

Canada was the leading exporter to the region, and supplied an average of 96 percent (by value) of the region's softwood lumber imports during 1967-76.

Wood Pulp

Wood pulp comprised 4 percent of the value of the South Pacific region's total forest product imports in 1976. The 36,000 tons (valued at $8.8 million) represented declines from the 1967 import levels by 31 percent in volume and 4 percent in nominal value. The region was the entrance for only 1 percent of all U.S. wood pulp imports in 1976.

Throughout 1967-76, Canada was practically the sole foreign source of pulp for the region. In only one year, 1967, did Canada account for less than 99 percent of the value of the region's total wood pulp imports. In that year, Portugal supplied 1 percent, and all other European countries together supplied 2 percent of the region's imports.

Hardwood Lumber

The South Pacific region accounted for 9 percent of U.S. 1976 hardwood lumber imports. The 16.3 million board-feet imported in that year (worth $8.5 million) represented a decline of 4 percent from the 1967 import volume. The nominal value rose by 153 percent during the period.

In 1976, 89 percent (by value) of the region's hardwood lumber imports originated in Asian countries, with Thailand, Taiwan, the Philippines, and Malaysia being the largest exporters.

Thailand grew rapidly as a lumber exporter during 1967-76. The 2 million board-feet exported to the South Pacific in 1976 was valued at $2.1 million, which represented increases over 1967 of 621 percent in volume and 1,031 percent in nominal value. Thailand was the source of one-fourth of the value of the region's 1976 hardwood lumber imports.

The Philippines exported 5 percent less (in volume) hardwood lumber in 1976 than in 1967. The 1976 exports to the South Pacific totaled 3.9 million board-feet (worth $.14 million), which was 17 percent of the value of the region's hardwood lumber imports.

Taiwan increased its exports of hardwood lumber to the region during 1967-76 by an average of 5 percent per year in volume, and 17 percent annually in nominal value. Taiwan's share of the region's imports grew from 11 percent to 18 percent (by value) during that period.

The volume of imports from Malaysia in 1976 was very close to the 1967 volume, but in the intervening years, the volume had been as much as three times the 1976 level. For six of the ten years, the value of hardwood lumber imports from Malaysia was higher than that from any other country. From 1969 through 1974, Malaysia was responsible for an average of 23 percent (by value) of the region's hardwood lumber imports, but in 1976 that share was just 14 percent.

Other Forest Products

All other forest products (pulpwood except chips, softwood logs, hardwood logs, wood veneer, reconstituted wood, pulpwood chips, other industrial roundwood, building board) contributed only 3 percent of the total value of the South Pacific's 1976 forest product imports.

Of the $7.9 million of these products imported in 1976, building board constituted $5.1 million, or 64 percent. The region, was, in fact, one of the largest and fastest growing U.S. importers of building board. The 1976 value, which represented 19 percent of all U.S. building board imports, was more than four and one-half times the 1967 import value. In volume, the region's imports increased by 140 percent over the 1967-76 period.

The South Pacific also considerably increased the value and volume of its wood veneer imports during 1967-76. The 1976 imports of 23.4 million square feet (worth $2.7 million) represented average annual increases of 5 percent in volume and 22 percent in nominal value. In that year, the region accounted for 4 percent of the value of all U.S. veneer imports.

EXPORTS
SOFTWOOD LOGS (2422)
SOUTH PACIFIC

VALUE--MILLION CURRENT DOLLARS

YEAR	MEXICO	EUROPE	REP KOR	JAPAN	OTH ASIA	OTHERS	TOTAL
1967	0.001	0.000	0.000	4.463	0.000	0.004	4.468
1968	0.001	0.002	0.810	17.988	0.007	0.231	19.038
1969	0.011	0.002	0.230	18.930	0.000	0.098	19.271
1970	0.030	0.009	1.722	19.524	0.000	0.000	21.285
1971	0.032	0.007	1.166	9.999	0.002	0.000	11.206
1972	0.028	0.000	1.145	9.093	0.000	0.002	10.268
1973	0.024	0.032	2.305	21.529	0.006	0.002	23.897
1974	0.015	0.012	2.807	21.432	0.188	0.002	24.455
1975	0.020	0.004	2.387	22.790	0.000	0.019	25.220
1976	0.045	0.005	4.064	31.613	0.035	0.001	35.765

QUANTITY--MILLION BF

YEAR	MEXICO	EUROPE	REP KOR	JAPAN	OTH ASIA	OTHERS	TOTAL
1967	0.024	0.000	0.000	55.182	0.000	0.007	55.214
1968	0.011	0.020	12.984	263.111	0.018	1.634	277.779
1969	0.140	0.003	4.165	247.921	0.000	0.905	253.134
1970	0.336	0.018	17.098	227.875	0.000	0.000	245.326
1971	0.174	0.019	7.442	129.126	0.003	0.000	136.763
1972	0.281	0.000	8.095	86.640	0.000	0.008	95.023
1973	0.214	0.061	13.733	130.419	0.006	0.008	144.441
1974	0.065	0.015	17.022	88.726	1.030	0.004	106.862
1975	0.137	0.005	12.329	107.423	0.000	0.034	119.927
1976	0.234	0.007	18.931	129.588	0.026	0.002	148.788

EXPORTS
HARDWOOD LOGS (2423)
SOUTH PACIFIC

VALUE--MILLION CURRENT DOLLARS

YEAR	MEXICO	FRANCE	ITALY	TAIWAN	JAPAN	OTHERS	TOTAL
1967	0.000	0.004	0.006	0.000	3.311	0.031	3.352
1968	0.003	0.008	0.000	0.000	3.991	0.072	4.074
1969	0.001	0.006	0.085	0.000	6.757	0.045	6.893
1970	0.001	0.003	0.014	0.000	3.233	0.043	3.294
1971	0.004	0.014	0.004	0.000	1.849	0.042	1.912
1972	0.006	0.017	0.000	0.046	2.072	0.192	2.333
1973	0.002	0.043	0.000	0.015	2.751	0.238	3.049
1974	0.021	0.024	0.000	0.000	1.627	0.212	1.884
1975	0.018	0.040	0.063	0.023	0.650	0.090	0.884
1976	0.082	0.089	0.086	0.052	1.277	0.030	1.617

QUANTITY--MILLION BF

YEAR	MEXICO	FRANCE	ITALY	TAIWAN	JAPAN	OTHERS	TOTAL
1967	0.000	0.003	0.015	0.000	3.574	0.052	3.643
1968	0.017	0.007	0.000	0.000	6.933	0.085	7.042
1969	0.009	0.003	0.104	0.000	10.552	0.050	10.718
1970	0.002	0.001	0.041	0.000	3.543	0.050	3.637
1971	0.015	0.016	0.005	0.000	1.908	0.040	1.984
1972	0.034	0.018	0.000	0.045	1.750	0.153	2.000
1973	0.011	0.068	0.000	0.043	1.813	0.213	2.148
1974	0.053	0.023	0.000	0.000	1.123	0.376	1.576
1975	0.035	0.031	0.050	0.017	4.821	0.055	5.009
1976	0.174	0.084	0.114	0.036	0.830	0.041	1.278

EXPORTS
SOFTWOOD LUMBER (2432)
SOUTH PACIFIC

VALUE--MILLION CURRENT DOLLARS

YEAR	MEXICO	NETHLNDS	W GERMNY	ASIA	OCEANIA	OTHERS	TOTAL
1967	2.558	1.837	2.631	1.152	1.836	5.477	15.489
1968	3.285	3.665	3.176	1.166	1.521	4.520	17.332
1969	3.521	3.056	2.639	1.378	1.779	4.481	16.853
1970	4.207	2.453	2.520	1.229	1.226	4.318	15.954
1971	4.890	1.778	2.649	1.149	0.633	2.351	13.449
1972	6.343	2.229	5.178	2.313	1.807	2.824	20.696
1973	8.461	4.932	8.727	1.524	6.062	6.298	36.003
1974	8.851	1.839	3.143	1.239	4.053	4.075	23.200
1975	8.844	1.628	3.678	1.844	2.148	3.799	21.941
1976	10.325	4.193	5.929	4.243	3.567	3.891	32.149

QUANTITY--MILLION BF

YEAR	MEXICO	NETHLNDS	W GERMNY	ASIA	OCEANIA	OTHERS	TOTAL
1967	32.043	11.944	17.534	18.451	14.973	45.416	140.361
1968	36.984	22.630	19.638	8.848	10.769	30.941	129.809
1969	37.424	14.620	11.891	11.504	10.410	21.434	107.283
1970	45.102	10.865	9.781	8.560	6.033	20.480	100.821
1971	41.029	8.080	11.448	4.630	2.752	15.508	83.448
1972	48.256	8.374	18.443	8.355	8.250	12.351	104.028
1973	51.093	12.749	22.456	7.008	17.522	22.582	133.410
1974	53.719	3.298	6.549	3.997	9.737	15.437	92.738
1975	58.420	3.323	6.801	5.940	5.070	10.954	90.507
1976	59.910	8.511	10.235	11.884	7.762	9.225	107.527

EXPORTS
HARDWOOD LUMBER (2433)
SOUTH PACIFIC

VALUE--MILLION CURRENT DOLLARS

YEAR	MEXICO	EUROPE	TAIWAN	JAPAN	OCEANIA	OTHERS	TOTAL
1967	0.224	0.009	0.000	0.362	0.022	0.017	0.634
1968	0.199	0.003	0.000	0.271	0.044	0.006	0.522
1969	0.590	0.010	0.000	1.120	0.025	0.090	1.834
1970	..189	0.029	0.000	3.650	0.048	0.066	4.982
1971	1.038	0.004	0.000	2.306	0.038	0.058	3.444
1972	1.603	0.000	0.039	11.135	0.060	0.076	12.913
1973	2.743	0.598	0.019	0.315	0.118	0.027	3.819
1974	2.634	0.065	0.011	0.029	0.118	0.061	2.917
1975	0.649	0.002	0.061	0.133	0.102	0.018	0.965
1976	0.610	0.050	0.165	0.207	0.059	0.070	1.159

QUANTITY--MILLION BF

YEAR	MEXICO	EUROPE	TAIWAN	JAPAN	OCEANIA	OTHERS	TOTAL
1967	1.048	0.035	0.000	2.089	0.065	0.052	3.289
1968	1.124	0.009	0.000	0.408	0.230	0.017	1.787
1969	3.564	0.023	0.000	2.425	0.089	0.361	6.461
1970	7.899	0.105	0.000	7.741	0.194	0.228	16.167
1971	6.486	0.038	0.000	4.890	0.183	0.128	11.725
1972	8.879	0.000	0.051	9.902	0.288	0.172	19.292
1973	13.131	1.489	0.022	0.463	0.493	0.032	15.630
1974	10.025	0.167	0.016	0.045	0.325	0.131	10.709
1975	2.362	0.010	0.114	0.239	0.140	0.024	2.888
1976	1.988	0.091	0.172	0.518	0.134	0.167	3.070

EXPORTS
WOOD VENEER (6311)
SOUTH PACIFIC

VALUE--MILLION CURRENT DOLLARS

YEAR	MEXICO	EUROPE	TAIWAN	JAPAN	OCEANIA	OTHERS	TOTAL
1967	0.009	0.084	0.000	0.008	0.001	0.002	0.105
1968	0.023	0.000	0.000	0.012	0.003	0.001	0.038
1969	0.003	0.023	0.000	0.039	0.002	0.000	0.067
1970	0.002	0.004	0.000	0.000	0.000	0.000	0.007
1971	0.004	0.000	0.000	0.011	0.002	0.000	0.017
1972	0.006	0.004	0.000	0.008	0.000	0.000	0.018
1973	0.009	0.004	0.000	0.000	0.000	0.000	0.013
1974	0.013	0.000	0.000	0.021	0.000	0.005	0.039
1975	0.031	0.000	0.007	0.000	0.016	0.000	0.053
1976	0.262	0.000	0.014	0.026	0.006	0.000	0.307

QUANTITY--MIL SQ FT

YEAR	MEXICO	EUROPE	TAIWAN	JAPAN	OCEANIA	OTHERS	TOTAL
1967	0.118	0.045	0.000	0.005	0.004	0.004	0.176
1968	0.307	0.000	0.000	0.009	0.007	0.006	0.329
1969	0.051	0.021	0.000	0.016	0.004	0.000	0.092
1970	0.022	0.009	0.000	0.000	0.000	0.000	0.031
1971	0.020	0.000	0.000	0.004	0.009	0.000	0.033
1972	0.037	0.007	0.000	0.006	0.000	0.000	0.051
1973	0.186	0.008	0.000	0.000	0.000	0.000	0.194
1974	0.101	0.000	0.000	0.096	0.000	0.019	0.216
1975	0.386	0.000	0.013	0.000	0.030	0.000	0.429
1976	2.876	0.000	0.005	0.013	0.002	0.000	2.896

EXPORTS
PLYWOOD (6312)
SOUTH PACIFIC

VALUE--MILLION CURRENT DOLLARS

YEAR	MEXICO	NETHLNDS	W GERMNY	OTH EURO	ASIA	OTHERS	TOTAL
1967	0.522	0.066	0.140	0.386	0.195	0.085	1.395
1968	0.600	0.055	0.145	0.544	0.091	0.075	1.510
1969	0.532	0.045	0.272	1.102	0.132	0.324	2.407
1970	0.408	0.075	0.127	0.308	0.264	0.135	1.317
1971	0.756	0.071	0.178	0.360	0.132	0.146	1.643
1972	1.061	0.110	0.316	0.608	0.727	0.382	3.204
1973	1.071	0.167	0.551	0.829	1.720	0.186	4.524
1974	1.861	0.140	0.271	0.464	0.309	0.648	3.692
1975	1.638	0.286	0.321	0.849	0.376	0.794	4.265
1976	1.765	0.697	0.577	0.454	0.478	0.853	4.824

QUANTITY--MIL SQ FT

YEAR	MEXICO	NETHLNDS	W GERMNY	OTH EURO	ASIA	OTHERS	TOTAL
1967	5.396	0.275	0.686	1.769	1.566	0.482	10.175
1968	5.494	0.199	0.515	2.131	0.459	0.398	9.196
1969	4.508	0.159	0.936	4.264	0.896	1.232	11.995
1970	3.680	0.255	0.440	1.092	1.897	0.829	8.192
1971	6.854	0.248	0.782	1.176	0.978	0.936	10.974
1972	7.796	0.340	0.966	2.235	1.846	1.942	15.124
1973	7.192	0.522	1.593	2.840	4.095	0.793	17.035
1974	12.282	0.330	0.840	0.830	1.433	2.087	17.802
1975	11.058	1.230	0.855	3.133	0.748	2.910	19.835
1976	9.835	2.959	2.584	0.855	1.421	2.753	20.407

EXPORTS
RECONSTITUTED WOOD (6314)
SOUTH PACIFIC

VALUE--MILLION CURRENT DOLLARS

YEAR	MEXICO	OTH N AM	EUROPE	ASIA	OCEANIA	OTHERS	TOTAL
1967	0.081	0.000	0.000	0.000	0.002	0.000	0.083
1968	0.076	0.000	0.000	0.004	0.000	0.000	0.080
1969	0.079	0.002	0.000	0.000	0.000	0.000	0.081
1970	0.066	0.000	0.000	0.001	0.000	0.000	0.067
1971	0.105	0.000	0.001	0.002	0.006	0.000	0.114
1972	0.110	0.000	0.000	0.001	0.011	0.000	0.122
1973	0.180	0.000	0.007	0.000	0.078	0.000	0.265
1974	0.453	0.003	0.009	0.004	0.653	0.000	1.122
1975	0.384	0.007	0.000	0.047	0.016	0.001	0.454
1976	1.163	0.000	0.000	0.148	0.015	0.000	1.326

QUANTITY--MIL SQ FT

YEAR	MEXICO	OTH N AM	EUROPE	ASIA	OCEANIA	OTHERS	TOTAL
1967	0.231	0.000	0.000	0.000	0.007	0.000	0.238
1968	0.377	0.000	0.000	0.007	0.000	0.000	0.385
1969	0.445	0.011	0.000	0.000	0.000	0.000	0.455
1970	0.376	0.000	0.000	0.008	0.000	0.000	0.384
1971	0.624	0.000	0.002	0.006	0.025	0.000	0.657
1972	0.593	0.000	0.000	0.000	0.091	0.000	0.684
1973	0.996	0.000	0.018	0.000	0.634	0.000	1.647
1974	3.017	0.017	0.020	0.006	4.701	0.000	7.761
1975	2.413	0.015	0.000	0.043	0.050	0.000	2.521
1976	6.917	0.000	0.000	0.145	0.037	0.000	7.098

EXPORTS
PULPWOOD CHIPS (631.8320)
SOUTH PACIFIC

VALUE--MILLION CURRENT DOLLARS

YEAR	MEXICO	GUATMALA	U KINGDM	W GERMNY	JAPAN	OTHERS	TOTAL
1967	0.000	0.000	0.001	0.001	0.000	0.000	0.001
1968	0.001	0.000	0.000	0.000	1.230	0.000	1.231
1969	0.000	0.000	0.000	0.000	1.266	0.000	1.266
1970	0.000	0.001	0.000	0.000	0.511	0.005	0.516
1971	0.000	0.000	0.000	0.000	4.637	0.000	4.637
1972	0.000	0.000	0.000	0.000	6.275	0.000	6.275
1973	0.000	0.000	0.000	0.000	9.014	0.000	9.014
1974	0.001	0.000	0.000	0.000	7.427	0.000	7.428
1975	0.004	0.000	0.000	0.000	7.465	0.000	7.469
1976	0.003	0.000	0.000	0.000	12.745	0.000	12.748

QUANTITY--1000 STN

YEAR	MEXICO	GUATMALA	U KINGDM	W GERMNY	JAPAN	OTHERS	TOTAL
1967	0.000	0.000	0.043	0.032	0.000	0.000	0.075
1968	0.041	0.000	0.000	0.000	98.145	0.000	98.186
1969	0.000	0.000	0.000	0.000	105.961	0.000	105.961
1970	0.000	0.022	0.000	0.000	48.000	0.019	48.041
1971	0.000	0.000	0.000	0.000	233.070	0.000	233.070
1972	0.000	0.000	0.000	0.000	253.401	0.000	253.401
1973	0.000	0.000	0.000	0.000	369.403	0.000	369.403
1974	0.041	0.000	0.000	0.000	242.017	0.000	242.058
1975	0.113	0.000	0.000	0.000	257.735	0.000	257.848
1976	0.052	0.000	0.000	0.000	366.678	0.000	366.730

EXPORTS
PULPWOOD (EXC CHIPS)(2421)
SOUTH PACIFIC

VALUE--MILLION CURRENT DOLLARS

YEAR	MEXICO	EL SLVDR	PHIL REP	JAPAN	OTH ASIA	OTHERS	TOTAL
1967	0.000	0.003	0.000	0.000	0.000	0.000	0.003
1968	0.000	0.000	0.000	0.000	0.000	0.000	0.000
1969	0.000	0.000	0.000	0.000	0.000	0.000	0.000
1970	0.000	0.000	0.000	0.000	0.000	0.000	0.000
1971	0.000	0.000	0.000	0.000	0.000	0.000	0.000
1972	0.000	0.000	0.000	0.000	0.000	0.000	0.000
1973	0.000	0.000	0.000	0.000	0.000	0.000	0.000
1974	0.000	0.000	0.000	0.791	0.000	0.000	0.791
1975	0.002	0.000	0.000	0.286	0.000	0.000	0.287
1976	0.001	0.000	0.001	0.286	0.000	0.000	0.287

QUANTITY--1000 CORDS

YEAR	MEXICO	EL SLVDR	PHIL REP	JAPAN	OTH ASIA	OTHERS	TOTAL
1967	0.000	0.100	0.000	0.000	0.000	0.000	0.100
1968	0.000	0.000	0.000	0.000	0.000	0.000	0.000
1969	0.000	0.000	0.000	0.000	0.000	0.000	0.000
1970	0.000	0.000	0.000	0.000	0.000	0.000	0.000
1971	0.000	0.000	0.000	0.000	0.000	0.000	0.000
1972	0.000	0.000	0.000	0.000	0.000	0.000	0.000
1973	0.000	0.000	0.000	0.000	0.000	0.000	0.000
1974	0.000	0.000	0.000	20.383	0.000	0.000	20.383
1975	0.084	0.000	0.000	7.817	0.000	0.000	7.901
1976	0.028	0.000	0.023	7.817	0.000	0.000	7.868

EXPORTS
WOOD PULP (2510)
SOUTH PACIFIC

VALUE--MILLION CURRENT DOLLARS

YEAR	N AMERIC	S AMERIC	EUROPE	ASIA	OCEANIA	OTHERS	TOTAL
1967	0.545	0.723	2.192	6.867	0.150	0.000	10.477
1968	0.018	0.636	4.309	8.018	0.000	0.117	13.098
1969	0.123	0.320	5.845	11.871	0.063	0.299	18.520
1970	0.000	1.017	14.869	18.225	1.946	1.572	37.629
1971	0.000	0.626	13.608	13.212	0.713	0.017	28.176
1972	0.000	0.919	16.319	14.597	1.490	4.238	37.563
1973	0.002	2.814	20.684	29.758	2.334	3.048	58.640
1974	1.725	4.165	42.469	59.877	4.921	7.072	120.228
1975	0.394	3.409	46.790	48.882	0.720	5.043	105.239
1976	0.002	3.832	38.121	37.255	0.000	6.106	85.315

QUANTITY--1000 STN

YEAR	N AMERIC	S AMERIC	EUROPE	ASIA	OCEANIA	OTHERS	TOTAL
1967	5.194	7.046	20.893	65.052	1.120	0.002	99.306
1968	0.157	6.255	38.826	74.008	0.000	0.879	120.124
1969	1.193	2.591	51.723	101.252	0.618	2.509	159.887
1970	0.000	7.938	100.679	128.411	12.406	10.763	260.198
1971	0.000	4.501	84.156	84.062	5.205	0.120	178.044
1972	0.000	6.724	105.652	96.699	9.906	30.279	249.260
1973	0.011	17.395	117.833	151.122	11.972	17.501	315.835
1974	4.588	11.027	125.275	184.259	11.808	16.577	353.533
1975	0.129	10.136	125.058	148.506	1.901	12.716	298.447
1976	0.007	12.759	117.380	111.236	0.000	26.212	267.594

EXPORTS
NEWSPRINT (6411)
SOUTH PACIFIC

VALUE--MILLION CURRENT DOLLARS

YEAR	MEXICO	OTH N AM	PHIL REP	OTH ASIA	OCEANIA	OTHERS	TOTAL
1967	0.204	0.000	0.008	0.004	0.000	0.000	0.216
1968	0.332	0.000	0.006	0.007	0.000	0.000	0.344
1969	0.267	0.040	0.084	0.039	0.002	0.103	0.535
1970	0.337	0.000	0.027	0.010	0.001	0.000	0.375
1971	0.380	0.000	0.001	0.011	0.000	0.027	0.419
1972	0.322	0.000	0.002	0.007	0.001	0.000	0.332
1973	0.351	0.000	0.000	0.001	0.003	0.001	0.356
1974	0.293	0.000	0.013	0.225	0.001	0.000	0.532
1975	0.359	0.000	0.002	0.127	0.002	0.000	0.491
1976	1.239	0.121	0.000	0.033	0.000	0.000	1.394

QUANTITY--1000 STN

YEAR	MEXICO	OTH N AM	PHIL REP	OTH ASIA	OCEANIA	OTHERS	TOTAL
1967	1.380	0.000	0.046	0.025	0.000	0.000	1.452
1968	2.330	0.000	0.030	0.035	0.001	0.000	2.396
1969	1.943	0.248	0.672	0.280	0.013	1.167	4.323
1970	2.410	0.000	0.235	0.042	0.004	0.000	2.691
1971	2.627	0.000	0.004	0.067	0.000	0.224	2.922
1972	2.191	0.000	0.014	0.053	0.002	0.000	2.260
1973	2.130	0.000	0.000	0.009	0.007	0.004	2.150
1974	1.134	0.000	0.147	0.958	0.005	0.000	2.244
1975	0.765	0.000	0.022	0.355	0.002	0.000	1.144
1976	3.130	0.403	0.000	0.179	0.000	0.000	3.712

EXPORTS
BUILDING BOARD (6416)
SOUTH PACIFIC

VALUE--MILLION CURRENT DOLLARS

YEAR	MEXICO	U KINGDM	OTH EURO	ASIA	OCEANIA	OTHERS	TOTAL
1967	0.124	0.001	0.130	0.113	0.009	0.030	0.408
1968	0.090	0.000	0.069	0.035	0.026	0.001	0.221
1969	0.081	0.000	0.110	0.130	0.020	0.003	0.344
1970	0.084	0.000	0.104	0.103	0.024	0.000	0.316
1971	0.098	0.000	0.035	0.038	0.026	0.008	0.205
1972	0.140	0.004	0.038	0.113	0.096	0.007	0.397
1973	0.176	0.127	0.024	0.050	0.030	0.049	0.456
1974	0.494	0.525	0.129	0.226	0.412	0.018	1.804
1975	0.565	0.103	0.025	0.099	0.125	0.009	0.927
1976	0.508	0.389	0.109	0.464	0.128	0.000	1.598

QUANTITY--1000 STN

YEAR	MEXICO	U KINGDM	OTH EURO	ASIA	OCEANIA	OTHERS	TOTAL
1967	0.294	0.002	0.403	0.174	0.024	0.168	1.063
1968	0.396	0.000	0.148	0.142	0.163	0.003	0.853
1969	0.322	0.000	0.142	0.338	0.057	0.010	0.870
1970	0.304	0.000	0.178	0.388	0.090	0.000	0.961
1971	0.513	0.000	0.108	0.159	0.125	0.029	0.934
1972	0.721	0.004	0.099	0.218	0.456	0.005	1.502
1973	0.669	0.289	0.088	0.187	0.074	0.115	1.423
1974	2.453	1.159	0.283	0.734	1.082	0.020	5.730
1975	2.260	0.216	0.076	0.490	0.322	0.015	3.379
1976	2.301	0.992	0.178	1.491	0.400	0.000	5.361

EXPORTS
PAPER & PAPERBOARD (6410)
SOUTH PACIFIC

VALUE--MILLION CURRENT DOLLARS

YEAR	MEXICO	PHIL REP	JAPAN	OTH ASIA	AUSTRALA	OTHERS	TOTAL
1967	2.158	1.110	2.491	1.591	0.814	1.711	9.874
1968	2.690	1.171	1.102	1.679	1.404	2.298	10.343
1969	2.794	0.833	1.168	2.472	2.415	2.997	12.679
1970	3.967	0.461	0.934	2.058	1.049	2.312	10.781
1971	3.851	0.375	0.781	1.471	0.634	2.043	9.155
1972	4.447	0.776	1.458	1.902	0.654	2.849	12.085
1973	3.745	0.636	3.245	4.367	0.874	3.538	16.404
1974	5.739	2.353	13.050	7.394	2.109	6.743	37.388
1975	6.246	1.477	4.531	5.512	1.322	3.533	22.620
1976	6.412	2.437	7.318	8.377	2.422	4.402	31.368

QUANTITY--1000 STN

YEAR	MEXICO	PHIL REP	JAPAN	OTH ASIA	AUSTRALA	OTHERS	TOTAL
1967	9.776	5.107	13.722	8.517	0.847	5.279	43.247
1968	13.433	3.862	2.520	7.980	1.436	10.714	39.945
1969	13.112	2.339	1.332	9.081	3.014	8.577	37.455
1970	20.870	1.388	0.635	9.098	1.042	5.262	38.295
1971	19.366	1.018	0.589	4.896	0.857	5.400	32.127
1972	23.424	1.680	0.982	5.284	0.627	6.774	38.772
1973	17.184	1.333	4.578	12.880	1.020	8.899	45.894
1974	18.520	6.360	27.961	21.872	3.007	10.781	88.501
1975	19.378	2.661	6.903	8.152	0.764	4.335	42.193
1976	19.787	4.698	10.926	13.794	1.354	3.993	54.553

IMPORTS
HARDWOOD LOGS (2423)
SOUTH PACIFIC

VALUE--MILLION CURRENT DOLLARS

YEAR	N AMERIC	INDONESA	OTH ASIA	LIBERIA	OTH AFRI	OTHERS	TOTAL
1967	0.006	0.000	0.000	0.001	0.000	0.000	0.008
1968	0.012	0.000	0.000	0.000	0.000	0.009	0.020
1969	0.005	0.000	0.000	0.000	0.000	0.000	0.006
1970	0.000	0.002	0.005	0.000	0.000	0.000	0.008
1971	0.024	0.002	0.001	0.000	0.000	0.000	0.026
1972	0.000	0.004	0.009	0.000	0.000	0.000	0.013
1973	0.000	0.025	0.008	0.145	0.000	0.000	0.178
1974	0.000	0.352	0.027	0.000	0.000	0.000	0.379
1975	0.002	0.003	0.011	0.000	0.011	0.002	0.028
1976	0.000	0.000	0.006	0.000	0.000	0.010	0.016

QUANTITY--MILLION BF

YEAR	N AMERIC	INDONESA	OTH ASIA	LIBERIA	OTH AFRI	OTHERS	TOTAL
1967	0.097	0.000	0.002	0.004	0.000	0.002	0.106
1968	0.009	0.000	0.000	0.000	0.000	0.050	0.059
1969	0.036	0.000	0.000	0.000	0.000	0.000	0.036
1970	0.000	0.003	0.027	0.000	0.000	0.000	0.031
1971	0.038	0.006	0.002	0.000	0.000	0.000	0.046
1972	0.000	0.005	0.011	0.000	0.000	0.000	0.015
1973	0.000	0.022	0.014	0.922	0.000	0.000	0.958
1974	0.000	1.727	0.093	0.000	0.000	0.000	1.820
1975	0.052	0.003	0.018	0.000	0.023	0.004	0.100
1976	0.000	0.000	0.008	0.000	0.000	0.009	0.017

IMPORTS
SOFTWOOD LUMBER (2432)
SOUTH PACIFIC

VALUE--MILLION CURRENT DOLLARS

YEAR	CANADA	BRAZIL	OTH S AM	EUROPE	ASIA	OTHERS	TOTAL
1967	1.942	0.016	0.001	0.001	0.027	0.006	1.993
1968	3.032	0.017	0.002	0.004	0.011	0.026	3.091
1969	4.154	0.016	0.005	0.001	0.019	0.177	4.373
1970	3.067	0.028	0.003	0.047	0.166	0.191	3.502
1971	3.246	0.019	0.000	0.022	0.033	0.041	3.361
1972	5.505	0.080	0.000	0.000	0.013	0.001	5.599
1973	10.220	0.023	0.000	0.209	0.153	0.021	10.626
1974	7.763	0.003	0.000	0.012	0.349	0.053	8.180
1975	3.481	0.021	0.016	0.000	0.055	0.021	3.593
1976	15.005	0.001	0.052	0.000	0.023	0.000	15.081

QUANTITY--MILLION BF

YEAR	CANADA	BRAZIL	OTH S AM	EUROPE	ASIA	OTHERS	TOTAL
1967	27.667	0.160	0.001	0.006	0.184	0.027	28.044
1968	31.966	0.140	0.003	0.000	0.055	0.109	32.272
1969	44.653	0.133	0.022	0.000	0.207	0.951	45.967
1970	38.134	0.232	0.009	0.886	1.459	1.288	42.008
1971	36.817	0.140	0.000	0.290	0.139	0.253	37.639
1972	54.539	0.507	0.000	0.005	0.050	0.013	55.114
1973	78.180	0.091	0.000	1.063	1.079	0.112	80.525
1974	52.027	0.017	0.000	0.049	0.976	0.150	53.219
1975	22.094	0.075	0.009	0.000	0.264	0.050	22.492
1976	82.363	0.006	0.014	0.000	0.030	0.003	82.415

IMPORTS
HARDWOOD LUMBER (2433)
SOUTH PACIFIC

VALUE--MILLION CURRENT DOLLARS

YEAR	THAILAND	MALAYSIA	PHIL REP	TAIWAN	OTH ASIA	OTHERS	TOTAL
1967	0.190	0.439	0.619	0.380	1.219	0.513	3.360
1968	0.444	0.761	0.701	0.590	1.054	0.497	4.047
1969	0.405	1.510	0.914	0.796	1.230	1.580	6.436
1970	0.300	0.945	0.481	0.445	0.782	1.080	4.033
1971	0.237	1.008	0.467	0.435	0.836	0.949	3.932
1972	0.591	2.109	0.748	1.560	1.821	1.453	8.281
1973	1.237	1.639	1.662	2.225	1.909	1.868	10.542
1974	2.991	4.129	2.100	1.312	2.393	2.884	15.810
1975	1.671	0.811	0.470	0.948	0.874	0.787	5.561
1976	2.148	1.200	1.418	1.553	1.194	0.989	8.502

QUANTITY--MILLION BF

YEAR	THAILAND	MALAYSIA	PHIL REP	TAIWAN	OTH ASIA	OTHERS	TOTAL
1967	0.281	3.211	4.051	2.324	5.772	1.464	17.103
1968	0.650	5.151	4.615	3.648	5.018	1.971	21.053
1969	0.637	10.087	6.163	4.768	5.408	9.419	36.482
1970	0.482	5.378	2.556	2.389	3.601	5.349	19.757
1971	0.379	5.799	2.381	2.363	3.758	6.093	20.774
1972	1.046	12.226	3.915	8.968	6.108	7.264	39.528
1973	1.354	7.585	8.172	8.951	4.638	6.592	37.293
1974	2.405	13.335	7.578	3.355	4.116	6.986	37.775
1975	1.616	3.016	1.534	2.751	1.389	1.159	11.466
1976	2.026	3.711	3.874	3.752	1.778	1.195	16.336

IMPORTS
WOOD VENEER (6311)
SOUTH PACIFIC

VALUE--MILLION CURRENT DOLLARS

YEAR	SINGAPOR	PHIL REP	TAIWAN	OTH ASIA	OCEANIA	OTHERS	TOTAL
1967	0.030	0.284	0.000	0.070	0.002	0.063	0.450
1968	0.029	0.541	0.000	0.050	0.001	0.155	0.776
1969	0.067	0.515	0.000	0.025	0.002	0.100	0.708
1970	0.007	0.158	0.000	0.036	0.000	0.122	0.323
1971	0.023	0.122	0.005	0.056	0.001	0.080	0.288
1972	0.157	0.424	0.067	0.178	0.359	0.121	1.304
1973	0.160	0.217	0.330	0.029	0.284	0.320	1.340
1974	0.308	0.327	0.360	0.251	0.203	0.187	1.636
1975	0.249	0.147	0.296	0.446	0.275	0.015	1.428
1976	0.213	0.532	0.728	1.140	0.063	0.050	2.726

QUANTITY--MIL SQ FT

YEAR	SINGAPOR	PHIL REP	TAIWAN	OTH ASIA	OCEANIA	OTHERS	TOTAL
1967	0.373	3.162	0.000	0.277	0.003	0.425	4.241
1968	0.269	3.225	0.000	0.248	0.001	2.621	6.364
1969	0.686	5.215	0.000	0.161	0.004	0.314	6.381
1970	0.091	1.841	0.000	0.360	0.000	0.149	2.440
1971	0.131	1.582	0.024	0.488	0.012	0.113	2.351
1972	1.222	5.365	0.267	2.164	1.834	0.106	10.958
1973	0.468	1.825	0.840	0.234	1.897	0.090	5.354
1974	0.902	2.010	0.719	0.888	0.551	0.085	5.155
1975	0.611	1.218	0.579	0.694	0.582	0.035	3.718
1976	0.406	2.959	1.604	1.136	0.085	0.034	6.224

IMPORTS
PLYWOOD (6312)
SOUTH PACIFIC

VALUE--MILLION CURRENT DOLLARS

YEAR	PHIL REP	KOR REP	TAIWAN	JAPAN	OTH ASIA	OTHERS	TOTAL
1967	4.714	6.770	4.177	12.609	0.391	0.932	29.592
1968	6.371	9.744	8.299	18.494	1.054	1.233	45.195
1969	4.880	17.502	8.083	18.538	1.880	3.377	54.259
1970	5.290	11.905	5.496	12.151	0.716	1.647	37.204
1971	3.525	14.352	4.995	11.430	0.309	1.145	35.756
1972	4.953	11.714	11.964	16.361	0.305	1.432	46.728
1973	6.705	20.668	11.478	15.575	0.361	2.229	57.016
1974	5.169	8.553	7.375	14.087	1.506	2.189	38.880
1975	3.170	14.562	6.805	12.205	0.631	1.049	38.423
1976	7.192	22.183	8.787	15.019	0.684	1.534	55.399

QUANTITY--MIL SQ FT

YEAR	PHIL REP	KOR REP	TAIWAN	JAPAN	OTH ASIA	OTHERS	TOTAL
1967	50.742	67.356	42.293	75.932	3.879	7.167	247.368
1968	58.596	83.494	75.562	91.084	10.547	6.767	326.049
1969	43.259	148.060	71.517	96.994	18.449	24.255	402.534
1970	52.357	98.924	46.710	70.550	7.481	10.619	286.641
1971	32.916	124.230	43.403	62.813	2.733	9.791	275.886
1972	49.373	97.556	103.272	72.125	2.780	7.742	332.847
1973	47.159	110.782	63.363	47.017	2.173	7.737	278.231
1974	30.333	45.923	37.424	38.049	8.543	6.062	166.333
1975	23.215	91.479	41.607	42.561	4.275	3.387	206.525
1976	45.010	117.164	41.538	41.810	3.336	6.470	255.327

IMPORTS
RECONSTITUTED WOOD (6314)
SOUTH PACIFIC

VALUE--MILLION CURRENT DOLLARS

YEAR	BRAZIL	FINLAND	OTH EURO	TAIWAN	JAPAN	OTHERS	TOTAL
1967	0.000	0.000	0.000	0.000	0.000	0.000	0.000
1968	0.000	0.000	0.001	0.005	0.002	0.000	0.008
1969	0.000	0.025	0.004	0.094	0.011	0.000	0.135
1970	0.000	0.000	0.003	0.099	0.007	0.000	0.110
1971	0.000	0.000	0.006	0.109	0.000	0.000	0.116
1972	0.000	0.005	0.000	0.022	0.000	0.001	0.027
1973	0.006	0.000	0.001	0.005	0.000	0.000	0.012
1974	0.000	0.000	0.000	0.000	0.000	0.000	0.000
1975	0.005	0.000	0.000	0.000	0.000	0.000	0.005
1976	0.000	0.005	0.000	0.020	0.005	0.001	0.031

QUANTITY--MIL SQ FT

YEAR	BRAZIL	FINLAND	OTH EURO	TAIWAN	JAPAN	OTHERS	TOTAL
1967	0.000	0.000	0.000	0.000	0.000	0.000	0.000
1968	0.000	0.000	0.011	0.002	0.003	0.000	0.016
1969	0.000	0.242	0.036	0.050	0.005	0.000	0.333
1970	0.000	0.000	0.015	0.065	0.003	0.000	0.082
1971	0.000	0.000	0.042	0.078	0.000	0.000	0.120
1972	0.000	0.020	0.000	0.022	0.000	0.001	0.043
1973	0.055	0.000	0.002	0.003	0.000	0.000	0.060
1974	0.000	0.000	0.000	0.000	0.000	0.000	0.000
1975	0.015	0.000	0.000	0.000	0.000	0.000	0.015
1976	0.000	0.025	0.000	0.019	0.002	0.005	0.050

IMPORTS
WOOD PULP (2510)
SOUTH PACIFIC

VALUE--MILLION CURRENT DOLLARS

YEAR	CANADA	PORTUGAL	OTH EURO	ASIA	AFRICA	OTHERS	TOTAL
1967	8.850	0.092	0.164	0.019	0.000	0.000	9.124
1968	8.612	0.054	0.006	0.009	0.000	0.000	8.681
1969	11.418	0.014	0.000	0.000	0.000	0.000	11.432
1970	9.925	0.012	0.000	0.000	0.000	0.000	9.936
1971	8.257	0.000	0.000	0.000	0.000	0.000	8.257
1972	10.668	0.000	0.002	0.000	0.000	0.000	10.671
1973	11.878	0.000	0.000	0.000	0.000	0.000	11.878
1974	14.680	0.000	0.000	0.007	0.000	0.000	14.686
1975	6.525	0.000	0.000	0.024	0.000	0.000	6.548
1976	8.802	0.000	0.000	0.000	0.000	0.000	8.802

QUANTITY--1000 STN

YEAR	CANADA	PORTUGAL	OTH EURO	ASIA	AFRICA	OTHERS	TOTAL
1967	113.166	1.004	1.709	0.194	0.000	0.000	116.073
1968	125.642	0.602	0.209	0.099	0.000	0.000	126.552
1969	161.308	0.148	0.000	0.000	0.000	0.000	161.456
1970	106.589	0.105	0.000	0.000	0.000	0.000	106.694
1971	81.379	0.000	0.000	0.000	0.000	0.000	81.379
1972	143.815	0.000	0.022	0.000	0.000	0.000	143.837
1973	106.804	0.000	0.000	0.000	0.000	0.000	106.804
1974	85.653	0.000	0.000	0.006	0.000	0.000	85.659
1975	22.815	0.000	0.000	0.047	0.000	0.000	22.862
1976	35.763	0.000	0.000	0.000	0.000	0.000	35.763

IMPORTS
NEWSPRINT (6411)
SOUTH PACIFIC

VALUE--MILLION CURRENT DOLLARS

YEAR	CANADA	SWEDEN	FINLAND	FRANCE	REP KOR	OTHERS	TOTAL
1967	73.206	0.001	6.615	0.000	0.000	0.000	79.822
1968	60.515	0.137	4.368	0.000	0.000	0.000	65.020
1969	63.424	0.176	5.616	0.000	0.000	0.001	69.218
1970	60.599	0.000	8.268	0.000	0.000	0.000	68.866
1971	49.210	0.000	7.174	0.000	0.000	0.000	56.383
1972	57.598	0.000	11.263	0.000	0.000	0.000	68.861
1973	81.498	0.000	12.833	0.000	0.000	0.000	94.332
1974	100.026	0.000	11.767	0.000	0.000	0.000	111.793
1975	103.233	0.000	4.404	0.005	0.002	0.000	107.644
1976	131.069	0.000	0.000	0.000	0.004	0.000	131.073

QUANTITY--1000 STN

YEAR	CANADA	SWEDEN	FINLAND	FRANCE	REP KOR	OTHERS	TOTAL
1967	620.901	0.010	60.537	0.000	0.000	0.000	681.448
1968	496.109	1.237	39.720	0.000	0.000	0.000	537.067
1969	492.205	1.706	48.712	0.000	0.000	0.008	542.631
1970	475.764	0.000	68.800	0.000	0.000	0.000	544.564
1971	400.853	0.000	57.082	0.000	0.000	0.000	457.935
1972	515.061	0.000	91.317	0.000	0.000	0.000	606.378
1973	537.341	0.000	93.460	0.000	0.000	0.000	630.801
1974	526.923	0.000	79.401	0.000	0.000	0.000	606.324
1975	429.943	0.000	22.028	0.003	0.010	0.000	451.985
1976	560.982	0.000	0.000	0.000	0.020	0.000	561.002

IMPORTS
BUILDING BOARD (6416)
SOUTH PACIFIC

VALUE--MILLION CURRENT DOLLARS

YEAR	BRAZIL	OTH S AM	SWEDEN	OTH EURO	REP KOR	OTHERS	TOTAL
1967	0.076	0.000	0.674	0.269	0.000	0.076	1.094
1968	0.023	0.000	1.307	0.509	0.000	0.183	2.023
1969	0.000	0.000	1.235	0.810	0.000	0.262	2.307
1970	0.105	0.000	1.362	0.570	0.000	0.376	2.413
1971	0.326	0.000	0.628	0.518	0.000	0.374	1.846
1972	0.638	0.000	1.126	1.634	0.000	1.003	4.401
1973	1.005	0.047	1.418	1.392	0.003	1.582	5.447
1974	1.857	0.103	1.455	1.359	0.087	1.820	6.682
1975	0.945	0.038	0.523	0.188	0.348	0.357	2.400
1976	2.020	0.453	0.942	0.220	0.971	0.464	5.070

QUANTITY--1000 STN

YEAR	BRAZIL	OTH S AM	SWEDEN	OTH EURO	REP KOR	OTHERS	TOTAL
1967	0.967	0.000	9.315	3.743	0.000	1.126	15.151
1968	0.213	0.000	19.802	7.457	0.000	2.575	30.047
1969	0.004	0.000	18.500	11.961	0.000	3.472	33.938
1970	1.436	0.000	18.356	7.246	0.000	5.202	32.240
1971	3.727	0.000	8.270	6.511	0.000	5.517	24.025
1972	7.479	0.000	14.750	21.666	0.000	12.356	56.251
1973	9.857	0.594	12.844	13.089	0.024	16.239	52.647
1974	12.472	1.149	9.896	9.343	0.492	17.522	50.875
1975	6.492	0.346	3.931	1.687	2.472	3.652	18.580
1976	13.019	3.835	7.055	2.290	5.893	4.305	36.396

IMPORTS
PAPER & PAPERBOARD (641C)
SOUTH PACIFIC

VALUE--MILLION CURRENT DOLLARS

YEAR	CANADA	SWEDEN	FINLAND	KOR REP	JAPAN	OTHERS	TOTAL
1967	2.362	0.422	0.244	0.168	1.178	0.326	4.700
1968	2.594	0.407	0.623	0.271	1.127	0.477	5.499
1969	1.241	0.526	1.635	0.296	1.656	0.650	6.003
1970	2.071	0.581	1.784	0.506	1.798	0.473	7.212
1971	2.340	0.442	1.374	0.334	1.445	0.381	6.315
1972	2.103	0.455	1.060	0.587	1.790	0.619	6.614
1973	2.913	0.825	1.657	0.861	2.653	1.084	9.993
1974	1.105	1.189	0.962	1.484	3.900	0.973	9.632
1975	0.443	0.787	0.264	1.315	2.729	0.934	6.472
1976	0.940	1.150	2.649	2.630	5.395	1.509	14.273

QUANTITY--1000 STN

YEAR	CANADA	SWEDEN	FINLAND	KOR REP	JAPAN	OTHERS	TOTAL
1967	15.619	1.974	1.765	0.071	3.897	0.852	24.199
1968	17.579	1.924	4.565	0.116	1.351	1.351	26.886
1969	8.703	2.338	10.062	0.122	5.072	1.908	28.144
1970	13.908	2.356	11.718	0.178	3.092	1.380	32.632
1971	17.657	1.699	8.664	0.117	2.408	0.960	31.506
1972	17.762	1.052	6.822	0.179	0.923	1.646	28.384
1973	18.855	2.604	9.568	0.251	1.147	3.534	35.959
1974	5.827	2.693	4.181	0.328	1.475	1.330	15.835
1975	1.557	1.646	0.630	0.276	0.831	0.865	5.805
1976	3.180	2.471	10.137	0.572	1.509	2.165	20.034

Appendix I

THE GREAT LAKES REGION

Exports

Regional Overview

Even though the Great Lakes region is the largest U.S. importer of forest products, it is also a significant exporter. In 1976 the $426.7 million worth of forest products shipped from the Great Lakes represented 11 percent of all U.S. forest product exports. This is within a few percentage points of the exports captured by each of the North Atlantic, South Atlantic, and Gulf regions.

The Great Lakes region grew steadily as an exporter during 1967-76. The average growth in the nominal value of forest product exports was 19 percent annually.

The most important export is paper and paperboard, which in 1976 was valued at $220.5 million, or 52 percent of the region's total forest product exports. The remainder of the region's pulp and paper exports, which includes wood pulp ($17.1 million) and building board ($14.5 million), represented 7 percent of the Great Lakes exports.

Of the solid wood products, the $68.1 million worth of softwood lumber was 16 percent of the region's total forest product exports, while hardwood lumber ($45.9 million), plywood ($24.5 million), hardwood logs ($14 million), and pulpwood except chips ($6.3 million) together represented 22 percent of the region's 1976 exports.

The outstanding characteristic of Great Lakes forest product export trade is that Canada was the destination for more than 90 percent (by value) of the region's exports throughout 1967-76. In 1976 the region shipped $402.6 million worth of forest products to Canada, or 94 percent of the region's total exports.

Paper and Paperboard

The Great Lakes was the fastest growing major paper and paperboard exporting region during 1967-76. The 459,000 tons (worth $220.5 million) of paper and paperboard exports in 1976 was 324 percent higher in volume and 408 percent higher in value than the 1967 exports. The equivalent average annual growth rates for the period are 17 percent in volume and 20 percent in value.

Nearly one-fifth of all U.S. exports of paper and paperboard was shipped from the Great Lakes region in 1976. Only the Gulf and North Atlantic regions were larger exporters in that year.

Canada's average share of the value of the Great Lakes exports was 96 percent during 1967-76. Various European countries were the markets for most of the remainder throughout the period.

Softwood Lumber

Softwood lumber was the region's second largest forest product export throughout 1967-76. In an average year, Great Lakes softwood lumber exports were 16 percent of the region's total forest product exports.

The Great Lakes was the second largest (in value terms) softwood lumber exporting region in 1976. In that year the region shipped 14 percent of the value and 13 percent of the volume of the total U.S. softwood lumber exports. The region's 1976 exports were 227.5 million board-feet (worth $68.1 million).

The region experienced good growth in softwood lumber exports from 1967 to 1976. The average annual growth in value was 18 percent, and in volume was 8 percent.

Canada was practically the only foreign market for Great Lakes softwood lumber throughout 1967-76. All other countries combined took less than 1 percent of the region's exports in every year.

Hardwood Lumber

More than half the 1976 U.S. exports of hardwood lumber was shipped from the Great Lakes region. The 133.7 million board-feet (worth $45.9 million) represented 57 percent of the quantity and 53 percent of the value of the country's total hardwood lumber exports.

The region's exports grew modestly during the 1967-76 period. The average growth rate in volume was 3 percent per year, and in nominal value it was 14 percent annually.

Canada is the principal market for Great Lakes hardwood lumber. In an average year, 86 percent (by value) of the region's hardwood lumber exports went to Canada. In 1976, 117.8 million board-feet (valued at $36.1 million) were exported from the Great Lakes to Canada.

European countries generally were the importers of the hardwood lumber not shipped to Canada.

In 1976 Europe-bound exports of 15.9 million board-feet (worth $9.8 million) represented 21 percent of the Great Lakes total hardwood lumber exports. The Netherlands and Belgium each imported about one-third of the hardwood lumber shipped to Europe.

Plywood

The Great Lakes was the second largest plywood exporting region in the United States in 1976; the 121.6 million square feet of exports (valued at $24.5 million) represented 12 percent of the quantity and 17 percent of the value of the country's total plywood exports.

The region's exports of plywood grew rapidly during 1967-76. By 1976 the value of plywood exports was more than 25 times the 1967 value, and volume had increased by more than 30 times. The average rates of growth for the period were 45 percent per year in value and 47 percent annually in volume.

In every year of the 1967-76 period, Canada was the destination for more than 99 percent of both the value and volume of the Great Lakes plywood exports.

Wood Pulp

The Great Lakes region was not a large exporter of wood pulp relative to other U.S. regions during 1967-76, but shipments from the Great Lakes were increasing more rapidly than were most other regions' pulp exports. The average annual export growth rates for pulp shipped from the region were 14 percent in quantity and 25 percent in value. The 1976 exports totaled 48,000 tons and were valued at $17.1 million.

Canada's share of the value of the region's pulp exports fell below 100 percent only in 1967 and 1968, when European countries were the destinations for 1 and 4 percent, respectively, of the Great Lakes total wood pulp exports.

Building Board

The largest building board exporting region throughout 1967-76 was the Great Lakes. In an average year, the region shipped 44 percent (by value) of all U.S. building board exports. In 1976 the 51,000 tons (worth $14.5 million) represented 46 percent of the country's building board exports.

The region substantially increased its building board exports from 1967 to 1976; the average export value growth rate was 21 percent annually, and the average rate of growth in volume was 19 percent per year.

There was only one year in which Canada did not import 97 percent to 99 percent of the value of the region's building board exports. In 1971 the Republic of South Africa imported 13 percent ($719,000) of the region's exports and Canada imported the remainder.

Hardwood Logs

Although the United States as a whole exported a smaller volume of hardwood logs in 1976 than it did in 1967, the Great Lakes region more than doubled its hardwood log export volume. The 26.2 million board-feet represented an increase of 153 percent over the 1967 volume. The equivalent annual growth rate is 11 percent. However, the nominal value of the region's exports increased at an average of only 8 percent per year.

The Great Lakes was the second largest hardwood log exporting region in 1976, when the $14 million it exported represented 22 percent of all U.S. hardwood log exports.

West Germany was the largest importer of the region's hardwood logs throughout 1967-76. In an average year, West Germany imported 46 percent (by value) of the Great Lakes' exports. In 1976 the European country imported 5.6 million board-feet (worth $6.3 million).

Canada, in 1976, imported 16.7 million board-feet of Great Lakes hardwood logs (valued at $4.7 million), which represented 34 percent of the region's total hardwood log exports.

European countries other than West Germany imported 22 percent of the value of the Great Lakes hardwood log exports in 1976. Italy was the largest importer among these countries; its 1976 imports totaled 1.6 million board-feet worth $1.2 million, or 9 percent of the region's total hardwood log exports.

Other Forest Products

Only 5 percent of the value of the region's 1976 forest product exports was embodied in the remaining commodities (pulpwood except chips, softwood logs, wood veneer, reconstituted wood, pulpwood chips, other industrial roundwood, and newsprint).

Forty percent of the value of 1976 exports of these commodities was in pulpwood except chips. The Great Lakes was the largest U.S. exporter of pulpwood in that year, shipping out 153,000 cords (worth $6.3 million), which represented 87 percent of all U.S. exports of non-chip pulpwood. Canada was the sole foreign importer of Great Lakes pulpwood throughout 1967-76.

The Great Lakes region was one of the largest reconstituted wood exporting regions in the United States throughout 1967-76. In 1976 the region shipped 21 million square feet (valued at $4.5 million), which was 24 percent of the value of all U.S. exports of reconstituted wood. The region exhibited phenomenal growth in its reconstituted wood exports, which increased in nominal value at an average rate of 36 percent and in volume at an average of 53 percent per year. Canada again was the only significant importer.

The Great Lakes dropped from its position as the leading exporter of wood veneer in 1967, when it shipped 55 percent of the value of all U.S. veneer exports, to the third largest exporter in 1976, when its share of U.S. exports was only 13 percent. The region actually increased its absolute volume of exports by 42 percent during the decade, but the nominal value increased only by 39 percent, which is equivalent to an annual growth of just 4 percent. Although European countries, particularly, West Germany, imported up to one-fourth of the region's total veneer exports in the early years of the period, by 1976 Canada was the destination for 95 percent of the value of the Great Lakes exports.

Imports

Regional Overview

The Great Lakes region is by far the largest forest product importing region in the United States. In 1976 the region imported $2.4 billion worth of forest products, which represented 46 percent of all U.S. forest product imports.

The region held its position as the dominant importer throughout 1967-76. In an average year, the Great Lakes was responsible for 44 percent of the value of all U.S. forest product imports.

The nominal value of the region's imports increased at an annual average rate of 12 percent, which is equal to the average import growth rate for the country as a whole.

In 1976 just four commodities--softwood lumber, wood pulp, newsprint, and paper and paperboard--constituted 98 percent of the value of the region's total forest product imports. Newsprint imports in 1976 were valued at $1 billion and represented 42 percent of the region's forest product imports. Wood pulp imports were worth $640.5 million (27 percent of the region's forest product imports); softwood lumber, $549.7 million (24 percent); and paper and paperboard, $115.3 million (5 percent).

The commodity mix in 1976 differed somewhat from 1967, when newsprint represented 54 percent (by value) of the region's forest product imports, wood pulp represented 21 percent, and softwood lumber represented 13 percent.

The Great Lakes region is almost exclusively dependent on Canada as a source of forest product. In 1976 shipments from Canada comprised more than 97 percent of the value of the region's total forest product imports.

Newsprint

Over the period extending from 1967 to 1976, newsprint became less important, relative to other forest products, as a Great Lakes import. The 1976 import volume of 3.7 million tons was a modest 11 percent greater than the 1967 imports. The nominal value of newsprint imports increased over the period by 122 percent, which is equivalent to an annual gain of 9 percent.

The Great Lakes region was the largest newsprint importer in the United States throughout 1967-76. In an average year, the region imported 55 percent (by value) of all U.S. newsprint imports. In 1976 the $1 billion of regional imports represented 57 percent of the U.S. total.

Canada was for practical purposes, the sole foreign source of the region's newsprint in every year of 1967-76; only occasionally during the early years of the period were shipments received from Sweden and Finland.

Wood Pulp

In an average year during 1967-76, more than half of all U.S. imports of wood pulp entered the Great Lakes region. The region's 1976 imports of 1.9 million tons (worth $640.5 million) represented 51 percent of both the volume and value of U.S. total wood pulp imports.

The region significantly increased its wood pulp imports from 1967 to 1976. The average import growth rate in value was 15 percent per year and, in volume, was 3 percent annually.

Canada supplied, on average, more than 99 percent of the value of the Great Lakes' pulp imports during the study period. Relatively small quantities were exported from Sweden and Finland, principally in the early years of the period.

Softwood Lumber

The Great Lakes region was the largest softwood lumber importing region in the United States throughout 1967-76. In 1976 the 3.2 billion board-feet (worth $549.7 million) represented 41 percent (by value) of all U.S. softwood lumber imports. That share was the highest taken by the region during the study period.

The region substantially increased its imports of softwood lumber from 1967 to 1976. In value, imports grew at an average rate of 19 percent per year and, in volume, at 8 percent annually.

Again, Canada was the only practical foreign supplier of softwood lumber to the region. Brazil, some European countries, and Taiwan shipped relatively small quantities to the Great Lakes at various times during the period.

Paper and Paperboard

Although paper and paperboard constituted only 5 percent (by value) of the region's 1976 forest product imports, that 5 percent was 53 percent of total U.S. paper and paperboard imports.

Not only was the region the largest U.S. importer, it was also one of the fastest growing importing regions. The 1976 imports of 328,000 tons were worth $115.3 million, which represented increases over the 1967 volume and value of 106 percent and 297 percent, respectively. The equivalent annual growth rates are 8 percent in volume and 13 percent in value.

Canada was the most important foreign source of paper and paperboard for the region. In an average year, 97 percent (by value) of the region's imports originated in Canada.

European countries, especially Finland, provided most of the paper and paperboard which was not imported from Canada. In 1976 the $2.4 million imported from Finland was 2 percent of the region's total paper and paperboard imports.

Reconstituted Wood

Until 1975-76, the Great Lakes region was not a large importer, relative to other regions, of reconstituted wood. In most years, the region imported between 3 percent and 10 percent of the value of all U.S. reconstituted wood imports, but, in 1975, that share grew to 33 percent, and in 1976, ballooned to 73 percent. In those two years the nominal value of imports increased by 551 percent, and the volume gained 16,512 percent over the 1974 volume and value.

Canada was usually the only important source of reconstituted wood for the region, but in several years West Germany and other European countries were responsible for significant portions of the region's imports. For example, in 1974 Canada shipped 76 percent (by value) of the region's imports, and West Germany provided the other 24 percent. But in 1975 and 1976 Canada provided more than 99 percent of the Great Lakes reconstituted wood imports.

Wood Veneer

In an average year during 1967-76, the Great Lakes region was the entrance for 47 percent (by value) of all U.S. imports of wood veneer. The 1976 imports of 716.5 million square feet (worth $32.9 million) represented 44 percent of the value of the country's total veneer imports.

The region's veneer imports declined slightly during 1967-76. The 1976 volume was 10 percent less than the 1967 volume, and the 1976 nominal value was 29 percent higher than the 1967 value. On an equivalent annual basis, the nominal value of veneer imports increased by just 3 percent per year during the study period.

Most of the region's imports were shipped from Canada. In an average year, 94 percent of the value of the region's imports were of Canadian origin. The remaining 6 percent was supplied largely by African nations. The Ivory Coast and Zaire were the most important African veneer exporters.

Other Forest Products

The remaining forest products (pulpwood except chips, softwood logs, hardwood logs, hardwood lumber, plywood, pulpwood chips, other industrial roundwood, and building board) in 1976 constituted only 1 percent (by value) of the region's total forest product imports. However, the Great Lakes was responsible for importing a substantial share of the total U.S. imports of several of these commodities.

In 1976, 22 percent (by volume) of the country's hardwood lumber imports entered the United States through the Great Lakes region. The 58.3 million board-feet imported by the region was, however, only 59 percent of the volume imported in 1967, when the region was responsible for 31 percent of the country's total hardwood lumber imports.

Canada was the dominant foreign supplier of hardwood lumber to the region throughout 1967-76. Its 1976 exports to the region represented more than 99 percent of both the volume and value of the Great Lakes total hardwood lumber imports.

The region's 1976 imports of pulpwood except chips represented 19 percent of the value of U.S. total imports of that commodity. Imports to the Great Lakes of pulpwood except chips had been declining throughout 1967-76, so that the 1976 volume and nominal value were only 10 percent and 22 percent of the respective 1967 import figures. During that period, Canada was the region's only source of pulpwood imports.

The Great Lakes' 1976 imports of pulpwood chips (worth $4.5 million) represented 19 percent of all U.S. pulpwood chip imports. The region grew substantially in chip imports during 1967-76. The volume of imports increased at an average annual rate of 19 percent, and nominal value increased at 36 percent per year. Canada was practically the sole supplier of pulpwood chips to the region in every year of the study period.

EXPORTS
SOFTWOOD LOGS (2422)
GREAT LAKES

VALUE--MILLION CURRENT DOLLARS

YEAR	CANADA	NETHLNDS	W GERMNY	SWITZLND	ITALY	OTHERS	TOTAL
1967	0.464	0.022	0.000	0.000	0.000	0.000	0.486
1968	0.634	0.000	0.000	0.000	0.011	0.000	0.644
1969	0.317	0.000	0.000	0.005	0.000	0.000	0.322
1970	0.253	0.000	0.000	0.000	0.000	0.000	0.253
1971	0.000	0.000	0.000	0.000	0.000	0.000	0.000
1972	0.251	0.000	0.000	0.008	0.007	0.000	0.266
1973	0.595	0.000	0.000	0.000	0.000	0.000	0.595
1974	0.933	0.000	0.014	0.000	0.001	0.000	0.948
1975	1.283	0.000	0.000	0.000	0.000	0.000	1.283
1976	2.130	0.000	0.000	0.000	0.000	0.000	2.130

QUANTITY--MILLION BF

YEAR	CANADA	NETHLNDS	W GERMNY	SWITZLND	ITALY	OTHERS	TOTAL
1967	8.524	0.040	0.000	0.000	0.000	0.000	8.564
1968	13.535	0.000	0.000	0.000	0.015	0.000	13.550
1969	4.729	0.000	0.000	0.009	0.000	0.000	4.738
1970	1.121	0.000	0.000	0.000	0.000	0.000	1.121
1971	0.000	0.000	0.000	0.000	0.000	0.000	0.000
1972	1.875	0.000	0.000	0.048	0.042	0.000	1.965
1973	3.315	0.000	0.000	0.000	0.000	0.000	3.315
1974	3.735	0.000	0.035	0.000	0.008	0.000	3.778
1975	5.273	0.000	0.000	0.000	0.000	0.000	5.273
1976	10.085	0.000	0.000	0.000	0.000	0.000	10.085

EXPORTS
HARDWOOD LOGS (2423)
GREAT LAKES

VALUE--MILLION CURRENT DOLLARS

YEAR	CANADA	W GERMNY	SWITZLND	ITALY	OTH EURO	OTHERS	TOTAL
1967	2.115	2.742	0.638	1.077	0.273	0.003	6.847
1968	1.986	3.979	1.271	0.967	0.562	0.031	8.796
1969	2.059	4.434	0.889	0.949	0.795	0.298	9.425
1970	1.549	5.192	1.243	0.731	0.482	0.000	9.196
1971	2.263	4.608	1.363	0.551	0.560	0.165	9.510
1972	2.617	5.969	1.347	0.925	0.993	0.037	11.889
1973	3.169	4.846	1.054	0.961	1.528	0.000	11.559
1974	3.602	3.775	0.368	0.875	1.445	0.000	10.066
1975	2.565	6.074	0.667	0.659	1.236	0.240	11.442
1976	4.698	6.269	0.551	1.203	1.291	0.000	14.011

QUANTITY--MILLION BF

YEAR	CANADA	W GERMNY	SWITZLND	ITALY	OTH EURO	OTHERS	TOTAL
1967	4.441	2.623	0.784	1.551	0.958	0.002	10.360
1968	3.300	3.362	1.222	1.223	0.763	0.028	9.898
1969	3.323	4.054	0.971	1.376	1.060	0.258	11.043
1970	2.664	4.534	1.215	0.828	0.817	0.000	10.058
1971	3.123	3.813	1.233	0.677	0.697	0.138	9.681
1972	7.677	6.394	1.507	1.062	0.972	0.038	17.650
1973	9.463	5.497	1.336	1.494	1.530	0.000	19.320
1974	10.735	4.079	0.528	1.256	1.353	0.000	17.950
1975	8.713	5.174	0.632	0.866	0.921	0.315	16.622
1976	16.675	5.609	0.749	1.590	1.607	0.000	26.231

EXPORTS
SOFTWOOD LUMBER (2432)
GREAT LAKES

VALUE--MILLION CURRENT DOLLARS

YEAR	CANADA	OTH N AM	S AMERIC	EUROPE	ASIA	OTHERS	TOTAL
1967	14.973	0.000	0.000	0.044	0.000	0.000	15.018
1968	17.072	0.000	0.000	0.026	0.000	0.000	17.098
1969	19.438	0.000	0.000	0.009	0.005	0.000	19.451
1970	16.331	0.000	0.000	0.026	0.000	0.013	16.369
1971	19.688	0.000	0.000	0.004	0.001	0.000	19.693
1972	29.253	0.000	0.000	0.083	0.013	0.041	29.390
1973	42.435	0.000	0.000	0.102	0.008	0.005	42.550
1974	45.551	0.000	0.000	0.033	0.000	0.000	45.583
1975	46.643	0.000	0.000	0.000	0.000	0.000	46.643
1976	67.681	0.006	0.000	0.115	0.339	0.000	68.142

QUANTITY--MILLION BF

YEAR	CANADA	OTH N AM	S AMERIC	EUROPE	ASIA	OTHERS	TOTAL
1967	115.826	0.000	0.000	0.294	0.000	0.000	116.120
1968	120.310	0.000	0.000	0.168	0.000	0.000	120.478
1969	121.324	0.000	0.000	0.047	0.023	0.000	121.394
1970	117.316	0.000	0.000	0.133	0.000	0.096	117.545
1971	117.527	0.000	0.000	0.014	0.004	0.000	117.546
1972	165.093	0.000	0.000	0.417	0.125	0.150	165.785
1973	212.103	0.000	0.000	0.575	0.039	0.014	212.731
1974	175.894	0.000	0.000	0.148	0.000	0.000	176.042
1975	189.900	0.000	0.000	0.000	0.000	0.000	189.900
1976	226.532	0.033	0.000	0.363	0.609	0.000	227.537

EXPORTS
HARDWOOD LUMBER (2433)
GREAT LAKES

VALUE--MILLION CURRENT DOLLARS

YEAR	CANADA	OTH N AM	S AMERIC	EUROPE	ASIA	OTHERS	TOTAL
1967	14.094	0.000	0.000	0.380	0.000	0.000	14.474
1968	9.699	0.001	0.000	0.591	0.015	0.000	10.306
1969	10.887	0.000	0.000	1.509	0.000	0.016	12.412
1970	8.906	0.002	0.000	1.666	0.066	0.018	10.658
1971	9.937	0.000	0.015	1.333	1.213	0.018	12.517
1972	17.453	0.000	0.000	0.909	0.667	0.047	19.075
1973	21.119	0.000	0.000	3.697	0.367	0.145	25.328
1974	23.533	0.000	0.000	4.348	0.031	0.055	27.966
1975	23.840	0.009	0.010	5.171	0.114	0.000	29.143
1976	36.101	0.000	0.000	9.766	0.055	0.000	45.923

QUANTITY--MILLION BF

YEAR	CANADA	OTH N AM	S AMERIC	EUROPE	ASIA	OTHERS	TOTAL
1967	101.541	0.000	0.000	1.480	0.000	0.000	103.021
1968	60.991	0.008	0.000	1.898	0.023	0.000	62.920
1969	57.748	0.000	0.000	3.791	0.000	0.040	61.580
1970	46.120	0.001	0.000	4.041	0.130	0.046	50.338
1971	62.105	0.000	0.006	3.359	3.396	0.048	68.915
1972	92.698	0.000	0.000	3.465	2.265	0.108	98.536
1973	141.265	0.000	0.000	7.103	0.570	0.333	149.272
1974	88.277	0.000	0.000	8.191	0.049	0.099	96.616
1975	193.581	0.003	0.007	7.749	0.154	0.000	201.494
1976	117.778	0.000	0.000	15.894	0.071	0.000	133.744

EXPORTS
WOOD VENEER (6311)
GREAT LAKES

VALUE--MILLION CURRENT DOLLARS

YEAR	CANADA	S AMERIC	W GERMNY	OTH EURO	ASIA	OTHERS	TOTAL
1967	3.468	0.000	0.605	0.144	0.000	0.000	4.217
1968	3.951	0.004	0.962	0.163	0.043	0.000	5.122
1969	4.351	0.000	0.920	0.385	0.021	0.002	5.679
1970	3.305	0.000	1.072	0.083	0.020	0.000	4.481
1971	2.725	0.000	0.912	0.104	0.080	0.012	3.833
1972	4.138	0.000	0.133	0.120	0.000	0.045	4.437
1973	5.692	0.000	0.423	0.361	0.005	0.215	6.696
1974	7.611	0.000	0.010	0.161	0.000	0.007	7.789
1975	8.586	0.000	0.000	0.316	0.044	0.000	8.946
1976	5.518	0.000	0.128	0.137	0.015	0.012	5.810

QUANTITY--MIL SQ FT

YEAR	CANADA	S AMERIC	W GERMNY	OTH EURO	ASIA	OTHERS	TOTAL
1967	65.584	0.000	0.442	0.237	0.000	0.000	66.263
1968	78.118	0.005	0.669	0.135	0.029	0.000	78.956
1969	92.908	0.000	0.706	0.371	0.015	0.001	94.002
1970	69.634	0.000	0.985	0.149	0.008	0.000	70.776
1971	57.622	0.000	0.999	0.088	0.097	0.023	58.828
1972	84.461	0.000	0.104	0.191	0.000	0.067	84.823
1973	98.380	0.000	0.562	0.308	0.003	0.259	99.512
1974	103.606	0.000	0.008	0.273	0.000	0.007	103.894
1975	147.187	0.000	0.000	0.797	0.074	0.000	148.058
1976	94.272	0.000	0.469	0.134	0.054	0.015	94.944

EXPORTS
PLYWOOD (6312)
GREAT LAKES

VALUE--MILLION CURRENT DOLLARS

YEAR	CANADA	S AMERIC	EUROPE	ASIA	AFRICA	OTHERS	TOTAL
1967	0.857	0.000	0.000	0.000	0.000	0.000	0.857
1968	0.999	0.000	0.006	0.000	0.000	0.000	1.005
1969	4.486	0.000	0.008	0.000	0.011	0.000	4.505
1970	1.034	0.000	0.002	0.000	0.000	0.000	1.037
1971	1.378	0.000	0.000	0.000	0.000	0.000	1.378
1972	8.579	0.000	0.082	0.000	0.000	0.000	8.661
1973	9.867	0.000	0.005	0.000	0.000	0.000	9.872
1974	32.845	0.000	0.011	0.000	0.000	0.000	32.856
1975	50.067	0.000	0.028	0.020	0.000	0.000	50.115
1976	24.537	0.000	0.000	0.000	0.000	0.000	24.537

QUANTITY--MIL SQ FT

YEAR	CANADA	S AMERIC	EUROPE	ASIA	AFRICA	OTHERS	TOTAL
1967	3.907	0.000	0.000	0.000	0.000	0.000	3.907
1968	7.110	0.000	0.017	0.000	0.000	0.000	7.127
1969	39.302	0.000	0.023	0.000	0.079	0.000	39.404
1970	54.655	0.000	0.002	0.000	0.000	0.000	54.657
1971	8.657	0.000	0.000	0.000	0.000	0.000	8.657
1972	56.456	0.000	0.100	0.000	0.000	0.000	56.556
1973	81.866	0.000	0.008	0.000	0.000	0.000	81.874
1974	190.530	0.000	0.026	0.000	0.000	0.000	190.556
1975	283.375	0.000	0.042	0.040	0.000	0.000	283.458
1976	121.632	0.000	0.000	0.000	0.000	0.000	121.632

EXPORTS
RECONSTITUTED WOOD (6314)
GREAT LAKES

VALUE--MILLION CURRENT DOLLARS

YEAR	CANADA	OTH N AM	NETHLNDS	W GERMNY	OTH EURO	OTHERS	TOTAL
1967	0.290	0.000	0.000	0.000	0.000	0.000	0.290
1968	0.277	0.000	0.000	0.000	0.000	0.000	0.277
1969	1.423	0.000	0.000	0.000	0.000	0.000	1.423
1970	0.903	0.000	0.000	0.000	0.000	0.000	0.903
1971	1.732	0.000	0.000	0.001	0.000	0.000	1.733
1972	3.502	0.000	0.000	0.000	0.000	0.000	3.502
1973	5.619	0.000	0.002	0.003	0.000	0.000	5.624
1974	8.552	0.000	0.000	0.000	0.000	0.000	8.552
1975	6.134	0.000	0.000	0.000	0.000	0.000	6.134
1976	4.516	0.000	0.000	0.000	0.000	0.000	4.516

QUANTITY--MIL SQ FT

YEAR	CANADA	OTH N AM	NETHLNDS	W GERMNY	OTH EURO	OTHERS	TOTAL
1967	0.452	0.000	0.000	0.000	0.000	0.000	0.452
1968	1.567	0.000	0.000	0.000	0.000	0.000	1.567
1969	11.146	0.000	0.000	0.000	0.000	0.000	11.146
1970	6.714	0.000	0.000	0.000	0.000	0.000	6.714
1971	12.495	0.000	0.000	0.003	0.000	0.000	12.498
1972	29.539	0.000	0.000	0.000	0.000	0.000	29.539
1973	42.548	0.000	0.000	0.002	0.000	0.000	42.551
1974	61.056	0.000	0.000	0.000	0.000	0.000	61.056
1975	40.541	0.000	0.000	0.000	0.000	0.000	40.541
1976	21.010	0.000	0.000	0.000	0.000	0.000	21.010

EXPORTS
PULPWOOD CHIPS (631.8320)
GREAT LAKES

VALUE--MILLION CURRENT DOLLARS

YEAR	CANADA	OTH N AM	EUROPE	JAPAN	OTH ASIA	OTHERS	TOTAL
1967	0.168	0.000	0.000	0.000	0.000	0.000	0.168
1968	0.226	0.000	0.000	0.000	0.000	0.000	0.226
1969	0.262	0.000	0.000	0.000	0.000	0.000	0.262
1970	0.016	0.000	0.000	0.000	0.000	0.000	0.016
1971	0.033	0.000	0.000	0.000	0.000	0.000	0.033
1972	0.024	0.000	0.000	0.000	0.000	0.000	0.024
1973	0.576	0.000	0.000	0.000	0.000	0.000	0.576
1974	0.058	0.000	0.000	0.000	0.000	0.000	0.058
1975	0.706	0.000	0.000	0.000	0.000	0.000	0.706
1976	1.468	0.000	0.000	0.000	0.000	0.000	1.468

QUANTITY--1000 STN

YEAR	CANADA	OTH N AM	EUROPE	JAPAN	OTH ASIA	OTHERS	TOTAL
1967	10.698	0.000	0.000	0.000	0.000	0.000	10.698
1968	14.736	0.000	0.000	0.000	0.000	0.000	14.736
1969	16.760	0.000	0.000	0.000	0.000	0.000	16.760
1970	0.648	0.000	0.000	0.000	0.000	0.000	0.648
1971	1.377	0.000	0.000	0.000	0.000	0.000	1.377
1972	1.007	0.000	0.000	0.000	0.000	0.000	1.007
1973	19.338	0.000	0.000	0.000	0.000	0.000	19.338
1974	2.202	0.000	0.000	0.000	0.000	0.000	2.202
1975	13.222	0.000	0.000	0.000	0.000	0.000	13.222
1976	31.374	0.000	0.000	0.000	0.000	0.000	31.374

EXPORTS
PULPWOOD (EXC CHIPS)(2421)
GREAT LAKES

VALUE--MILLION CURRENT DOLLARS

YEAR	CANADA	OTH N AM	S AMERIC	BELGIUM	OTH EURO	OTHERS	TOTAL
1967	1.346	0.000	0.000	0.000	0.000	0.000	1.346
1968	1.447	0.000	0.000	0.000	0.000	0.000	1.447
1969	0.801	0.000	0.000	0.003	0.000	0.000	0.803
1970	0.774	0.000	0.000	0.000	0.000	0.000	0.774
1971	0.000	0.000	0.000	0.000	0.000	0.000	0.000
1972	2.412	0.000	0.000	0.000	0.000	0.000	2.412
1973	1.297	0.000	0.000	0.000	0.000	0.000	1.297
1974	3.785	0.000	0.000	0.000	0.000	0.000	3.785
1975	4.179	0.000	0.000	0.000	0.000	0.000	4.179
1976	6.310	0.000	0.000	0.000	0.000	0.000	6.310

QUANTITY--1000 CORDS

YEAR	CANADA	OTH N AM	S AMERIC	BELGIUM	OTH EURO	OTHERS	TOTAL
1967	48.634	0.000	0.000	0.000	0.000	0.000	48.634
1968	53.574	0.000	0.000	0.000	0.000	0.000	53.574
1969	28.250	0.000	0.000	0.073	0.000	0.000	28.323
1970	26.007	0.000	0.000	0.000	0.000	0.000	26.007
1971	0.000	0.000	0.000	0.000	0.000	0.000	0.000
1972	80.870	0.000	0.000	0.000	0.000	0.000	80.870
1973	43.843	0.000	0.000	0.000	0.000	0.000	43.843
1974	111.500	0.000	0.000	0.000	0.000	0.000	111.500
1975	125.245	0.000	0.000	0.000	0.000	0.000	125.245
1976	152.630	0.000	0.000	0.000	0.000	0.000	152.630

EXPORTS
WOOD PULP (2510)
GREAT LAKES

VALUE--MILLION CURRENT DOLLARS

YEAR	CANADA	OTH N AM	S AMERIC	EUROPE	ASIA	OTHERS	TOTAL
1967	2.275	0.000	0.000	0.019	0.001	0.000	2.294
1968	3.447	0.001	0.000	0.160	0.000	0.000	3.608
1969	2.370	0.000	0.000	0.000	0.001	0.000	2.371
1970	6.586	0.000	0.000	0.000	0.000	0.000	6.586
1971	7.175	0.000	0.000	0.000	0.000	0.000	7.175
1972	7.341	0.000	0.000	0.000	0.000	0.000	7.341
1973	6.988	0.000	0.000	0.030	0.000	0.000	7.018
1974	10.316	0.000	0.000	0.000	0.000	0.000	10.316
1975	9.662	0.000	0.000	0.000	0.000	0.000	9.662
1976	17.056	0.000	0.000	0.002	0.000	0.000	17.058

QUANTITY--1000 STN

YEAR	CANADA	OTH N AM	S AMERIC	EUROPE	ASIA	OTHERS	TOTAL
1967	14.723	0.000	0.000	0.059	0.002	0.000	14.784
1968	22.453	0.005	0.000	0.952	0.000	0.000	23.409
1969	15.037	0.000	0.000	0.000	0.003	0.000	15.040
1970	45.581	0.000	0.000	0.000	0.000	0.000	45.581
1971	47.129	0.000	0.000	0.000	0.000	0.000	47.129
1972	52.314	0.000	0.000	0.000	0.000	0.000	52.314
1973	55.270	0.000	0.000	0.175	0.000	0.000	55.445
1974	43.045	0.000	0.000	0.000	0.000	0.000	43.045
1975	29.947	0.000	0.000	0.000	0.000	0.000	29.947
1976	48.128	0.000	0.000	0.004	0.000	0.000	48.132

EXPORTS
NEWSPRINT (6411)
GREAT LAKES

VALUE--MILLION CURRENT DOLLARS

YEAR	CANADA	VENEZUEL	PERU	LEBANON	JAPAN	OTHERS	TOTAL
1967	0.129	0.000	0.000	0.000	0.000	0.005	0.134
1968	0.149	0.000	0.056	0.000	0.000	0.000	0.205
1969	0.242	0.000	0.000	0.000	0.000	0.000	0.242
1970	0.196	0.000	0.000	0.000	0.000	0.000	0.196
1971	0.179	0.000	0.000	0.000	0.001	0.000	0.179
1972	0.639	0.000	0.000	0.000	0.000	0.031	0.671
1973	0.263	0.000	0.000	0.000	0.000	0.000	0.263
1974	0.609	0.034	0.000	0.521	0.000	0.140	1.304
1975	1.610	0.000	0.157	0.000	0.000	0.000	1.768
1976	1.683	0.034	0.000	0.000	0.000	0.000	1.717

QUANTITY--1000 STN

YEAR	CANADA	VENEZUEL	PERU	LEBANON	JAPAN	OTHERS	TOTAL
1967	0.517	0.000	0.000	0.000	0.000	0.057	0.574
1968	0.593	0.000	0.448	0.000	0.000	0.000	1.041
1969	1.392	0.000	0.000	0.000	0.000	0.000	1.392
1970	0.820	0.000	0.000	0.000	0.000	0.000	0.820
1971	1.047	0.000	0.000	0.000	0.002	0.000	1.049
1972	3.706	0.000	0.000	0.000	0.000	0.063	3.769
1973	1.524	0.000	0.000	0.000	0.000	0.000	1.524
1974	2.664	0.114	0.000	1.343	0.000	0.464	4.585
1975	3.106	0.000	0.461	0.000	0.000	0.000	3.568
1976	3.191	0.100	0.000	0.000	0.000	0.000	3.291

EXPORTS
BUILDING BOARD (6416)
GREAT LAKES

VALUE--MILLION CURRENT DOLLARS

YEAR	CANADA	OTH N AM	EUROPE	ASIA	AFRICA	OTHERS	TOTAL
1967	2.655	0.001	0.038	0.000	0.000	0.000	2.694
1968	3.048	0.004	0.033	0.000	0.050	0.000	3.136
1969	4.521	0.002	0.011	0.000	0.046	0.000	4.581
1970	4.501	0.003	0.020	0.005	0.022	0.000	4.550
1971	4.845	0.030	0.009	0.001	0.719	0.000	5.604
1972	6.210	0.000	0.076	0.000	0.022	0.000	6.308
1973	12.964	0.001	0.138	0.000	0.200	0.000	13.303
1974	11.205	0.003	0.180	0.000	0.022	0.000	11.409
1975	11.776	0.000	0.075	0.000	0.000	0.000	11.851
1976	14.320	0.000	0.073	0.112	0.000	0.000	14.505

QUANTITY--1000 STN

YEAR	CANADA	OTH N AM	EUROPE	ASIA	AFRICA	OTHERS	TOTAL
1967	10.734	0.003	0.077	0.000	0.000	0.000	10.814
1968	14.718	0.007	0.084	0.000	0.151	0.000	14.960
1969	21.552	0.005	0.038	0.000	0.147	0.000	21.742
1970	24.849	0.004	0.053	0.011	0.063	0.000	24.980
1971	26.068	0.129	0.032	0.003	1.920	0.000	28.153
1972	33.303	0.000	0.182	0.000	0.069	0.000	33.553
1973	66.053	0.002	0.228	0.000	0.489	0.000	66.771
1974	45.819	0.004	0.495	0.000	0.028	0.000	46.346
1975	44.780	0.000	0.093	0.000	0.000	0.000	44.873
1976	50.613	0.000	0.109	0.298	0.000	0.000	51.021

EXPORTS
PAPER & PAPERBOARD (6410)
GREAT LAKES

VALUE--MILLION CURRENT DOLLARS

YEAR	CANADA	OTH N AM	EUROPE	ASIA	AFRICA	OTHERS	TOTAL
1967	43.059	0.046	0.197	0.008	0.026	0.102	43.438
1968	46.274	0.096	0.667	0.058	0.182	0.289	47.565
1969	49.892	0.044	4.667	0.036	0.074	0.301	55.014
1970	49.091	0.139	2.771	0.010	0.072	0.005	52.088
1971	61.313	0.108	4.438	0.169	0.500	0.130	66.657
1972	70.180	0.191	3.542	0.097	0.208	0.015	74.233
1973	84.014	0.095	4.386	0.038	0.390	0.003	88.927
1974	125.268	0.081	0.928	0.583	0.049	0.113	127.022
1975	169.564	0.020	0.989	0.003	0.015	0.118	170.710
1976	216.627	0.112	3.425	0.007	0.276	0.061	220.508

QUANTITY--1000 STN

YEAR	CANADA	OTH N AM	EUROPE	ASIA	AFRICA	OTHERS	TOTAL
1967	107.314	0.060	0.200	0.015	0.103	0.420	108.112
1968	141.710	0.305	0.827	0.204	0.339	1.421	144.805
1969	146.335	0.218	7.066	0.306	0.157	1.448	155.530
1970	129.638	1.011	3.226	0.010	0.144	0.005	134.034
1971	171.941	0.032	4.670	0.304	0.711	0.137	177.795
1972	195.080	0.047	4.043	0.129	0.287	0.018	199.605
1973	220.769	0.040	5.127	0.066	0.970	0.001	226.973
1974	305.637	0.015	1.190	0.790	0.073	0.338	308.043
1975	403.844	0.005	1.024	0.008	0.022	0.225	405.127
1976	454.544	0.136	3.331	0.020	0.440	0.085	458.557

IMPORTS
SOFTWOOD LOGS (2422)
GREAT LAKES

VALUE--MILLION CURRENT DOLLARS

YEAR	CANADA	OTH N AM	S AMERIC	U KINGDM	ASIA	OTHERS	TOTAL
1967	0.001	0.000	0.000	0.000	0.000	0.000	0.001
1968	0.000	0.000	0.000	0.000	0.000	0.000	0.000
1969	0.000	0.000	0.000	0.000	0.000	0.000	0.001
1970	0.013	0.000	0.000	0.000	0.000	0.000	0.013
1971	0.005	0.000	0.000	0.000	0.000	0.000	0.005
1972	0.005	0.000	0.000	0.000	0.000	0.000	0.005
1973	0.006	0.000	0.000	0.000	0.000	0.000	0.006
1974	0.002	0.000	0.000	0.000	0.000	0.000	0.002
1975	0.004	0.000	0.000	0.000	0.000	0.000	0.004
1976	0.000	0.000	0.000	0.000	0.000	0.000	0.000

QUANTITY--MILLION BF

YEAR	CANADA	OTH N AM	S AMERIC	U KINGDM	ASIA	OTHERS	TOTAL
1967	0.186	0.000	0.000	0.000	0.000	0.000	0.186
1968	0.000	0.000	0.000	0.000	0.000	0.000	0.000
1969	0.004	0.000	0.000	0.000	0.000	0.000	0.004
1970	0.135	0.000	0.000	0.000	0.000	0.000	0.135
1971	0.044	0.000	0.000	0.000	0.000	0.000	0.044
1972	0.169	0.000	0.000	0.000	0.000	0.000	0.169
1973	0.044	0.000	0.000	0.000	0.000	0.000	0.044
1974	0.012	0.000	0.000	0.000	0.000	0.000	0.012
1975	0.019	0.000	0.000	0.000	0.000	0.000	0.019
1976	0.000	0.000	0.000	0.000	0.000	0.000	0.000

IMPORTS
HARDWOOD LOGS (2423)
GREAT LAKES

VALUE--MILLION CURRENT DOLLARS

YEAR	CANADA	BRAZIL	W GERMNY	ASIA	IVRY CST	OTHERS	TOTAL
1967	1.052	0.008	0.000	0.000	0.000	0.000	1.060
1968	1.099	0.000	0.000	0.000	0.000	0.000	1.099
1969	1.315	0.000	0.001	0.000	0.000	0.000	1.316
1970	0.974	0.000	0.000	0.000	0.000	0.000	0.974
1971	0.903	0.000	0.000	0.000	0.000	0.000	0.903
1972	0.972	0.000	0.000	0.000	0.000	0.000	0.972
1973	0.703	0.000	0.000	0.000	0.012	0.000	0.715
1974	0.551	0.000	0.000	0.000	0.000	0.000	0.551
1975	0.261	0.000	0.000	0.000	0.000	0.000	0.261
1976	0.218	0.000	0.000	0.000	0.000	0.000	0.218

QUANTITY--MILLION BF

YEAR	CANADA	BRAZIL	W GERMNY	ASIA	IVRY CST	OTHERS	TOTAL
1967	5.443	0.004	0.000	0.000	0.000	0.000	5.447
1968	5.333	0.000	0.000	0.000	0.000	0.000	5.333
1969	6.298	0.000	0.000	0.000	0.000	0.000	6.298
1970	4.711	0.000	0.000	0.000	0.000	0.000	4.711
1971	3.969	0.000	0.000	0.000	0.000	0.000	3.969
1972	4.342	0.000	0.000	0.000	0.000	0.000	4.342
1973	3.654	0.000	0.000	0.000	0.021	0.000	3.675
1974	2.969	0.000	0.000	0.000	0.000	0.000	2.969
1975	1.121	0.000	0.000	0.000	0.000	0.000	1.121
1976	1.059	0.000	0.000	0.000	0.000	0.000	1.059

IMPORTS
SOFTWOOD LUMBER (2432)
GREAT LAKES

VALUE--MILLION CURRENT DOLLARS

YEAR	CANADA	BRAZIL	W GERMNY	OTH EURO	TAIWAN	OTHERS	TOTAL
1967	112.935	0.000	0.000	0.000	0.001	0.000	112.936
1968	175.757	0.000	0.004	0.000	0.039	0.001	175.801
1969	195.669	0.000	0.001	0.000	0.000	0.008	195.678
1970	167.610	0.000	0.000	0.000	0.023	0.000	167.633
1971	265.242	0.000	0.000	0.000	0.024	0.021	265.286
1972	410.982	0.000	0.000	0.001	0.010	0.000	410.993
1973	545.036	0.000	0.000	0.005	0.000	0.008	545.049
1974	347.082	0.007	0.000	0.003	0.060	0.001	347.153
1975	284.382	0.000	0.000	0.000	0.000	0.000	284.382
1976	549.724	0.000	0.004	0.000	0.000	0.000	549.729

QUANTITY--MILLION BF

YEAR	CANADA	BRAZIL	W GERMNY	OTH EURO	TAIWAN	OTHERS	TOTAL
1967	1618.985	0.000	0.000	0.000	0.005	0.000	1618.990
1968	2002.747	0.000	0.003	0.000	0.222	0.013	2002.985
1969	2075.529	0.000	0.000	0.000	0.000	0.076	2075.605
1970	2168.498	0.000	0.000	0.000	0.098	0.000	2168.596
1971	2721.367	0.000	0.000	0.000	0.083	0.203	2721.653
1972	3312.245	0.000	0.000	0.000	0.034	0.000	3312.279
1973	3569.033	0.000	0.000	0.095	0.000	0.052	3569.180
1974	2630.398	0.017	0.000	0.000	0.126	0.000	2630.541
1975	2295.542	0.000	0.000	0.000	0.000	0.000	2295.542
1976	3234.655	0.000	0.005	0.000	0.000	0.000	3234.660

IMPORTS
HARDWOOD LUMBER (2433)
GREAT LAKES

VALUE--MILLION CURRENT DOLLARS

YEAR	CANADA	OTH N AM	EUROPE	MALAYSIA	OTH ASIA	OTHERS	TOTAL
1967	21.456	0.000	0.096	0.009	0.047	0.057	21.664
1968	20.667	0.000	0.040	0.036	0.042	0.025	20.809
1969	23.412	0.000	0.037	0.027	0.055	0.000	23.531
1970	17.622	0.012	0.013	0.085	0.028	0.008	17.767
1971	19.471	0.005	0.004	0.196	0.065	0.051	19.792
1972	24.069	0.000	0.091	0.234	0.128	0.059	24.601
1973	28.410	0.238	0.131	0.023	0.136	0.042	28.979
1974	23.578	0.000	0.021	0.007	0.036	0.004	23.646
1975	14.463	0.000	0.093	0.031	0.008	0.000	14.597
1976	19.523	0.000	0.075	0.025	0.000	0.000	19.624

QUANTITY--MILLION BF

YEAR	CANADA	OTH N AM	EUROPE	MALAYSIA	OTH ASIA	OTHERS	TOTAL
1967	97.854	0.000	0.375	0.065	0.181	0.226	98.701
1968	92.695	0.000	0.149	0.142	0.167	0.146	93.299
1969	111.010	0.000	0.161	0.098	0.169	0.000	111.437
1970	82.778	0.015	0.030	0.247	0.106	0.034	83.211
1971	84.367	0.030	0.002	0.674	0.312	0.264	85.650
1972	88.343	0.000	0.145	0.733	0.483	0.203	89.907
1973	93.642	0.636	0.107	0.075	0.409	0.331	95.200
1974	72.349	0.000	0.003	0.013	0.049	0.012	72.426
1975	44.966	0.000	0.023	0.100	0.033	0.000	45.122
1976	58.240	0.000	0.010	0.078	0.000	0.000	58.328

IMPORTS
WOOD VENEER (6311)
GREAT LAKES

VALUE--MILLION CURRENT DOLLARS

YEAR	EUROPE	ASIA	IVRY CST	ZAIRE	OTH AFRI	OTHERS	TOTAL
1967	0.161	0.201	0.062	1.745	0.327	23.043	25.539
1968	0.304	0.309	0.000	2.065	0.000	23.407	26.084
1969	0.085	0.176	0.000	0.536	0.000	19.184	19.980
1970	0.075	0.044	0.168	1.020	0.734	16.666	18.707
1971	0.072	0.019	0.507	0.972	0.000	20.088	21.658
1972	0.213	0.299	0.660	0.246	0.075	27.186	28.679
1973	0.037	0.006	0.559	0.228	0.000	32.661	33.491
1974	0.043	0.000	0.000	0.000	0.000	31.411	31.454
1975	0.044	0.035	0.991	0.387	0.000	24.255	25.712
1976	0.204	0.000	0.010	0.000	0.000	32.692	32.906

QUANTITY--MIL SQ FT

YEAR	EUROPE	ASIA	IVRY CST	ZAIRE	OTH AFRI	OTHERS	TOTAL
1967	0.140	2.462	0.537	16.594	2.180	712.345	734.260
1968	0.378	4.674	0.000	19.696	0.000	745.100	769.848
1969	0.125	1.352	0.000	4.893	0.000	634.481	640.852
1970	0.150	0.743	1.418	9.777	8.530	589.387	610.004
1971	0.432	0.284	2.936	8.703	0.000	714.763	727.117
1972	0.199	3.927	5.386	2.125	0.140	851.170	862.846
1973	0.056	0.004	4.066	1.693	0.000	822.659	828.468
1974	0.020	0.000	0.000	0.000	0.000	651.259	651.279
1975	0.020	0.000	7.271	2.696	0.000	520.064	530.051
1976	0.084	0.000	0.068	0.000	0.000	715.877	716.029

IMPORTS
PLYWOOD (6312)
GREAT LAKES

VALUE--MILLION CURRENT DOLLARS

YEAR	PHIL REP	KOR REP	TAIWAN	JAPAN	OTH ASIA	OTHERS	TOTAL
1967	0.083	0.000	0.767	1.902	0.016	5.746	8.515
1968	0.041	0.000	1.021	2.653	0.500	6.222	10.437
1969	0.068	0.020	2.145	3.644	1.628	7.110	14.614
1970	2.447	0.680	2.014	1.603	1.208	3.995	11.947
1971	2.917	0.000	2.978	1.400	3.470	7.101	17.867
1972	3.483	0.866	2.703	1.406	3.282	8.339	20.080
1973	2.260	0.013	1.994	1.549	0.345	7.740	13.900
1974	0.000	3.749	0.825	0.084	0.006	5.897	10.561
1975	2.553	3.762	0.020	0.077	0.384	7.334	14.130
1976	0.000	0.240	0.079	0.000	0.000	8.205	8.524

QUANTITY--MIL SQ FT

YEAR	PHIL REP	KOR REP	TAIWAN	JAPAN	OTH ASIA	OTHERS	TOTAL
1967	0.870	0.003	7.107	12.892	0.159	50.847	71.876
1968	0.401	0.000	8.924	14.306	4.649	51.330	79.610
1969	0.576	0.165	19.511	20.908	15.896	54.001	111.057
1970	26.384	7.085	18.734	8.909	11.393	31.785	104.289
1971	28.308	0.000	24.783	7.715	35.637	57.035	153.478
1972	32.539	7.404	24.243	6.323	34.313	575.974	680.797
1973	13.990	0.086	8.671	3.695	1.495	53.222	81.159
1974	0.000	28.179	4.588	0.182	0.151	32.814	65.915
1975	17.580	22.843	0.090	0.386	2.940	45.149	88.988
1976	0.000	1.138	0.379	0.000	0.000	46.579	48.096

IMPORTS
RECONSTITUTED WOOD (6314)
GREAT LAKES

VALUE--MILLION CURRENT DOLLARS

YEAR	CANADA	W GERMNY	OTH EURO	TAIWAN	OTH ASIA	OTHERS	TOTAL
1967	0.053	0.000	0.000	0.000	0.000	0.000	0.053
1968	0.014	0.000	0.000	0.000	0.000	0.000	0.014
1969	0.021	0.011	0.051	0.000	0.000	0.000	0.084
1970	0.112	0.010	0.000	0.000	0.000	0.000	0.122
1971	0.178	0.005	0.000	0.001	0.000	0.000	0.185
1972	0.162	0.005	0.000	0.000	0.000	0.000	0.166
1973	0.429	0.007	0.000	0.000	0.000	0.000	0.436
1974	0.078	0.024	0.000	0.000	0.000	0.000	0.102
1975	1.415	0.000	0.000	0.000	0.000	0.000	1.415
1976	9.168	0.041	0.000	0.000	0.000	0.001	9.210

QUANTITY--MIL SQ FT

YEAR	CANADA	W GERMNY	OTH EURO	TAIWAN	OTH ASIA	OTHERS	TOTAL
1967	0.011	0.000	0.000	0.000	0.000	0.000	0.011
1968	0.067	0.000	0.000	0.000	0.000	0.000	0.067
1969	0.080	0.015	0.473	0.000	0.000	0.000	0.568
1970	0.652	0.014	0.000	0.000	0.000	0.000	0.666
1971	1.053	0.008	0.000	0.002	0.000	0.000	1.062
1972	0.860	0.005	0.000	0.000	0.000	0.000	0.865
1973	2.245	0.007	0.000	0.000	0.000	0.000	2.252
1974	0.292	0.017	0.000	0.000	0.000	0.000	0.309
1975	7.484	0.000	0.000	0.000	0.000	0.000	7.484
1976	51.300	0.030	0.000	0.000	0.000	0.002	51.332

IMPORTS
PULPWOOD CHIPS (631.8320)
GREAT LAKES

VALUE--MILLION CURRENT DOLLARS

YEAR	CANADA	PANAMA	VENEZUEL	W GERMNY	AUSTRALA	OTHERS	TOTAL
1967	0.273	0.000	0.000	0.000	0.000	0.000	0.273
1968	0.119	0.000	0.000	0.000	0.000	0.000	0.119
1969	0.252	0.000	0.000	0.000	0.000	0.000	0.252
1970	0.521	0.000	0.000	0.000	0.001	0.000	0.522
1971	0.729	0.002	0.000	0.000	0.000	0.000	0.731
1972	1.316	0.000	0.000	0.000	0.000	0.000	1.317
1973	2.097	0.000	0.000	0.000	0.000	0.000	2.097
1974	3.000	0.000	0.000	0.000	0.000	0.000	3.000
1975	2.298	0.000	0.000	0.000	0.000	0.000	2.298
1976	4.465	0.000	0.000	0.001	0.000	0.000	4.466

QUANTITY--1000 STN

YEAR	CANADA	PANAMA	VENEZUEL	W GERMNY	AUSTRALA	OTHERS	TOTAL
1967	23.924	0.000	0.000	0.000	0.000	0.000	23.924
1968	12.234	0.000	0.000	0.000	0.000	0.000	12.234
1969	22.216	0.000	0.000	0.000	0.000	0.000	22.216
1970	31.418	0.000	0.000	0.000	0.080	0.000	31.498
1971	32.051	0.071	0.000	0.000	0.000	0.000	32.122
1972	60.592	0.000	0.029	0.000	0.000	0.000	60.621
1973	102.225	0.000	0.000	0.000	0.000	0.000	102.225
1974	124.557	0.000	0.000	0.000	0.000	0.000	124.557
1975	65.055	0.000	0.000	0.000	0.000	0.000	65.055
1976	115.686	0.000	0.000	0.001	0.000	0.000	115.687

IMPORTS
PULPWOOD (EXC CHIPS)(2421)
GREAT LAKES

VALUE--MILLION CURRENT DOLLARS

YEAR	CANADA	OTH N AM	S AMERIC	EUROPE	ASIA	OTHERS	TOTAL
1967	9.791	0.000	0.000	0.000	0.000	0.000	9.791
1968	8.911	0.000	0.000	0.000	0.000	0.000	8.911
1969	6.599	0.001	0.000	0.000	0.000	0.000	6.600
1970	4.815	0.000	0.000	0.000	0.000	0.000	4.815
1971	3.930	0.000	0.000	0.000	0.000	0.000	3.930
1972	2.854	0.000	0.000	0.000	0.000	0.000	2.854
1973	2.975	0.000	0.000	0.000	0.000	0.000	2.975
1974	2.946	0.000	0.000	0.000	0.000	0.000	2.946
1975	1.415	0.000	0.000	0.000	0.000	0.000	1.415
1976	2.175	0.000	0.000	0.000	0.000	0.000	2.175

QUANTITY--1000 CORDS

YEAR	CANADA	OTH N AM	S AMERIC	EUROPE	ASIA	OTHERS	TOTAL
1967	392.054	0.000	0.000	0.000	0.000	0.000	392.054
1968	359.990	0.000	0.000	0.000	0.000	0.000	359.990
1969	266.649	0.040	0.000	0.000	0.000	0.000	266.689
1970	194.451	0.000	0.000	0.000	0.000	0.000	194.451
1971	145.229	0.000	0.000	0.000	0.000	0.000	145.229
1972	88.648	0.000	0.000	0.000	0.000	0.000	88.648
1973	94.396	0.000	0.000	0.000	0.000	0.000	94.396
1974	98.226	0.000	0.000	0.000	0.000	0.000	98.226
1975	41.834	0.000	0.000	0.000	0.000	0.000	41.834
1976	37.766	0.000	0.000	0.000	0.000	0.000	37.766

IMPORTS
WOOD PULP (2510)
GREAT LAKES

VALUE--MILLION CURRENT DOLLARS

YEAR	CANADA	SWEDEN	FINLAND	OTH EURO	ASIA	OTHERS	TOTAL
1967	171.932	2.110	1.777	0.120	0.000	0.000	175.938
1968	207.092	1.137	2.610	0.207	0.000	0.000	211.046
1969	246.994	0.468	1.546	0.163	0.000	0.043	249.213
1970	251.797	0.142	0.391	0.000	0.000	0.000	252.331
1971	221.757	0.001	0.892	0.000	0.000	0.065	222.714
1972	241.350	0.369	0.114	0.000	0.000	0.000	241.833
1973	312.608	2.668	1.008	0.502	0.000	0.020	316.806
1974	546.551	0.000	0.000	0.000	0.000	0.019	546.569
1975	510.328	4.206	0.730	0.000	0.000	0.258	515.522
1976	640.482	0.019	0.000	0.003	0.000	0.000	640.504

QUANTITY--1000 STN

YEAR	CANADA	SWEDEN	FINLAND	OTH EURO	ASIA	OTHERS	TOTAL
1967	1378.698	17.678	20.322	1.173	0.000	0.000	1417.872
1968	1685.692	9.889	34.759	1.139	0.000	0.000	1731.479
1969	1986.093	3.660	17.327	0.896	0.000	0.295	2008.270
1970	1846.431	0.827	5.174	0.000	0.000	0.000	1852.432
1971	1776.715	0.003	9.810	0.000	0.000	0.486	1787.014
1972	1812.208	2.627	2.513	0.000	0.000	0.000	1817.348
1973	1901.261	9.766	8.007	3.774	0.000	0.104	1922.912
1974	2030.328	0.000	0.000	0.000	0.000	0.062	2030.390
1975	1565.225	12.065	2.310	0.001	0.000	0.756	1580.357
1976	1885.459	0.043	0.000	0.009	0.000	0.000	1885.511

IMPORTS
NEWSPRINT (6411)
GREAT LAKES

VALUE--MILLION CURRENT DOLLARS

YEAR	CANADA	SWEDEN	FINLAND	OTH EURO	ASIA	OTHERS	TOTAL
1967	449.661	0.000	0.055	0.000	0.000	0.000	449.717
1968	449.262	0.000	0.000	0.000	0.000	0.000	449.262
1969	516.692	0.047	0.000	0.000	0.000	0.000	516.739
1970	502.354	0.000	0.000	0.000	0.000	0.000	502.354
1971	549.355	0.000	0.000	0.000	0.000	0.000	549.355
1972	577.218	0.000	0.000	0.000	0.000	0.000	577.218
1973	663.370	0.000	0.000	0.000	0.000	0.000	663.370
1974	864.249	0.000	0.000	0.000	0.000	0.000	864.249
1975	834.423	0.000	0.000	0.000	0.000	0.000	834.423
1976	1000.461	0.000	0.000	0.000	0.000	0.000	1000.461

QUANTITY--1000 STN

YEAR	CANADA	SWEDEN	FINLAND	OTH EURO	ASIA	OTHERS	TOTAL
1967	3318.083	0.000	0.510	0.000	0.000	0.000	3318.593
1968	3257.139	0.000	0.000	0.000	0.000	0.000	3257.139
1969	3632.346	0.364	0.000	0.000	0.000	0.000	3632.710
1970	3516.726	0.000	0.000	0.000	0.000	0.000	3516.726
1971	3618.866	0.000	0.000	0.000	0.000	0.000	3618.866
1972	3721.346	0.000	0.000	0.000	0.000	0.000	3721.346
1973	4064.261	0.000	0.000	0.000	0.000	0.000	4064.261
1974	4222.174	0.000	0.000	0.000	0.000	0.000	4222.174
1975	3432.714	0.000	0.000	0.000	0.000	0.000	3432.714
1976	3672.241	0.000	0.000	0.000	0.000	0.000	3672.241

IMPORTS
BUILDING BOARD (6416)
GREAT LAKES

VALUE--MILLION CURRENT DOLLARS

YEAR	CANADA	BRAZIL	EUROPE	ASIA	AFRICA	OTHERS	TOTAL
1967	1.963	0.256	0.313	0.004	0.000	0.000	2.535
1968	3.414	0.040	0.250	0.000	0.158	0.024	3.886
1969	4.782	0.051	0.343	0.000	0.000	0.001	5.176
1970	4.096	0.050	0.113	0.000	0.000	0.000	4.258
1971	4.642	0.118	0.155	0.000	0.000	0.014	4.930
1972	8.501	0.119	0.577	0.000	0.254	0.096	9.547
1973	10.183	0.080	0.336	0.597	0.096	0.112	11.404
1974	7.370	0.139	0.259	0.000	0.000	0.092	7.860
1975	2.983	0.675	0.059	0.000	0.000	0.031	3.748
1976	2.603	1.262	0.096	0.058	0.000	0.000	4.019

QUANTITY--1000 STN

YEAR	CANADA	BRAZIL	EUROPE	ASIA	AFRICA	OTHERS	TOTAL
1967	16.876	2.678	3.953	0.005	0.000	0.000	23.512
1968	31.906	0.359	3.019	0.000	1.894	0.289	37.467
1969	44.972	0.636	4.158	0.000	0.000	0.003	49.769
1970	33.340	0.682	1.146	0.000	0.000	0.000	35.168
1971	41.551	1.477	1.855	0.000	0.000	0.328	45.211
1972	74.098	1.312	5.556	0.000	3.759	1.438	86.164
1973	77.804	0.650	4.098	6.910	1.136	1.262	91.861
1974	41.894	0.847	1.717	0.000	0.000	0.753	45.210
1975	18.699	4.356	0.410	0.000	0.000	0.251	23.716
1976	13.560	7.580	0.797	0.340	0.000	0.000	22.277

IMPORTS
PAPER & PAPERBOARD (6410)
GREAT LAKES

VALUE--MILLION CURRENT DOLLARS

YEAR	CANADA	FINLAND	U KINGDM	OTH EURO	ASIA	OTHERS	TOTAL
1967	28.294	0.427	0.043	0.264	0.000	0.000	29.029
1968	28.226	0.460	0.129	0.235	0.001	0.003	29.054
1969	38.886	0.524	0.182	0.526	0.000	0.000	40.118
1970	45.546	0.577	0.107	0.625	0.001	0.000	46.854
1971	48.922	0.932	0.199	0.300	0.009	0.000	50.362
1972	50.511	0.009	0.241	0.314	0.009	0.000	51.084
1973	75.119	0.170	0.213	0.392	0.012	0.005	75.911
1974	112.956	0.000	0.187	1.066	0.003	0.002	114.213
1975	69.436	4.043	0.475	0.243	0.003	0.000	74.199
1976	111.965	2.363	0.615	0.354	0.006	0.000	115.304

QUANTITY--1000 STN

YEAR	CANADA	FINLAND	U KINGDM	OTH EURO	ASIA	OTHERS	TOTAL
1967	155.506	2.315	0.052	1.022	0.000	0.000	158.895
1968	143.380	2.246	0.226	0.739	0.000	0.003	146.593
1969	196.352	3.123	0.323	1.643	0.000	0.000	201.441
1970	225.413	4.006	0.180	0.872	0.002	0.000	230.471
1971	257.557	7.153	0.333	0.711	0.006	0.000	265.761
1972	257.560	0.030	0.325	0.514	0.006	0.000	258.434
1973	342.091	1.308	0.193	0.776	0.004	0.006	344.378
1974	367.658	0.000	0.125	2.320	0.003	0.004	370.110
1975	225.205	13.387	0.261	0.149	0.001	0.000	239.003
1976	319.440	7.943	0.382	0.274	0.001	0.000	328.041

Appendix J

THE NORTH CENTRAL REGION

Exports

Regional Overview

Relative to other U.S. regions, the North Central is an insignificant exporter of forest products. The region is landlocked, and borders on Canada's western provinces. More than 99 percent of the value of the region's 1976 forest product exports were shipped to Canada.

The region's 1976 exports totaled $58.6 million, which was an increase of 879 percent over the nominal value of the 1967 exports. The average growth rate for the period was 29 percent per year.

The North Central region predominantly exports solid-wood products. Softwood lumber constituted 28 percent (by value) of the region's 1976 forest product exports, while plywood accounted for 15 percent, reconstituted wood for 10 percent, and hardwood lumber for 6 percent. All solid-wood products together represented 61 percent of the region's 1976 exports.

Of the pulp and paper products, paper and paperboard was the most important, accounting for 27 percent of the value of the region's 1976 forest product exports. Wood pulp and building board each constituted about 6 percent of the region's exports.

Because Canada was nearly the only importer of North Central forest products, very little mention will be made of the country structure of trade in the commodity analyses below.

Softwood Lumber. The North Central region was the fastest growing softwood-lumber-exporting region in the United States during 1967-1976. The region's 1976 exports of 64 million board-feet (valued at $16.6 million) represented increases over the 1967 levels of 425 percent in volume and 839 percent in value. The average volume and value growth rates are, respectively, 17 percent and 27 percent per year.

During the period, the region doubled its share of total U.S. softwood lumber exports from 2 percent (by value) in 1967 to 4 percent in 1976.

Paper and Paperboard. Paper and paperboard exports from the North Central region grew steadily and substantially from 1967 to 1976. The average growth rate, in volume, was 21 percent per year and, in nominal value, was 26 percent annually. The 1976 exports amounted to 32,000 tons (worth $15.9 million) and represented 1 percent of all U.S. paper and paperboard exports.

Plywood. Exports of plywood from the region in 1976 totaled 51.8 million square feet (worth $8.9 million), which was 6 percent of the value of all U.S. plywood exports.

The North Central region expanded its plywood export considerably from 1967 to 1976. In volume, exports increased by an average of 64 percent per year, and in nominal value by 72 percent annually.

Reconstituted Wood. Over the course of the 1967-76 decade, the North Central region grew to become the largest reconstituted wood-exporting region in the United States. The 1976 exports of 44.1 million square feet (worth $6.1 million) represented 36 percent of the volume and 33 percent of the value of all U.S. reconstituted wood exports.

The average export growth rates for the period were nothing less than spectacular--96 percent per year in volume and 62 percent in nominal value. The region's share of the value of U.S. reconstituted wood exports grew from 8 percent in 1967 to 33 percent in 1976.

Wood Pulp. The North Central was a rapidly expanding exporter of wood pulp during 1975-76, but its share of U.S. pulp exports, as compared with other regions, was insignificant. The 1976 exports of 14,000 tons were valued at $3.2 million, which represented gains over 1967 of 2,422 percent in volume and 7,833 percent in value. The equivalent average annual growth rates are 43 percent in volume and 63 percent in nominal value.

Building Board. Building board exports from the North Central region grew substantially from 1967 to 1976. In 1976 exports totaled 11,100 tons (worth $3.1 million), which was an increase over the 1967 exports of 1,111 percent in volume and 834 percent in nominal value. The region's share of the value of U.S. building board exports grew from 4 percent in 1967 to 10 percent in 1976.

Hardwood Lumber. The North Central region's 1976 exports of hardwood lumber were 8.6 million board-feet (worth $3.3 million). That value represented 4 percent of all U.S. hardwood lumber exports.

The region's growth in exports was better than that for the country as a whole. In an average year, exports increased by 9 percent in volume and 16 percent in nominal value.

In only one year did Canada not import all of the region's hardwood lumber exports. In 1971 Japan imported 5 percent, and France 1 percent, of the value of the North Central's exports.

Other Forest Products. The region's exports of the remaining forest products (pulpwood except chips, softwood logs, hardwood logs, wood veneer, pulpwood chips, other industrial roundwood, and newsprint) totaled only $1.4 million in 1976.

Although the nominal value of exports of these "other" products increased by 136 percent from 1967 to 1976, the share which these products constituted of the region's total forest product exports declined from 10 percent in 1967 to 2 percent in 1976. This is caused largely by the decline in export of pulpwood except chips, which comprised 8 percent of the value of the region's forest product exports in 1967, but less than 1 percent in 1976.

Imports

Regional Overview

In an average year during 1967-76 the North Central region imported 9 percent of the value of all U.S. forest product imports.

The 1976 imports of $454.9 million represented a 214 percent increase over the 1967 imports (in nominal dollars), equivalent to an annual growth rate of 14 percent.

More than 98 percent of the value of the region's 1976 imports was contained in just three commodities--softwood lumber, wood pulp, and newsprint. The $227.9 million of softwood lumber imports represented half of the region's total forest product imports; wood pulp ($153.5 million) represented 34 percent and newsprint ($66.2 million) 15 percent of the region's imports.

Like the region's export trade, the North Central's import trade involves only one partner--Canada. More than 99 percent of the region's imports originated in Canada in every year of the decade 1967-76.

Softwood Lumber. The North Central region was a large importer of softwood lumber throughout 1967-76. In an average year, the region imported 18 percent of the value of all U.S. softwood lumber imports. The 1976 imports totaled 1.2 billion board-feet (worth $227.9 million).

The average growth rate for the 1967-76 period was 3 percent per year in volume, and 15 percent annually in nominal value. In volume terms, the region's imports peaked in 1972-73 when imports average 1.7 billion board-feet.

Wood Pulp. In 1976 the North Central was the third largest wood-pulp-importing region in the United States, accounting for 12 percent of the value of all U.S. pulp imports.

The region's 1976 imports of 431,000 tons (value at $153.5 million), represented a 100 percent increase over the 1967 import volume and a 459 percent increase over the 1967 nominal value. The equivalent annual growth rates are 8 percent in volume and 21 percent in nominal value. The rapid growth was also reflected in the region's share of U.S. pulp imports, which grew from 8 percent (by value) in 1967 to 12 percent in 1976.

Newsprint. Newsprint imports to the North Central region declined by 39 percent from 1967 to 1976. In the same period, the nominal value increased by 22 percent, of by just 2 percent per year.

The 1976 imports of 252,000 tons (worth $66.2 million) represented 4 percent of the value of all U.S. newsprint imports.

Other Forest Products. As mentioned above, all other forest products combined (pulpwood except chips, softwood and hardwood logs, hardwood lumber, veneer, plywood, reconstituted wood, pulpwood chips, other industrial roundwood, building board, paper and paperboard) constitued less than 2 percent of the region's 1976 imports.

In 1976 imports of these "other" forest products totaled $7.2 million. Of these, paper and paperboard were the most important commodities, accounting for $4.3 million followed by reconstituted wood ($1.5 million) and building board ($1.1 million).

EXPORTS
SOFTWOOD LOGS (2422)
NORTH CENTRAL

VALUE--MILLION CURRENT DOLLARS

YEAR	CANADA	OTH N AM	S AMERIC	EUROPE	ASIA	OTHERS	TOTAL
1967	0.021	0.000	0.000	0.000	0.000	0.000	0.021
1968	0.126	0.000	0.000	0.000	0.000	0.000	0.126
1969	0.019	0.000	0.000	0.000	0.000	0.000	0.019
1970	0.011	0.000	0.000	0.000	0.000	0.000	0.011
1971	0.000	0.000	0.000	0.000	0.000	0.000	0.000
1972	0.057	0.000	0.000	0.000	0.000	0.000	0.057
1973	0.089	0.000	0.000	0.000	0.000	0.000	0.089
1974	0.146	0.000	0.000	0.000	0.000	0.000	0.146
1975	0.247	0.000	0.000	0.000	0.000	0.000	0.247
1976	0.241	0.000	0.000	0.000	0.000	0.000	0.241

QUANTITY--MILLION BF

YEAR	CANADA	OTH N AM	S AMERIC	EUROPE	ASIA	OTHERS	TOTAL
1967	0.397	0.000	0.000	0.000	0.000	0.000	0.397
1968	1.516	0.000	0.000	0.000	0.000	0.000	1.516
1969	0.228	0.000	0.000	0.000	0.000	0.000	0.228
1970	0.096	0.000	0.000	0.000	0.000	0.000	0.096
1971	0.000	0.000	0.000	0.000	0.000	0.000	0.000
1972	3.272	0.000	0.000	0.000	0.000	0.000	3.272
1973	0.523	0.000	0.000	0.000	0.000	0.000	0.523
1974	0.626	0.000	0.000	0.000	0.000	0.000	0.626
1975	1.027	0.000	0.000	0.000	0.000	0.000	1.027
1976	1.062	0.000	0.000	0.000	0.000	0.000	1.062

EXPORTS
HARDWOOD LOGS (2423)
NORTH CENTRAL

VALUE--MILLION CURRENT DOLLARS

YEAR	CANADA	OTH N AM	S AMERIC	EUROPE	ASIA	OTHERS	TOTAL
1967	0.009	0.000	0.000	0.000	0.000	0.000	0.009
1968	0.020	0.000	0.000	0.000	0.000	0.000	0.020
1969	0.035	0.000	0.000	0.000	0.000	0.000	0.035
1970	0.003	0.000	0.000	0.000	0.000	0.000	0.003
1971	0.032	0.000	0.000	0.000	0.000	0.000	0.032
1972	0.024	0.000	0.000	0.000	0.000	0.000	0.024
1973	0.121	0.000	0.000	0.000	0.000	0.000	0.121
1974	0.044	0.000	0.000	0.000	0.000	0.000	0.044
1975	0.071	0.000	0.000	0.000	0.000	0.000	0.071
1976	0.160	0.000	0.000	0.000	0.000	0.000	0.160

QUANTITY--MILLION BF

YEAR	CANADA	OTH N AM	S AMERIC	EUROPE	ASIA	OTHERS	TOTAL
1967	0.031	0.000	0.000	0.000	0.000	0.000	0.031
1968	0.148	0.000	0.000	0.000	0.000	0.000	0.148
1969	0.090	0.000	0.000	0.000	0.000	0.000	0.090
1970	0.075	0.000	0.000	0.000	0.000	0.000	0.075
1971	0.080	0.000	0.000	0.000	0.000	0.000	0.080
1972	0.090	0.000	0.000	0.000	0.000	0.000	0.090
1973	0.524	0.000	0.000	0.000	0.000	0.000	0.524
1974	0.165	0.000	0.000	0.000	0.000	0.000	0.165
1975	0.266	0.000	0.000	0.000	0.000	0.000	0.266
1976	0.501	0.000	0.000	0.000	0.000	0.000	0.501

EXPORTS
SOFTWOOD LUMBER (2432)
NORTH CENTRAL

VALUE--MILLION CURRENT DOLLARS

YEAR	CANADA	U KINGDM	OTH EURO	JAPAN	AUSTRALA	OTHERS	TOTAL
1967	1.770	0.000	0.000	0.000	0.000	0.000	1.770
1968	2.153	0.000	0.000	0.000	0.000	0.000	2.153
1969	2.569	0.000	0.006	0.000	0.000	0.000	2.575
1970	1.759	0.000	0.000	0.000	0.000	0.000	1.759
1971	2.229	0.000	0.000	0.000	0.002	0.000	2.231
1972	3.420	0.000	0.000	0.000	0.000	0.000	3.420
1973	6.032	0.030	0.035	0.003	0.014	0.000	6.113
1974	7.589	0.026	0.000	0.000	0.000	0.000	7.615
1975	11.262	0.000	0.000	0.000	0.000	0.000	11.262
1976	16.624	0.000	0.000	0.000	0.000	0.000	16.624

QUANTITY--MILLION BF

YEAR	CANADA	U KINGDM	OTH EURO	JAPAN	AUSTRALA	OTHERS	TOTAL
1967	15.063	0.000	0.000	0.000	0.000	0.000	15.063
1968	15.681	0.000	0.000	0.000	0.000	0.000	15.681
1969	16.854	0.000	0.024	0.000	0.000	0.000	16.878
1970	14.191	0.000	0.000	0.000	0.000	0.000	14.191
1971	15.752	0.000	0.000	0.000	0.013	0.000	15.765
1972	20.946	0.000	0.000	0.000	0.000	0.000	20.946
1973	54.828	0.216	0.273	0.012	0.110	0.000	55.439
1974	35.670	0.130	0.000	0.000	0.000	0.000	35.800
1975	54.797	0.000	0.000	0.000	0.000	0.000	54.797
1976	64.003	0.000	0.000	0.000	0.000	0.000	64.003

EXPORTS
HARDWOOD LUMBER (2433)
NORTH CENTRAL

VALUE--MILLION CURRENT DOLLARS

YEAR	CANADA	FRANCE	OTH EURO	JAPAN	OTH ASIA	OTHERS	TOTAL
1967	0.828	0.000	0.000	0.000	0.000	0.000	0.828
1968	1.045	0.000	0.000	0.000	0.000	0.000	1.045
1969	1.213	0.000	0.000	0.000	0.000	0.000	1.213
1970	0.799	0.000	0.000	0.000	0.000	0.000	0.799
1971	0.902	0.005	0.000	0.049	0.000	0.000	0.956
1972	1.550	0.000	0.000	0.000	0.000	0.000	1.550
1973	2.316	0.000	0.000	0.000	0.000	0.000	2.316
1974	2.838	0.000	0.000	0.000	0.000	0.000	2.838
1975	2.455	0.000	0.000	0.000	0.000	0.000	2.455
1976	3.250	0.000	0.000	0.000	0.000	0.000	3.250

QUANTITY--MILLION BF

YEAR	CANADA	FRANCE	OTH EURO	JAPAN	OTH ASIA	OTHERS	TOTAL
1967	3.938	0.000	0.000	0.000	0.000	0.000	3.938
1968	5.658	0.000	0.000	0.000	0.000	0.000	5.658
1969	6.016	0.000	0.000	0.000	0.000	0.000	6.016
1970	3.709	0.000	0.000	0.000	0.000	0.000	3.709
1971	4.388	0.012	0.000	0.137	0.000	0.000	4.537
1972	6.098	0.000	0.000	0.000	0.000	0.000	6.098
1973	8.424	0.000	0.000	0.000	0.000	0.000	8.424
1974	8.362	0.000	0.000	0.000	0.000	0.000	8.362
1975	7.714	0.000	0.000	0.000	0.000	0.000	7.714
1976	8.575	0.000	0.000	0.000	0.000	0.000	8.575

EXPORTS
WOOD VENEER (6311)
NORTH CENTRAL

VALUE--MILLION CURRENT DOLLARS

YEAR	CANADA	OTH N AM	S AMERIC	EUROPE	ASIA	OTHERS	TOTAL
1967	0.050	0.000	0.000	0.000	0.000	0.000	0.050
1968	0.077	0.000	0.000	0.000	0.000	0.000	0.077
1969	0.019	0.000	0.000	0.000	0.000	0.000	0.019
1970	0.071	0.000	0.000	0.000	0.000	0.000	0.071
1971	0.126	0.000	0.000	0.000	0.000	0.000	0.126
1972	0.043	0.000	0.000	0.000	0.000	0.000	0.043
1973	0.225	0.000	0.000	0.000	0.000	0.000	0.225
1974	0.295	0.000	0.000	0.000	0.000	0.000	0.295
1975	0.366	0.000	0.000	0.000	0.000	0.000	0.366
1976	0.528	0.000	0.000	0.000	0.000	0.000	0.528

QUANTITY--MIL SQ FT

YEAR	CANADA	OTH N AM	S AMERIC	EUROPE	ASIA	OTHERS	TOTAL
1967	1.193	0.000	0.000	0.000	0.000	0.000	1.193
1968	2.444	0.000	0.000	0.000	0.000	0.000	2.444
1969	0.280	0.000	0.000	0.000	0.000	0.000	0.280
1970	0.888	0.000	0.000	0.000	0.000	0.000	0.888
1971	4.320	0.000	0.000	0.000	0.000	0.000	4.320
1972	1.235	0.000	0.000	0.000	0.000	0.000	1.235
1973	3.655	0.000	0.000	0.000	0.000	0.000	3.655
1974	4.102	0.000	0.000	0.000	0.000	0.000	4.102
1975	3.254	0.000	0.000	0.000	0.000	0.000	3.254
1976	5.305	0.000	0.000	0.000	0.000	0.000	5.305

EXPORTS
PLYWOOD (6312)
NORTH CENTRAL

VALUE--MILLION CURRENT DOLLARS

YEAR	CANADA	OTH N AM	S AMERIC	EUROPE	ASIA	OTHERS	TOTAL
1967	0.329	0.000	0.000	0.000	0.000	0.000	0.329
1968	0.674	0.000	0.000	0.000	0.000	0.000	0.674
1969	1.502	0.000	0.000	0.000	0.000	0.000	1.502
1970	0.629	0.000	0.000	0.000	0.000	0.000	0.629
1971	0.622	0.000	0.000	0.000	0.000	0.000	0.622
1972	1.922	0.000	0.000	0.000	0.000	0.000	1.922
1973	2.874	0.000	0.000	0.000	0.000	0.000	2.874
1974	9.841	0.000	0.000	0.000	0.000	0.000	9.841
1975	13.831	0.000	0.000	0.000	0.000	0.000	13.831
1976	8.884	0.000	0.000	0.000	0.000	0.000	8.884

QUANTITY--MIL SQ FT

YEAR	CANADA	OTH N AM	S AMERIC	EUROPE	ASIA	OTHERS	TOTAL
1967	0.615	0.000	0.000	0.000	0.000	0.000	0.615
1968	3.148	0.000	0.000	0.000	0.000	0.000	3.148
1969	8.666	0.000	0.000	0.000	0.000	0.000	8.666
1970	7.554	0.000	0.000	0.000	0.000	0.000	7.554
1971	5.509	0.000	0.000	0.000	0.000	0.000	5.509
1972	18.710	0.000	0.000	0.000	0.000	0.000	18.710
1973	23.836	0.000	0.000	0.000	0.000	0.000	23.836
1974	68.359	0.000	0.000	0.000	0.000	0.000	68.359
1975	92.061	0.000	0.000	0.000	0.000	0.000	92.061
1976	51.764	0.000	0.000	0.000	0.000	0.000	51.764

EXPORTS
RECONSTITUTED WOOD (6314)
NORTH CENTRAL

VALUE--MILLION CURRENT DOLLARS

YEAR	CANADA	OTH N AM	S AMERIC	EUROPE	ASIA	OTHERS	TOTAL
1967	0.079	0.000	0.000	0.000	0.000	0.000	0.079
1968	0.057	0.000	0.000	0.000	0.000	0.000	0.057
1969	0.153	0.000	0.000	0.000	0.000	0.000	0.153
1970	0.224	0.000	0.000	0.000	0.000	0.000	0.224
1971	0.945	0.000	0.000	0.000	0.000	0.000	0.945
1972	2.099	0.000	0.000	0.000	0.000	0.000	2.099
1973	3.219	0.000	0.000	0.000	0.000	0.000	3.219
1974	3.306	0.000	0.000	0.000	0.000	0.000	3.306
1975	4.093	0.000	0.000	0.000	0.000	0.000	4.093
1976	6.073	0.000	0.000	0.000	0.000	0.000	6.073

QUANTITY--MIL SQ FT

YEAR	CANADA	OTH N AM	S AMERIC	EUROPE	ASIA	OTHERS	TOTAL
1967	0.103	0.000	0.000	0.000	0.000	0.000	0.103
1968	0.571	0.000	0.000	0.000	0.000	0.000	0.571
1969	0.997	0.000	0.000	0.000	0.000	0.000	0.997
1970	1.401	0.000	0.000	0.000	0.000	0.000	1.401
1971	7.370	0.000	0.000	0.000	0.000	0.000	7.370
1972	16.632	0.000	0.000	0.000	0.000	0.000	16.632
1973	24.771	0.000	0.000	0.000	0.000	0.000	24.771
1974	22.896	0.000	0.000	0.000	0.000	0.000	22.896
1975	29.478	0.000	0.000	0.000	0.000	0.000	29.478
1976	44.084	0.000	0.000	0.000	0.000	0.000	44.084

EXPORTS
PULPWOOD (EXC CHIPS)(2421)
NORTH CENTRAL

VALUE--MILLION CURRENT DOLLARS

YEAR	CANADA	OTH N AM	S AMERIC	EUROPE	ASIA	OTHERS	TOTAL
1967	0.505	0.000	0.000	0.000	0.000	0.000	0.505
1968	0.381	0.000	0.000	0.000	0.000	0.000	0.381
1969	0.252	0.000	0.000	0.000	0.000	0.000	0.252
1970	0.475	0.000	0.000	0.000	0.000	0.000	0.475
1971	0.000	0.000	0.000	0.000	0.000	0.000	0.000
1972	0.264	0.000	0.000	0.000	0.000	0.000	0.264
1973	0.245	0.000	0.000	0.000	0.000	0.000	0.245
1974	0.321	0.000	0.000	0.000	0.000	0.000	0.321
1975	0.758	0.000	0.000	0.000	0.000	0.000	0.758
1976	0.348	0.000	0.000	0.000	0.000	0.000	0.348

QUANTITY--1000 CORDS

YEAR	CANADA	OTH N AM	S AMERIC	EUROPE	ASIA	OTHERS	TOTAL
1967	22.120	0.000	0.000	0.000	0.000	0.000	22.120
1968	15.769	0.000	0.000	0.000	0.000	0.000	15.769
1969	10.784	0.000	0.000	0.000	0.000	0.000	10.784
1970	20.153	0.000	0.000	0.000	0.000	0.000	20.153
1971	0.000	0.000	0.000	0.000	0.000	0.000	0.000
1972	11.373	0.000	0.000	0.000	0.000	0.000	11.373
1973	10.584	0.000	0.000	0.000	0.000	0.000	10.584
1974	11.519	0.000	0.000	0.000	0.000	0.000	11.519
1975	32.647	0.000	0.000	0.000	0.000	0.000	32.647
1976	6.520	0.000	0.000	0.000	0.000	0.000	6.520

EXPORTS
WOOD PULP (2510)
NORTH CENTRAL

VALUE--MILLION CURRENT DOLLARS

YEAR	CANADA	OTH N AM	S AMERIC	EUROPE	ASIA	OTHERS	TOTAL
1967	0.040	0.000	0.000	0.000	0.000	0.000	0.040
1968	0.005	0.000	0.000	0.000	0.000	0.000	0.005
1969	0.011	0.000	0.000	0.000	0.000	0.000	0.011
1970	0.067	0.000	0.000	0.000	0.000	0.000	0.067
1971	0.109	0.000	0.000	0.000	0.000	0.000	0.109
1972	0.005	0.000	0.000	0.000	0.000	0.000	0.005
1973	0.195	0.000	0.000	0.000	0.000	0.000	0.195
1974	1.933	0.000	0.000	0.000	0.000	0.000	1.933
1975	3.785	0.000	0.000	0.000	0.000	0.000	3.785
1976	3.201	0.000	0.000	0.000	0.000	0.000	3.201

QUANTITY--1000 STN

YEAR	CANADA	OTH N AM	S AMERIC	EUROPE	ASIA	OTHERS	TOTAL
1967	0.547	0.000	0.000	0.000	0.000	0.000	0.547
1968	0.017	0.000	0.000	0.000	0.000	0.000	0.017
1969	0.066	0.000	0.000	0.000	0.000	0.000	0.066
1970	0.405	0.000	0.000	0.000	0.000	0.000	0.405
1971	0.653	0.000	0.000	0.000	0.000	0.000	0.653
1972	0.063	0.000	0.000	0.000	0.000	0.000	0.063
1973	3.140	0.000	0.000	0.000	0.000	0.000	3.140
1974	9.829	0.000	0.000	0.000	0.000	0.000	9.829
1975	14.015	0.000	0.000	0.000	0.000	0.000	14.015
1976	13.796	0.000	0.000	0.000	0.000	0.000	13.796

EXPORTS
BUILDING BOARD (6416)
NORTH CENTRAL

VALUE--MILLION CURRENT DOLLARS

YEAR	CANADA	OTH N AM	S AMERIC	EUROPE	ASIA	OTHERS	TOTAL
1967	0.337	0.000	0.000	0.000	0.000	0.000	0.337
1968	0.204	0.000	0.000	0.000	0.000	0.000	0.204
1969	0.506	0.000	0.000	0.000	0.000	0.000	0.506
1970	0.370	0.000	0.000	0.000	0.000	0.000	0.370
1971	0.805	0.000	0.000	0.000	0.000	0.000	0.805
1972	0.890	0.000	0.000	0.000	0.000	0.000	0.890
1973	2.031	0.000	0.000	0.000	0.000	0.000	2.031
1974	2.028	0.000	0.000	0.000	0.000	0.000	2.028
1975	2.724	0.000	0.000	0.000	0.000	0.000	2.724
1976	3.145	0.000	0.000	0.000	0.000	0.000	3.145

QUANTITY--1000 STN

YEAR	CANADA	OTH N AM	S AMERIC	EUROPE	ASIA	OTHERS	TOTAL
1967	0.919	0.000	0.000	0.000	0.000	0.000	0.919
1968	0.828	0.000	0.000	0.000	0.000	0.000	0.828
1969	1.902	0.000	0.000	0.000	0.000	0.000	1.902
1970	1.739	0.000	0.000	0.000	0.000	0.000	1.739
1971	3.384	0.000	0.000	0.000	0.000	0.000	3.384
1972	4.243	0.000	0.000	0.000	0.000	0.000	4.243
1973	7.919	0.000	0.000	0.000	0.000	0.000	7.919
1974	7.272	0.000	0.000	0.000	0.000	0.000	7.272
1975	9.502	0.000	0.000	0.000	0.000	0.000	9.502
1976	11.134	0.000	0.000	0.000	0.000	0.000	11.134

EXPORTS
PAPER & PAPERBOARD (6410)
NORTH CENTRAL

VALUE--MILLION CURRENT DOLLARS

YEAR	CANADA	MEXICO	OTH N AM	EUROPE	ASIA	OTHERS	TOTAL
1967	2.006	0.000	0.000	0.000	0.000	0.000	2.006
1968	1.915	0.002	0.000	0.000	0.000	0.000	1.917
1969	2.578	0.000	0.000	0.000	0.000	0.000	2.578
1970	2.801	0.000	0.000	0.000	0.000	0.000	2.801
1971	3.470	0.000	0.000	0.000	0.000	0.000	3.470
1972	3.873	0.000	0.000	0.000	0.000	0.000	3.873
1973	4.818	0.001	0.000	0.000	0.000	0.000	4.818
1974	7.319	0.000	0.000	0.000	0.000	0.000	7.319
1975	12.102	0.000	0.000	0.000	0.000	0.000	12.102
1976	15.946	0.001	0.000	0.000	0.000	0.000	15.947

QUANTITY--1000 STN

YEAR	CANADA	MEXICO	OTH N AM	EUROPE	ASIA	OTHERS	TOTAL
1967	5.834	0.000	0.000	0.000	0.000	0.000	5.834
1968	5.378	0.000	0.000	0.000	0.000	0.000	5.378
1969	7.789	0.000	0.000	0.000	0.000	0.000	7.789
1970	10.132	0.000	0.000	0.000	0.000	0.000	10.132
1971	12.476	0.000	0.000	0.000	0.000	0.000	12.476
1972	12.905	0.000	0.000	0.000	0.000	0.000	12.905
1973	17.624	0.000	0.000	0.000	0.000	0.000	17.624
1974	18.954	0.000	0.000	0.000	0.000	0.000	18.954
1975	30.769	0.000	0.000	0.000	0.000	0.000	30.769
1976	32.423	0.003	0.000	0.000	0.000	0.000	32.426

IMPORTS
SOFTWOOD LUMBER (2432)
NORTH CENTRAL

VALUE--MILLION CURRENT DOLLARS

YEAR	CANADA	MEXICO	OTH N AM	EUROPE	ASIA	OTHERS	TOTAL
1967	62.311	0.000	0.000	0.000	0.000	0.000	62.311
1968	89.982	0.004	0.000	0.000	0.000	0.000	89.986
1969	100.625	0.000	0.000	0.000	0.000	0.000	100.625
1970	83.735	0.000	0.000	0.000	0.000	0.000	83.735
1971	139.010	0.000	0.000	0.000	0.000	0.000	139.010
1972	217.647	0.018	0.000	0.000	0.000	0.000	217.665
1973	253.175	0.008	0.000	0.000	0.000	0.000	253.183
1974	153.501	0.009	0.000	0.000	0.000	0.000	153.510
1975	108.531	0.000	0.000	0.000	0.000	0.000	108.531
1976	227.917	0.000	0.000	0.000	0.000	0.000	227.917

QUANTITY--MILLION BF

YEAR	CANADA	MEXICO	OTH N AM	EUROPE	ASIA	OTHERS	TOTAL
1967	900.643	0.000	0.000	0.000	0.000	0.000	900.643
1968	1017.895	0.046	0.000	0.000	0.000	0.000	1017.941
1969	1076.151	0.000	0.000	0.000	0.000	0.000	1076.151
1970	1091.773	0.000	0.000	0.000	0.000	0.000	1091.773
1971	1416.928	0.000	0.000	0.000	0.000	0.000	1416.928
1972	1768.427	0.091	0.000	0.000	0.000	0.000	1768.518
1973	1619.539	0.064	0.000	0.000	0.000	0.000	1619.603
1974	1119.216	0.070	0.000	0.000	0.000	0.000	1119.286
1975	815.528	0.000	0.000	0.000	0.000	0.000	815.528
1976	1221.315	0.000	0.000	0.000	0.000	0.000	1221.315

IMPORTS
HARDWOOD LUMBER (2433)
NORTH CENTRAL

VALUE--MILLION CURRENT DOLLARS

YEAR	CANADA	OTH N AM	S AMERIC	EUROPE	ASIA	OTHERS	TOTAL
1967	0.114	0.000	0.000	0.000	0.000	0.000	0.114
1968	0.120	0.000	0.000	0.000	0.000	0.000	0.120
1969	0.181	0.000	0.000	0.000	0.000	0.000	0.181
1970	0.167	0.000	0.000	0.000	0.000	0.000	0.167
1971	0.254	0.000	0.000	0.000	0.000	0.000	0.254
1972	1.023	0.000	0.000	0.000	0.000	0.000	1.023
1973	1.795	0.000	0.000	0.000	0.000	0.000	1.795
1974	0.607	0.000	0.000	0.000	0.000	0.000	0.607
1975	0.121	0.000	0.000	0.000	0.000	0.000	0.121
1976	0.197	0.000	0.000	0.000	0.000	0.000	0.197

QUANTITY--MILLION BF

YEAR	CANADA	OTH N AM	S AMERIC	EUROPE	ASIA	OTHERS	TOTAL
1967	1.282	0.000	0.000	0.000	0.000	0.000	1.282
1968	1.132	0.000	0.000	0.000	0.000	0.000	1.132
1969	1.460	0.000	0.000	0.000	0.000	0.000	1.460
1970	1.417	0.000	0.000	0.000	0.000	0.000	1.417
1971	2.530	0.000	0.000	0.000	0.000	0.000	2.530
1972	10.015	0.000	0.000	0.000	0.000	0.000	10.015
1973	16.677	0.000	0.000	0.000	0.000	0.000	16.677
1974	7.053	0.000	0.000	0.000	0.000	0.000	7.053
1975	0.642	0.000	0.000	0.000	0.000	0.000	0.642
1976	1.106	0.000	0.000	0.000	0.000	0.000	1.106

IMPORTS
PLYWOOD (6312)
NORTH CENTRAL

VALUE--MILLION CURRENT DOLLARS

YEAR	CANADA	MEXICO	OTH N AM	EUROPE	ASIA	OTHERS	TOTAL
1967	0.000	0.000	0.000	0.000	0.000	0.000	0.000
1968	0.065	0.000	0.000	0.000	0.000	0.000	0.065
1969	0.237	0.000	0.000	0.000	0.000	0.000	0.237
1970	0.002	0.000	0.000	0.000	0.000	0.000	0.002
1971	0.001	0.000	0.000	0.000	0.000	0.000	0.001
1972	0.211	0.000	0.000	0.000	0.000	0.000	0.211
1973	0.728	0.000	0.000	0.000	0.000	0.000	0.728
1974	0.002	0.000	0.000	0.000	0.000	0.000	0.002
1975	0.002	0.000	0.000	0.000	0.000	0.000	0.002
1976	0.012	0.000	0.000	0.000	0.000	0.000	0.012

QUANTITY--MIL SQ FT

YEAR	CANADA	MEXICO	OTH N AM	EUROPE	ASIA	OTHERS	TOTAL
1967	0.006	0.000	0.000	0.000	0.000	0.000	0.006
1968	0.411	0.000	0.000	0.000	0.000	0.000	0.411
1969	0.985	0.000	0.000	0.000	0.000	0.000	0.985
1970	0.099	0.000	0.000	0.000	0.000	0.000	0.099
1971	0.001	0.000	0.000	0.000	0.000	0.000	0.001
1972	1.009	0.000	0.000	0.000	0.000	0.000	1.009
1973	2.290	0.000	0.000	0.000	0.000	0.000	2.290
1974	0.011	0.000	0.000	0.000	0.000	0.000	0.011
1975	0.010	0.000	0.000	0.000	0.000	0.000	0.010
1976	0.049	0.000	0.000	0.000	0.000	0.000	0.049

IMPORTS
RECONSTITUTED WOOD (6314)
NORTH CENTRAL

VALUE--MILLION CURRENT DOLLARS

YEAR	CANADA	OTH N AM	S AMERIC	EUROPE	ASIA	OTHERS	TOTAL
1967	0.000	0.000	0.000	0.000	0.000	0.000	0.000
1968	0.002	0.000	0.000	0.000	0.000	0.000	0.002
1969	0.007	0.000	0.000	0.000	0.000	0.000	0.007
1970	0.153	0.000	0.000	0.000	0.000	0.000	0.153
1971	0.981	0.000	0.000	0.000	0.000	0.000	0.981
1972	1.241	0.000	0.000	0.000	0.000	0.000	1.241
1973	0.553	0.000	0.000	0.000	0.000	0.000	0.553
1974	0.000	0.000	0.000	0.000	0.000	0.000	0.000
1975	1.190	0.000	0.000	0.000	0.000	0.000	1.190
1976	1.511	0.000	0.000	0.000	0.000	0.000	1.511

QUANTITY--MIL SQ FT

YEAR	CANADA	OTH N AM	S AMERIC	EUROPE	ASIA	OTHERS	TOTAL
1967	0.000	0.000	0.000	0.000	0.000	0.000	0.000
1968	0.002	0.000	0.000	0.000	0.000	0.000	0.002
1969	0.043	0.000	0.000	0.000	0.000	0.000	0.043
1970	0.919	0.000	0.000	0.000	0.000	0.000	0.919
1971	6.357	0.000	0.000	0.000	0.000	0.000	6.357
1972	7.204	0.000	0.000	0.000	0.000	0.000	7.204
1973	3.090	0.000	0.000	0.000	0.000	0.000	3.090
1974	0.001	0.000	0.000	0.000	0.000	0.000	0.001
1975	5.780	0.000	0.000	0.000	0.000	0.000	5.780
1976	6.002	0.000	0.000	0.000	0.000	0.000	6.002

IMPORTS
WOOD PULP (2510)
NORTH CENTRAL

VALUE--MILLION CURRENT DOLLARS

YEAR	CANADA	MEXICO	OTH N AM	EUROPE	ASIA	OTHERS	TOTAL
1967	27.437	0.000	0.000	0.000	0.000	0.000	27.437
1968	29.674	0.000	0.000	0.000	0.000	0.000	29.674
1969	40.817	0.007	0.000	0.000	0.000	0.000	40.824
1970	39.743	0.000	0.000	0.000	0.000	0.000	39.743
1971	42.937	0.061	0.000	0.000	0.000	0.000	42.998
1972	49.122	0.000	0.000	0.000	0.000	0.000	49.122
1973	70.816	0.000	0.000	0.000	0.000	0.000	70.816
1974	138.010	0.000	0.000	0.000	0.000	0.000	138.010
1975	119.164	0.000	0.000	0.000	0.000	0.000	119.164
1976	153.476	0.000	0.000	0.000	0.000	0.000	153.476

QUANTITY--1000 STN

YEAR	CANADA	MEXICO	OTH N AM	EUROPE	ASIA	OTHERS	TOTAL
1967	215.447	0.000	0.000	0.000	0.000	0.000	215.447
1968	266.141	0.000	0.000	0.000	0.000	0.000	266.141
1969	349.176	0.057	0.000	0.000	0.000	0.000	349.233
1970	294.032	0.000	0.000	0.000	0.000	0.000	294.032
1971	321.632	0.411	0.000	0.000	0.000	0.000	322.043
1972	392.690	0.000	0.000	0.000	0.000	0.000	392.690
1973	427.888	0.000	0.000	0.000	0.000	0.000	427.888
1974	498.107	0.000	0.000	0.000	0.000	0.000	498.107
1975	346.508	0.000	0.000	0.000	0.000	0.000	346.508
1976	430.733	0.000	0.000	0.000	0.000	0.000	430.733

IMPORTS
NEWSPRINT (6411)
NORTH CENTRAL

VALUE--MILLION CURRENT DOLLARS

YEAR	CANADA	OTH N AM	S AMERIC	EUROPE	ASIA	OTHERS	TOTAL
1967	54.178	0.000	0.000	0.000	0.000	0.000	54.178
1968	51.971	0.000	0.000	0.000	0.000	0.000	51.971
1969	49.308	0.000	0.000	0.000	0.000	0.000	49.308
1970	48.887	0.000	0.000	0.000	0.000	0.000	48.887
1971	51.424	0.000	0.000	0.000	0.000	0.000	51.424
1972	53.868	0.000	0.000	0.000	0.000	0.000	53.868
1973	52.277	0.000	0.000	0.000	0.000	0.000	52.277
1974	76.567	0.000	0.000	0.000	0.000	0.000	76.567
1975	63.909	0.000	0.000	0.000	0.000	0.000	63.909
1976	66.245	0.000	0.000	0.000	0.000	0.000	66.245

QUANTITY--1000 STN

YEAR	CANADA	OTH N AM	S AMERIC	EUROPE	ASIA	OTHERS	TOTAL
1967	415.682	0.000	0.000	0.000	0.000	0.000	415.682
1968	400.952	0.000	0.000	0.000	0.000	0.000	400.952
1969	376.398	0.000	0.000	0.000	0.000	0.000	376.398
1970	366.463	0.000	0.000	0.000	0.000	0.000	366.463
1971	369.442	0.000	0.000	0.000	0.000	0.000	369.442
1972	374.759	0.000	0.000	0.000	0.000	0.000	374.759
1973	342.668	0.000	0.000	0.000	0.000	0.000	342.668
1974	407.012	0.000	0.000	0.000	0.000	0.000	407.012
1975	276.352	0.000	0.000	0.000	0.000	0.000	276.352
1976	251.940	0.000	0.000	0.000	0.000	0.000	251.940

IMPORTS
BUILDING BOARD (6416)
NORTH CENTRAL

VALUE--MILLION CURRENT DOLLARS

YEAR	CANADA	OTH N AM	S AMERIC	EUROPE	ASIA	OTHERS	TOTAL
1967	0.200	0.000	0.000	0.000	0.000	0.000	0.200
1968	0.266	0.000	0.000	0.000	0.000	0.000	0.266
1969	0.269	0.000	0.000	0.000	0.000	0.000	0.269
1970	0.263	0.000	0.000	0.000	0.000	0.000	0.263
1971	0.312	0.000	0.000	0.000	0.000	0.000	0.312
1972	0.497	0.000	0.000	0.000	0.000	0.000	0.497
1973	1.154	0.000	0.000	0.000	0.000	0.000	1.154
1974	0.877	0.000	0.000	0.000	0.000	0.000	0.877
1975	0.633	0.000	0.000	0.000	0.000	0.000	0.633
1976	1.147	0.000	0.000	0.000	0.000	0.000	1.147

QUANTITY--1000 STN

YEAR	CANADA	OTH N AM	S AMERIC	EUROPE	ASIA	OTHERS	TOTAL
1967	2.464	0.000	0.000	0.000	0.000	0.000	2.464
1968	3.267	0.000	0.000	0.000	0.000	0.000	3.267
1969	3.585	0.000	0.000	0.000	0.000	0.000	3.585
1970	3.390	0.000	0.000	0.000	0.000	0.000	3.390
1971	3.789	0.000	0.000	0.000	0.000	0.000	3.789
1972	5.171	0.000	0.000	0.000	0.000	0.000	5.171
1973	11.339	0.000	0.000	0.000	0.000	0.000	11.339
1974	7.199	0.000	0.000	0.000	0.000	0.000	7.199
1975	5.175	0.000	0.000	0.000	0.000	0.000	5.175
1976	8.064	0.000	0.000	0.000	0.000	0.000	8.064

IMPORTS
PAPER & PAPERBOARD (6410)
NORTH CENTRAL

VALUE--MILLION CURRENT DOLLARS

YEAR	CANADA	OTH N AM	S AMERIC	EUROPE	ASIA	OTHERS	TOTAL
1967	0.575	0.000	0.000	0.000	0.000	0.000	0.575
1968	0.719	0.000	0.000	0.000	0.000	0.000	0.719
1969	1.597	0.000	0.000	0.000	0.000	0.000	1.597
1970	1.835	0.000	0.000	0.000	0.000	0.000	1.835
1971	1.244	0.000	0.000	0.000	0.000	0.000	1.244
1972	1.643	0.000	0.000	0.000	0.000	0.000	1.643
1973	4.570	0.000	0.000	0.000	0.000	0.000	4.570
1974	5.989	0.000	0.000	0.000	0.000	0.000	5.989
1975	3.537	0.000	0.000	0.000	0.000	0.000	3.537
1976	4.322	0.000	0.000	0.000	0.000	0.000	4.322

QUANTITY--1000 STN

YEAR	CANADA	OTH N AM	S AMERIC	EUROPE	ASIA	OTHERS	TOTAL
1967	3.302	0.000	0.000	0.000	0.000	0.000	3.302
1968	4.239	0.000	0.000	0.000	0.000	0.000	4.239
1969	8.688	0.000	0.000	0.000	0.000	0.000	8.688
1970	10.102	0.000	0.000	0.000	0.000	0.000	10.102
1971	6.990	0.000	0.000	0.000	0.000	0.000	6.990
1972	7.749	0.000	0.000	0.000	0.000	0.000	7.749
1973	19.068	0.000	0.000	0.000	0.000	0.000	19.068
1974	17.928	0.000	0.000	0.000	0.000	0.000	17.928
1975	11.307	0.000	0.000	0.000	0.000	0.000	11.307
1976	13.889	0.000	0.000	0.000	0.000	0.000	13.889

Appendix K

THE SOUTH CENTRAL REGION

Exports

Regional Overview

The South Central region is similar to the North Central in that it is landlocked, and borders on just one country--in this case, Mexico. In 1976 only $8,862 out of the region's exports of $15.6 million went to countries outside of Mexico.

The region's exports in 1976 represented less than one-half of 1 percent of the value of all U.S. forest product exports.

From 1967 to 1976, exports from the South Central region grew at about the same rate as did exports from the United States as a whole--on average, exports increased by 18 percent per year in nominal value.

The commodity mix of the region's exports favored pulp and paper products. Paper and paperboard accounted for 45 percent of the value of the region's 1976 exports, and wood pulp and newsprint each accounted for about 10 percent. Solid wood products together constituted 36 percent of the 1976 exports. Softwood lumber was the most important of these, and represented 22 percent of the region's forest product exports.

Paper and Paperboard. The South Central region was the fastest growing paper and paperboard exporter during 1967-76. The 1976 exports of 21,000 tons valued at $7.1 million represented increases above the 1967 levels of 8,098 percent in volume and 5,814 percent in nominal value. The equivalent average annual growth rates are 63 percent in volume and 57 percent in nominal value.

Compared with other regions, the South Central is a very small exporter of paper and paperboard. In 1976 less than 1 percent of all U.S. exports of paper and paperboard were shipped from the South Central region.

Softwood Lumber. Softwood lumber exports from the region were somewhat erratic during 1967-76, but generally increased over the course of the period. The 1976 exports of 12 million board-feet were valued at $3.4 million, which represented increases over the 1967 levels of 41 percent in volume and 362 percent in nominal value.

Throughout 1967-76, the region's exports were about 1 percent (by value) of all U.S. exports of softwood lumber.

Wood Pulp. The region's 1976 exports of wood pulp amounted to 4,525 tons (worth $1.6 million). The region experienced slow pulp-export growth during the decade, on the order of 1 percent per year in volume and 9 percent annually in nominal value.

Although, in 1976, wood pulp represented 10 percent (by value) of the South Central's forest product exports, it accounted for less than one-half of 1 percent of all U.S. pulp exports.

Newsprint. Newsprint was an important commodity in 1967, when the South Central region shipped 15 percent of the value of all U.S. newsprint exports. That share declined to 3 percent by 1976, as the absolute amount of exports declined by 68 percent in volume and 20 percent in nominal value. The 1976 exports totaled 3,906 tons (growth $1.4 million). In that year, newsprint constituted 9 percent of the value of the region's total forest product exports.

Plywood. Although the South central was not an important plywood-exporting region, its exports were rapidly expanding over the period from

1967-76. The 1976 shipments of 4.2 million square feet were worth $869,000, which represented increases over the 1967 levels of 640 percent in volume and 1,227 percent in nominal value. The equivalent annual growth rates are 25 percent in volume and 33 percent in nominal value. In 1976 the region's exports were only 1 percent (value) of all U.S. plywood exports.

Other Forest Products. The remaining forest products (pulpwood except chips, softwood and hardwood logs, hardwood lumber, wood veneer, reconstituted wood, pulpwood chips, other industrial roundwood, and building board) constituted just 8 percent of the value of the South Central region's 1976 forest product exports. The 1976 value of exports of these "other" products was $1.3 million, which represented a 1,266 percent increase over the 1967 value. Much of the increase was due to the export of $512,000 worth of pulpwood chips in 1976. Chips had been exported from the region in only one other year during 1967-76, and at a much lower level.

Imports

Regional Overview

Only four commodities were consistently, or even frequently, imported by the South Central region. Softwood lumber, paper and paperboard, plywood, and newsprint constituted more than 99 percent of the region's forest product imports, which, in 1976, totaled only $1.3 million. That figure represents about two one-hundredths of 1 percent of all U.S. 1976 forest product imports.

The region's imports increased by only 28 percent in nominal value from 1967 to 1976, which is equivalent to an average annual gain of just 3 percent.

Mexico and Canada were the only foreign sources of forest products for the region during 1967-76. Canada was an important supplier of softwood lumber and newsprint. Overall, Mexico provided about 80 percent of the value of the region's 1976 imports, and Canada provided the remaining 20 percent.

Softwood Lumber

Softwood lumber constituted nearly one-half of the value of the region's 1976 forest product imports, but less than 1 percent of U.S. softwood lumber imports. The 1976 imports totaled 2.5 million board-feet (worth $638,000).

The region decreased the volume of its softwood lumber imports to about two-thirds of their 1967 level. The nominal value increased by just 13 percent, or only 1 percent per year, from 1967 to 1976.

Both Mexico and Canada were somewhat erratic in their shipment of softwood lumber to the region. Mexico shipped 3 million board-feet in 1967, steadily increased to 4.5 million board-feet in 1971, and then exploded in 1972-73, averaging 17.8 million board-feet. But by 1976, imports from Mexico were down to 1.2 million board-feet, less than half their 1967 level. Imports from Canada were similarly unstable, but seemed to be generally increasing, especially in 1973-76.

In 1976 Mexico supplied two-thirds of the value of the region's softwood lumber imports, and Canada supplied the remainder. During 1967-73, Canada's average share of the region's imports had been just 6 percent.

Paper and Paperboard. The South Central's 1976 imports of paper and paperboard totaled 3,392 tons (worth $419,000). The region did not import paper and paperboard in 1967, but from 1968 to 1976 the region's imports grew at an average annual rate of 65 percent in volume and 56 percent in nominal value.

Mexico was the dominant exporter of paper and paperboard to the South Central region. In 1976 Mexico provided 97 percent of the value of the region's imports, and Canada provided the remaining 3 percent.

Plywood. Plywood constituted 16 percent (by value) of the region's 1976 forest product imports, down from 27 percent in 1967. Plywood's share of the region's imports dropped largely because of the decline in the absolute amount of plywood imports. The 1976 imports of 604,000 square feet (worth $202,000) represented losses of 64 percent in volume and 25 percent in nominal value.

Canada sporadically shipped plywood to the region, but Mexico was the principal source and always accounted for more than 98 percent of the value of the South Central region's plywood imports.

Newsprint. Although newsprint constituted 3 percent of the region's forest product imports in 1976, it had not been imported at all in the previous five years. The 1976 imports, which were valued at just $35,000, were all shipped from Canada.

EXPORTS
SOFTWOOD LUMBER (2432)
SOUTH CENTRAL

VALUE--MILLION CURRENT DOLLARS

YEAR	MEXICO	NICARAGA	COLOMBIA	U KINGDM	W GERMNY	OTHERS	TOTAL
1967	0.739	0.001	0.000	0.000	0.000	0.000	0.740
1968	1.730	0.000	0.000	0.000	0.000	0.000	1.730
1969	1.156	0.000	0.000	0.000	0.000	0.000	1.156
1970	1.730	0.000	0.000	0.000	0.000	0.000	1.730
1971	1.663	0.000	0.000	0.000	0.000	0.000	1.663
1972	1.338	0.000	0.000	0.004	0.008	0.000	1.351
1973	2.489	0.000	0.000	0.000	0.000	0.000	2.489
1974	1.826	0.000	0.000	0.000	0.000	0.000	1.826
1975	2.500	0.000	0.002	0.000	0.000	0.000	2.502
1976	3.422	0.000	0.000	0.000	0.000	0.000	3.422

QUANTITY--MILLION BF

YEAR	MEXICO	NICARAGA	COLOMBIA	U KINGDM	W GERMNY	OTHERS	TOTAL
1967	8.477	0.013	0.000	0.000	0.000	0.000	8.490
1968	13.532	0.000	0.000	0.000	0.000	0.000	13.532
1969	9.607	0.000	0.000	0.000	0.000	0.000	9.607
1970	14.260	0.000	0.000	0.000	0.000	0.000	14.260
1971	13.187	0.000	0.000	0.000	0.000	0.000	13.187
1972	8.744	0.000	0.000	0.061	0.026	0.000	8.831
1973	15.148	0.000	0.000	0.000	0.000	0.000	15.148
1974	7.681	0.000	0.000	0.000	0.000	0.000	7.681
1975	9.338	0.000	0.013	0.000	0.000	0.000	9.351
1976	11.967	0.000	0.000	0.000	0.000	0.000	11.967

EXPORTS
HARDWOOD LUMBER (2433)
SOUTH CENTRAL

VALUE--MILLION CURRENT DOLLARS

YEAR	MEXICO	OTH N AM	S AMERIC	EUROPE	ASIA	OTHERS	TOTAL
1967	0.092	0.000	0.000	0.000	0.000	0.000	0.092
1968	0.096	0.000	0.000	0.000	0.000	0.000	0.096
1969	0.085	0.000	0.000	0.000	0.000	0.000	0.085
1970	0.083	0.000	0.000	0.000	0.000	0.000	0.083
1971	0.065	0.000	0.000	0.000	0.000	0.000	0.065
1972	0.064	0.000	0.000	0.000	0.000	0.000	0.064
1973	0.110	0.000	0.000	0.000	0.000	0.000	0.110
1974	0.135	0.000	0.000	0.000	0.000	0.000	0.135
1975	0.145	0.000	0.000	0.000	0.000	0.000	0.145
1976	0.458	0.000	0.000	0.000	0.000	0.000	0.458

QUANTITY--MILLION BF

YEAR	MEXICO	OTH N AM	S AMERIC	EUROPE	ASIA	OTHERS	TOTAL
1967	0.459	0.000	0.000	0.000	0.000	0.000	0.459
1968	0.521	0.000	0.000	0.000	0.000	0.000	0.521
1969	0.308	0.000	0.000	0.000	0.000	0.000	0.308
1970	0.385	0.000	0.000	0.000	0.000	0.000	0.385
1971	0.283	0.000	0.000	0.000	0.000	0.000	0.283
1972	0.235	0.000	0.000	0.000	0.000	0.000	0.235
1973	0.333	0.000	0.000	0.000	0.000	0.000	0.333
1974	0.592	0.000	0.000	0.000	0.000	0.000	0.592
1975	0.433	0.000	0.000	0.000	0.000	0.000	0.433
1976	1.502	0.000	0.000	0.000	0.000	0.000	1.502

EXPORTS
PLYWOOD (6312)
SOUTH CENTRAL

VALUE--MILLION CURRENT DOLLARS

YEAR	CANADA	MEXICO	OTH N AM	S AMERIC	EUROPE	OTHERS	TOTAL
1967	0.000	0.066	0.000	0.000	0.000	0.000	0.066
1968	0.000	0.073	0.000	0.000	0.000	0.000	0.073
1969	0.000	0.099	0.000	0.000	0.000	0.000	0.099
1970	0.000	0.205	0.000	0.000	0.000	0.000	0.205
1971	0.000	0.110	0.000	0.000	0.000	0.000	0.110
1972	0.000	0.127	0.000	0.000	0.000	0.000	0.127
1973	0.000	0.258	0.000	0.000	0.000	0.000	0.258
1974	0.000	0.574	0.000	0.000	0.000	0.000	0.574
1975	0.000	0.773	0.000	0.000	0.000	0.000	0.773
1976	0.003	0.866	0.000	0.000	0.000	0.000	0.869

QUANTITY--MIL SQ FT

YEAR	CANADA	MEXICO	OTH N AM	S AMERIC	EUROPE	OTHERS	TOTAL
1967	0.000	0.573	0.000	0.000	0.000	0.000	0.573
1968	0.000	0.484	0.000	0.000	0.000	0.000	0.484
1969	0.000	0.716	0.000	0.000	0.000	0.000	0.716
1970	0.000	1.705	0.000	0.000	0.000	0.000	1.705
1971	0.000	0.985	0.000	0.000	0.000	0.000	0.985
1972	0.000	1.038	0.000	0.000	0.000	0.000	1.038
1973	0.000	1.670	0.000	0.000	0.000	0.000	1.670
1974	0.000	3.328	0.000	0.000	0.000	0.000	3.328
1975	0.000	4.174	0.000	0.000	0.000	0.000	4.174
1976	0.001	4.243	0.000	0.000	0.000	0.000	4.244

EXPORTS
RECONSTITUTED WOOD (6314)
SOUTH CENTRAL

VALUE--MILLION CURRENT DOLLARS

YEAR	MEXICO	OTH N AM	S AMERIC	EUROPE	ASIA	OTHERS	TOTAL
1967	0.001	0.000	0.000	0.000	0.000	0.000	0.001
1968	0.001	0.000	0.000	0.000	0.000	0.000	0.001
1969	0.010	0.000	0.000	0.000	0.000	0.000	0.010
1970	0.001	0.000	0.000	0.000	0.000	0.000	0.001
1971	0.003	0.000	0.000	0.000	0.000	0.000	0.003
1972	0.008	0.000	0.000	0.000	0.000	0.000	0.008
1973	0.297	0.000	0.000	0.000	0.000	0.000	0.297
1974	1.050	0.000	0.000	0.000	0.000	0.000	1.050
1975	0.187	0.000	0.000	0.000	0.000	0.000	0.187
1976	0.238	0.000	0.000	0.000	0.000	0.000	0.238

QUANTITY--MIL SQ FT

YEAR	MEXICO	OTH N AM	S AMERIC	EUROPE	ASIA	OTHERS	TOTAL
1967	0.001	0.000	0.000	0.000	0.000	0.000	0.001
1968	0.003	0.000	0.000	0.000	0.000	0.000	0.003
1969	0.083	0.000	0.000	0.000	0.000	0.000	0.083
1970	0.003	0.000	0.000	0.000	0.000	0.000	0.003
1971	0.027	0.000	0.000	0.000	0.000	0.000	0.027
1972	0.060	0.000	0.000	0.000	0.000	0.000	0.060
1973	1.401	0.000	0.000	0.000	0.000	0.000	1.401
1974	9.522	0.000	0.000	0.000	0.000	0.000	9.522
1975	1.458	0.000	0.000	0.000	0.000	0.000	1.458
1976	1.651	0.000	0.000	0.000	0.000	0.000	1.651

EXPORTS
PULPWOOD CHIPS (631.8320)
SOUTH CENTRAL

VALUE--MILLION CURRENT DOLLARS

YEAR	MEXICO	OTH N AM	S AMERIC	EUROPE	JAPAN	OTHERS	TOTAL
1967	0.000	0.000	0.000	0.000	0.000	0.000	0.000
1968	0.000	0.000	0.000	0.000	0.000	0.000	0.000
1969	0.000	0.000	0.000	0.000	0.000	0.000	0.000
1970	0.002	0.000	0.000	0.000	0.000	0.000	0.002
1971	0.000	0.000	0.000	0.000	0.000	0.000	0.000
1972	0.000	0.000	0.000	0.000	0.000	0.000	0.000
1973	0.000	0.000	0.000	0.000	0.000	0.000	0.000
1974	0.000	0.000	0.000	0.000	0.000	0.000	0.000
1975	0.000	0.000	0.000	0.000	0.000	0.000	0.000
1976	0.000	0.000	0.000	0.000	0.512	0.000	0.512

QUANTITY--1000 STN

YEAR	MEXICO	OTH N AM	S AMERIC	EUROPE	JAPAN	OTHERS	TOTAL
1967	0.000	0.000	0.000	0.000	0.000	0.000	0.000
1968	0.000	0.000	0.000	0.000	0.000	0.000	0.000
1969	0.000	0.000	0.000	0.000	0.000	0.000	0.000
1970	0.083	0.000	0.000	0.000	0.000	0.000	0.083
1971	0.000	0.000	0.000	0.000	0.000	0.000	0.000
1972	0.000	0.000	0.000	0.000	0.000	0.000	0.000
1973	0.000	0.000	0.000	0.000	0.000	0.000	0.000
1974	0.000	0.000	0.000	0.000	0.000	0.000	0.000
1975	0.000	0.000	0.000	0.000	0.000	0.000	0.000
1976	0.000	0.000	0.000	0.000	11.396	0.000	11.396

EXPORTS
WOOD PULP (2510)
SOUTH CENTRAL

VALUE--MILLION CURRENT DOLLARS

YEAR	MEXICO	OTH N AM	S AMERIC	EUROPE	ASIA	OTHERS	TOTAL
1967	0.724	0.000	0.000	0.000	0.000	0.000	0.724
1968	0.813	0.000	0.000	0.000	0.000	0.000	0.813
1969	1.053	0.000	0.000	0.000	0.000	0.000	1.053
1970	0.990	0.000	0.000	0.000	0.000	0.000	0.990
1971	1.043	0.000	0.000	0.000	0.000	0.000	1.043
1972	1.097	0.000	0.000	0.000	0.000	0.000	1.097
1973	1.710	0.000	0.000	0.000	0.000	0.000	1.710
1974	1.888	0.000	0.000	0.000	0.000	0.000	1.888
1975	2.569	0.000	0.000	0.000	0.000	0.000	2.569
1976	1.553	0.000	0.000	0.000	0.000	0.000	1.553

QUANTITY--1000 STN

YEAR	MEXICO	OTH N AM	S AMERIC	EUROPE	ASIA	OTHERS	TOTAL
1967	4.048	0.000	0.000	0.000	0.000	0.000	4.048
1968	4.423	0.000	0.000	0.000	0.000	0.000	4.423
1969	5.974	0.000	0.000	0.000	0.000	0.000	5.974
1970	5.476	0.000	0.000	0.000	0.000	0.000	5.476
1971	5.174	0.000	0.000	0.000	0.000	0.000	5.174
1972	5.270	0.000	0.000	0.000	0.000	0.000	5.270
1973	10.370	0.000	0.000	0.000	0.000	0.000	10.370
1974	6.337	0.000	0.000	0.000	0.000	0.000	6.337
1975	6.856	0.000	0.000	0.000	0.000	0.000	6.856
1976	4.525	0.000	0.000	0.000	0.000	0.000	4.525

EXPORTS
NEWSPRINT (6411)
SOUTH CENTRAL

VALUE--MILLION CURRENT DOLLARS

YEAR	MEXICO	OTH N AM	S AMERIC	EUROPE	ASIA	OTHERS	TOTAL
1967	1.711	0.000	0.000	0.000	0.000	0.000	1.711
1968	1.681	0.000	0.000	0.000	0.000	0.000	1.681
1969	0.433	0.000	0.000	0.000	0.000	0.000	0.433
1970	1.802	0.000	0.000	0.000	0.000	0.000	1.802
1971	0.623	0.000	0.000	0.000	0.000	0.000	0.623
1972	0.444	0.000	0.000	0.000	0.000	0.000	0.444
1973	0.285	0.000	0.000	0.000	0.000	0.000	0.285
1974	1.126	0.000	0.000	0.000	0.000	0.000	1.126
1975	1.126	0.000	0.000	0.000	0.000	0.000	1.126
1976	1.369	0.000	0.000	0.000	0.000	0.000	1.369

QUANTITY--1000 STN

YEAR	MEXICO	OTH N AM	S AMERIC	EUROPE	ASIA	OTHERS	TOTAL
1967	12.311	0.000	0.000	0.000	0.000	0.000	12.311
1968	12.191	0.000	0.000	0.000	0.000	0.000	12.191
1969	3.057	0.000	0.000	0.000	0.000	0.000	3.057
1970	12.951	0.000	0.000	0.000	0.000	0.000	12.951
1971	4.023	0.000	0.000	0.000	0.000	0.000	4.023
1972	2.817	0.000	0.000	0.000	0.000	0.000	2.817
1973	1.709	0.000	0.000	0.000	0.000	0.000	1.709
1974	5.584	0.000	0.000	0.000	0.000	0.000	5.584
1975	4.049	0.000	0.000	0.000	0.000	0.000	4.049
1976	3.906	0.000	0.000	0.000	0.000	0.000	3.906

EXPORTS
PAPER & PAPERBOARD (6410)
SOUTH CENTRAL

VALUE--MILLION CURRENT DOLLARS

YEAR	CANADA	MEXICO	OTH N AM	S AMERIC	EUROPE	OTHERS	TOTAL
1967	0.000	0.105	0.000	0.000	0.000	0.000	0.105
1968	0.000	0.249	0.000	0.000	0.000	0.000	0.249
1969	0.002	0.194	0.000	0.000	0.000	0.000	0.197
1970	0.000	1.153	0.000	0.000	0.000	0.000	1.153
1971	0.000	4.172	0.000	0.000	0.000	0.000	4.172
1972	0.008	5.487	0.000	0.000	0.000	0.000	5.495
1973	0.000	4.023	0.000	0.000	0.000	0.000	4.023
1974	0.008	3.713	0.000	0.000	0.000	0.000	3.721
1975	0.010	5.939	0.000	0.000	0.000	0.000	5.949
1976	0.006	7.115	0.000	0.000	0.000	0.000	7.120

QUANTITY--1000 STN

YEAR	CANADA	MEXICO	OTH N AM	S AMERIC	EUROPE	OTHERS	TOTAL
1967	0.000	0.256	0.000	0.000	0.000	0.000	0.256
1968	0.000	0.605	0.000	0.000	0.000	0.000	0.605
1969	0.000	0.504	0.000	0.000	0.000	0.000	0.505
1970	0.000	1.375	0.000	0.000	0.000	0.000	1.375
1971	0.000	6.802	0.000	0.000	0.000	0.000	6.802
1972	0.003	14.606	0.000	0.000	0.000	0.000	14.610
1973	0.000	16.311	0.000	0.000	0.000	0.000	16.311
1974	0.027	13.609	0.000	0.000	0.000	0.000	13.636
1975	0.008	19.033	0.000	0.000	0.000	0.000	19.041
1976	0.012	21.362	0.000	0.000	0.000	0.000	21.374

IMPORTS
SOFTWOOD LUMBER (2432)
SOUTH CENTRAL

VALUE--MILLION CURRENT DOLLARS

YEAR	CANADA	MEXICO	OTH N AM	S AMERIC	EUROPE	OTHERS	TOTAL
1967	0.041	0.522	0.000	0.000	0.000	0.000	0.563
1968	0.102	0.439	0.000	0.000	0.000	0.000	0.540
1969	0.018	0.786	0.000	0.000	0.000	0.000	0.803
1970	0.004	0.775	0.000	0.000	0.000	0.000	0.780
1971	0.087	0.674	0.000	0.000	0.000	0.000	0.761
1972	0.000	1.993	0.000	0.000	0.000	0.000	1.993
1973	0.078	2.673	0.000	0.000	0.000	0.000	2.752
1974	0.144	0.501	0.000	0.000	0.000	0.000	0.645
1975	0.099	0.191	0.000	0.000	0.000	0.000	0.290
1976	0.211	0.428	0.000	0.000	0.000	0.000	0.638

QUANTITY--MILLION BF

YEAR	CANADA	MEXICO	OTH N AM	S AMERIC	EUROPE	OTHERS	TOTAL
1967	0.632	3.045	0.000	0.000	0.000	0.000	3.677
1968	1.308	2.659	0.000	0.000	0.000	0.000	3.967
1969	0.172	4.518	0.000	0.000	0.000	0.000	4.690
1970	0.059	4.014	0.000	0.000	0.000	0.000	4.073
1971	0.929	4.524	0.000	0.000	0.000	0.000	5.453
1972	0.000	18.481	0.000	0.000	0.000	0.000	18.481
1973	0.401	17.079	0.000	0.000	0.000	0.000	17.480
1974	1.214	2.714	0.000	0.000	0.000	0.000	3.928
1975	1.045	0.450	0.000	0.000	0.000	0.000	1.495
1976	1.214	1.241	0.000	0.000	0.000	0.000	2.455

IMPORTS
PLYWOOD (6312)
SOUTH CENTRAL

VALUE--MILLION CURRENT DOLLARS

YEAR	CANADA	MEXICO	OTH N AM	S AMERIC	EUROPE	OTHERS	TOTAL
1967	0.000	0.271	0.000	0.000	0.000	0.000	0.271
1968	0.000	0.217	0.000	0.000	0.000	0.000	0.217
1969	0.000	0.202	0.000	0.000	0.000	0.000	0.202
1970	0.000	0.118	0.000	0.000	0.000	0.000	0.118
1971	0.000	0.042	0.000	0.000	0.000	0.000	0.042
1972	0.000	0.080	0.000	0.000	0.000	0.000	0.080
1973	0.000	0.104	0.000	0.000	0.000	0.000	0.104
1974	0.001	0.063	0.000	0.000	0.000	0.000	0.064
1975	0.000	0.001	0.000	0.000	0.000	0.000	0.001
1976	0.001	0.201	0.000	0.000	0.000	0.000	0.202

QUANTITY--MIL SQ FT

YEAR	CANADA	MEXICO	OTH N AM	S AMERIC	EUROPE	OTHERS	TOTAL
1967	0.000	1.676	0.000	0.000	0.000	0.000	1.676
1968	0.000	1.482	0.000	0.000	0.000	0.000	1.482
1969	0.000	2.130	0.000	0.000	0.000	0.000	2.130
1970	0.000	0.789	0.000	0.000	0.000	0.000	0.789
1971	0.000	0.234	0.000	0.000	0.000	0.000	0.234
1972	0.000	0.439	0.000	0.000	0.000	0.000	0.439
1973	0.000	0.563	0.000	0.000	0.000	0.000	0.563
1974	0.000	0.250	0.000	0.000	0.000	0.000	0.250
1975	0.000	0.009	0.000	0.000	0.000	0.000	0.009
1976	0.003	0.601	0.000	0.000	0.000	0.000	0.604

IMPORTS
PAPER & PAPERBOARD (6410)
SOUTH CENTRAL

VALUE--MILLION CURRENT DOLLARS

YEAR	CANADA	MEXICO	OTH N AM	S AMERIC	EUROPE	OTHERS	TOTAL
1967	0.000	0.000	0.000	0.000	0.000	0.000	0.000
1968	0.012	0.000	0.000	0.000	0.000	0.000	0.012
1969	0.000	0.001	0.000	0.000	0.000	0.000	0.001
1970	0.001	0.000	0.000	0.000	0.000	0.000	0.001
1971	0.000	0.000	0.000	0.000	0.000	0.000	0.000
1972	0.000	0.013	0.000	0.000	0.000	0.000	0.013
1973	0.002	0.003	0.000	0.000	0.000	0.000	0.005
1974	0.000	0.106	0.000	0.000	0.000	0.000	0.106
1975	0.000	0.284	0.000	0.000	0.000	0.000	0.284
1976	0.014	0.405	0.000	0.000	0.000	0.000	0.419

QUANTITY--1000 STN

YEAR	CANADA	MEXICO	OTH N AM	S AMERIC	EUROPE	OTHERS	TOTAL
1967	0.000	0.000	0.000	0.000	0.000	0.000	0.000
1968	0.062	0.000	0.000	0.000	0.000	0.000	0.062
1969	0.000	0.000	0.000	0.000	0.000	0.000	0.000
1970	0.007	0.000	0.000	0.000	0.000	0.000	0.007
1971	0.000	0.000	0.000	0.000	0.000	0.000	0.000
1972	0.000	0.005	0.000	0.000	0.000	0.000	0.005
1973	0.001	0.031	0.000	0.000	0.000	0.000	0.032
1974	0.000	0.775	0.000	0.000	0.000	0.000	0.775
1975	0.000	1.997	0.000	0.000	0.000	0.000	1.997
1976	0.047	3.345	0.000	0.000	0.000	0.000	3.392

Appendix L

ALASKA

Exports

Regional Overview

Alaska consistently exported only four forest products during 1967-76. The 1976 exports of softwood logs ($5.9 million), softwood lumber ($67.1 million), pulpwood chips ($4.1 million), and wood pulp ($66.9 million) totaled $143.9 million, which was practically 100 percent of the region's forest product exports.

From 1967 to 1976 Alaska steadily increased the amount of forest products it exported. The average annual growth rate in the nominal export value was 15 percent.

The 1976 exports represented just 4 percent of the value of the entire U.S. forest products exports.

For all of Alaska's major export commodities, Japan was the principal market. In 1976 Japan was the destination for 83 percent of the value of the region's total forest product exports.

Softwood Lumber

Alaska was a major U.S. exporter of softwood lumber throughout 1967-76. In an average year, the region shipped 17 percent (by value) of all U.S. softwood lumber exports. Alaska's 1976 exports of 401.9 million board-feet (worth $67.1 million) represented 22 percent of the volume, and 14 percent of the value of the country's softwood lumber exports.

Export growth during this period was good. The average annual growth rate was 20 percent in nominal value, and 5 percent in volume.

Japan was the only significant market for Alaskan softwood lumber during 1967-76. In every year the decade, Japan imported more than 99 percent of the value of the region's softwood lumber exports.

Wood Pulp

In 1976 Alaska exported 198,000 tons of wood pulp valued at $66.9 million, which represented 8 percent of all U.S. wood pulp exports.

Wood pulp exports from Alaska increased at a relatively slow rate from 1967-76. The average annual gain in nominal value was 11 percent, and in volume was less than 1 percent.

In an average year, Asian countries imported 95 percent of the value of Alaska's wood pulp exports. In 1976 Japan was the largest importer followed by Korea, Taiwan, and India.

The volume of Japanese imports declined by 11 percent from 1967 to 1976, and the nominal value rose by 89 percent, equivalent to an annual increase of just 7 percent. During the period, Japan's share of the value of Alaska's wood pulp exports fell from 87 percent to 64 percent.

Korea had been steadily increasing its imports of Alaskan pulp through most of the 1967-75 period, and in 1976, more than doubled the amount which it had imported in the previous year. That large jump allowed Korea to double its share of Alaska's exports, from 7 percent in 1975 to 14 percent in 1976. India's imports of Alaskan wood pulp dropped drastically from 1967 to 1976. Its 1976 imports of 4,680 tons (worth $1.6 million) represented declines from the 1967 levels of 77 percent in volume and 48 percent nominal value. In 1967 India imported 12 percent (by value) of the region's wood pulp exports, but by 1976 its share was only 2 percent.

Taiwan, which did not import pulp from Alaska in 1967, accounted for 7 percent of the value of the region's 1976 pulp exports.

Softwood Logs

From 1967 to 1973, Alaska's softwood log exports were increasing in volume at a rate of 9 percent per year, but fell off in each of the next three years, so that by 1976 the region was exporting a smaller volume of logs than it had been in 1967. At the same time, the nominal value was 97 percent higher than the 1967 level, equivalent to an average growth rate of 8 percent per year.

Throughout the study period, Alaska shipped 1 percent to 2 percent of the value of all U.S. softwood log exports.

In an average year, Japan was the destination for 95 percent of the value of the region's softwood log exports. Canada was usually the market for the logs not bound for Japan, and Korea imported Alaskan logs infrequently. In 1976 Japan was responsible for 89 percent, Canada for 8 percent, and Korea for 2 percent of the value of Alaska's softwood log exports.

Pulpwood Chips

Alaska began exporting pulpwood chips in 1970, and steadily increased its exports through 1976. In 1976 Alaska shipped 107,000 tons (worth $1.4 million), which represented 3 percent of all U.S. chip exports.

The average growth in the region's chip exports from 1970 to 1976 was 48 percent per year in volume and 57 percent annually in nominal value.

From 1970 to 1976, with the exception of one year, Japan consumed 100 percent of Alaska's chip exports. In 1974 Canada imported $110,000 worth of chips, which was 11 percent of Alaska's wood chip exports.

Imports

Regional Overview

Only six forest products were imported by Alaska with any regularity during 1967-76. Softwood logs, softwood lumber, hardwood lumber, plywood, newsprint, and other industrial roundwood constituted the entire set of forest product imports in 1976, and their combined import value was $1.9 million. That amount represented less than one-twenty-fifth of 1 percent of the value of all U.S. forest product imports.

The nominal value of Alaska's imports increased steadily from 1967 to 1976. The 1976 value was 433 percent greater than the 1967 value, which is equivalent to an average annual gain of 20 percent.

Newsprint constituted almost half ($928,000) of the region's imports in 1976. Softwood lumber imports in that year totaled $837,000 and made up 43 percent of the region's forest product imports. The remaining four commodities together were valued at just $182,000 or 9 percent of the region's forest product imports.

Newsprint

During 1967-76, Alaska rapidly increased its import of newsprint, while the rest of the country generally diminished its dependence on foreign newsprint. The 1976 Alaskan imports of 3,527 tons (worth $928,000) resulted from a decade of tremendous growth, with average annual gains of 60 percent in volume and 75 percent in nominal value. Despite this growth, Alaska's 1976 imports were still less than one-tenth of 1 percent of all U.S. newsprint imports.

Canada has been the only foreign country to ever export newsprint to Alaska.

Softwood Lumber

Alaska was a steady importer of softwood lumber throughout 1967-76. In the early years of the period, softwood lumber accounted for nearly half the value of the region's total forest product imports, but in the 1970s that share averaged 31 percent.

Alaska's softwood lumber imports grew at an average annual rate of 10 percent in volume and 20 percent in nominal value. The 1976 imports totaled 4.6 million board-feet (valued at $837,000).

With the exception of two years in which Japan shipped small quantities to Alaska, Canada was the region's sole foreign source of softwood lumber.

Softwood Logs

Softwood log imports to Alaska were somewhat sporadic in 1967-76. In three years there were no imports, and in another four years the value of log imports was less than $10,000. The peak year was 1970, when the region imported 2 million board-feet (worth $139,000). The 1976 imports amounted to 302,000 board-feet (valued at $75,000).

Canada was the only exporter of softwood logs to Alaska during the 1967-76 period.

Hardwood Lumber

Like softwood log imports, Alaska's hardwood lumber imports were erratic and small, relative to other regions' imports. In only three years did the region's hardwood lumber imports exceed $10,000. The 1976 imports totaled 63,000 board-feet (valued at $23,000).

In the early years of 1967-76, and in the years when imports were greatest, Japan and Malaysia supplied the bulk of Alaska's hardwood lumber imports. Canada was the only exporter to the region in 1974, when it first began hardwood lumber export to Alaska, and in 1975. In 1976 Malaysia supplied 83 percent of the value of the region's hardwood lumber imports, and Canada supplied the remainder.

EXPORTS
SOFTWOOD LOGS (2422)
ALASKA

VALUE--MILLION CURRENT DOLLARS

YEAR	CANADA	EUROPE	KOR REP	JAPAN	OTH ASIA	OTHERS	TOTAL
1967	0.177	0.000	0.000	2.804	0.000	0.000	2.980
1968	0.000	0.000	0.000	3.336	0.000	0.000	3.336
1969	0.031	0.000	0.000	3.762	0.000	0.000	3.793
1970	0.301	0.000	0.000	6.170	0.000	0.000	6.471
1971	0.208	0.000	0.046	4.710	0.000	0.000	4.965
1972	0.244	0.000	0.000	8.145	0.000	0.000	8.388
1973	0.003	0.000	0.000	17.770	0.000	0.000	17.773
1974	0.270	0.000	0.802	7.351	0.000	0.000	8.422
1975	0.370	0.000	0.000	8.709	0.000	0.000	9.079
1976	0.487	0.000	0.145	5.251	0.000	0.000	5.884

QUANTITY--MILLION BF

YEAR	CANADA	EUROPE	KOR REP	JAPAN	OTH ASIA	OTHERS	TOTAL
1967	2.108	0.000	0.000	33.766	0.000	0.000	35.874
1968	0.000	0.000	0.000	48.820	0.000	0.000	48.820
1969	1.000	0.000	0.000	41.644	0.000	0.000	42.644
1970	3.948	0.000	0.000	51.909	0.000	0.000	55.857
1971	2.593	0.000	1.469	46.715	0.000	0.000	50.777
1972	3.956	0.000	0.000	66.692	0.000	0.000	70.648
1973	0.014	0.000	0.000	78.853	0.000	0.000	78.867
1974	1.936	0.000	5.218	35.647	0.000	0.000	42.801
1975	4.700	0.000	0.000	35.674	0.000	0.000	40.374
1976	4.548	0.000	1.382	24.273	0.000	0.000	30.204

EXPORTS
HARDWOOD LOGS (2423)
ALASKA

VALUE--MILLION CURRENT DOLLARS

YEAR	CANADA	S AMERIC	EUROPE	JAPAN	OTH ASIA	OTHERS	TOTAL
1967	0.000	0.000	0.000	0.008	0.000	0.000	0.008
1968	0.000	0.000	0.000	0.027	0.000	0.000	0.027
1969	0.000	0.000	0.000	0.016	0.000	0.000	0.016
1970	0.000	0.000	0.000	0.000	0.000	0.000	0.000
1971	0.208	0.000	0.000	0.000	0.000	0.000	0.208
1972	0.000	0.000	0.000	0.000	0.000	0.000	0.000
1973	0.000	0.000	0.000	0.236	0.000	0.000	0.236
1974	0.000	0.000	0.000	0.060	0.000	0.000	0.060
1975	0.000	0.000	0.000	0.000	0.000	0.000	0.000
1976	0.000	0.000	0.000	0.000	0.000	0.000	0.000

QUANTITY--MILLION BF

YEAR	CANADA	S AMERIC	EUROPE	JAPAN	OTH ASIA	OTHERS	TOTAL
1967	0.000	0.000	0.000	0.115	0.000	0.000	0.115
1968	0.000	0.000	0.000	0.477	0.000	0.000	0.477
1969	0.000	0.000	0.000	0.309	0.000	0.000	0.309
1970	0.000	0.000	0.000	0.000	0.000	0.000	0.000
1971	2.593	0.000	0.000	0.000	0.000	0.000	2.593
1972	0.000	0.000	0.000	0.000	0.000	0.000	0.000
1973	0.000	0.000	0.000	1.855	0.000	0.000	1.855
1974	0.000	0.000	0.000	2.470	0.000	0.000	2.470
1975	0.000	0.000	0.000	0.000	0.000	0.000	0.000
1976	0.000	0.000	0.000	0.000	0.000	0.000	0.000

EXPORTS
SOFTWOOD LUMBER (2432)
ALASKA

VALUE--MILLION CURRENT DOLLARS

YEAR	CANADA	ITALY	KOR REP	JAPAN	AUSTRALA	OTHERS	TOTAL
1967	0.000	0.000	0.004	12.998	0.000	0.000	13.002
1968	0.000	0.000	0.000	17.285	0.150	0.000	17.435
1969	0.000	0.000	0.000	24.672	0.050	0.000	24.722
1970	0.010	0.000	0.000	29.818	0.000	0.000	29.828
1971	0.166	0.000	0.000	24.340	0.000	0.000	24.507
1972	0.334	0.000	0.000	37.774	0.000	0.000	38.108
1973	0.051	0.000	0.132	81.990	0.000	0.000	82.173
1974	0.075	0.000	0.062	79.747	0.000	0.000	79.884
1975	0.058	0.000	0.000	67.721	0.000	0.000	67.779
1976	0.079	0.034	0.143	66.800	0.000	0.000	67.056

QUANTITY--MILLION BF

YEAR	CANADA	ITALY	KOR REP	JAPAN	AUSTRALA	OTHERS	TOTAL
1967	0.000	0.000	0.050	261.332	0.000	0.000	261.382
1968	0.000	0.000	0.000	316.419	2.306	0.000	318.725
1969	0.000	0.000	0.000	363.569	0.533	0.000	364.102
1970	0.200	0.000	0.000	447.377	0.000	0.000	447.577
1971	1.440	0.000	0.000	304.552	0.000	0.000	305.992
1972	3.258	0.000	0.000	440.231	0.000	0.000	443.489
1973	0.322	0.000	0.783	570.225	0.000	0.000	571.330
1974	0.421	0.000	0.413	517.285	0.000	0.000	518.119
1975	0.331	0.000	0.000	445.186	0.000	0.000	445.517
1976	0.396	0.159	0.363	400.975	0.000	0.000	401.894

EXPORTS
HARDWOOD LUMBER (2433)
ALASKA

VALUE--MILLION CURRENT DOLLARS

YEAR	CANADA	S AMERIC	EUROPE	JAPAN	OTH ASIA	OTHERS	TOTAL
1967	0.000	0.000	0.000	0.000	0.000	0.000	0.000
1968	0.004	0.000	0.000	0.000	0.000	0.000	0.004
1969	0.000	0.000	0.000	0.000	0.000	0.000	0.000
1970	0.000	0.000	0.000	0.199	0.000	0.000	0.199
1971	0.000	0.000	0.000	0.105	0.000	0.000	0.105
1972	0.000	0.000	0.000	0.000	0.000	0.000	0.000
1973	0.003	0.000	0.000	0.000	0.000	0.000	0.003
1974	0.000	0.000	0.000	0.000	0.000	0.000	0.000
1975	0.000	0.000	0.000	0.000	0.000	0.000	0.000
1976	0.000	0.000	0.000	0.000	0.000	0.000	0.000

QUANTITY--MILLION BF

YEAR	CANADA	S AMERIC	EUROPE	JAPAN	OTH ASIA	OTHERS	TOTAL
1967	0.000	0.000	0.000	0.000	0.000	0.000	0.000
1968	0.013	0.000	0.000	0.000	0.000	0.000	0.013
1969	0.000	0.000	0.000	0.000	0.000	0.000	0.000
1970	0.000	0.000	0.000	2.824	0.000	0.000	2.824
1971	0.000	0.000	0.000	1.498	0.000	0.000	1.498
1972	0.000	0.000	0.000	0.000	0.000	0.000	0.000
1973	0.022	0.000	0.000	0.000	0.000	0.000	0.022
1974	0.000	0.000	0.000	0.000	0.000	0.000	0.000
1975	0.000	0.000	0.000	0.000	0.000	0.000	0.000
1976	0.000	0.000	0.000	0.000	0.000	0.000	0.000

EXPORTS
PULPWOOD CHIPS (631.8320)
ALASKA

VALUE--MILLION CURRENT DOLLARS

YEAR	CANADA	OTH N AM	S AMERIC	JAPAN	OTH ASIA	OTHERS	TOTAL
1967	0.000	0.000	0.000	0.000	0.000	0.000	0.000
1968	0.000	0.000	0.000	0.000	0.000	0.000	0.000
1969	0.000	0.000	0.000	0.000	0.000	0.000	0.000
1970	0.000	0.000	0.000	0.270	0.000	0.000	0.270
1971	0.000	0.000	0.000	0.830	0.000	0.000	0.830
1972	0.000	0.000	0.000	0.580	0.000	0.000	0.580
1973	0.000	0.000	0.000	0.000	0.000	0.000	0.000
1974	0.110	0.000	0.000	0.897	0.000	0.000	1.007
1975	0.000	0.000	0.000	1.572	0.000	0.000	1.572
1976	0.000	0.000	0.000	4.078	0.000	0.000	4.078

QUANTITY--1000 STN

YEAR	CANADA	OTH N AM	S AMERIC	JAPAN	OTH ASIA	OTHERS	TOTAL
1967	0.000	0.000	0.000	0.000	0.000	0.000	0.000
1968	0.000	0.000	0.000	0.000	0.000	0.000	0.000
1969	0.000	0.000	0.000	0.000	0.000	0.000	0.000
1970	0.000	0.000	0.000	10.450	0.000	0.000	10.450
1971	0.000	0.000	0.000	27.800	0.000	0.000	27.800
1972	0.000	0.000	0.000	20.185	0.000	0.000	20.185
1973	0.000	0.000	0.000	0.000	0.000	0.000	0.000
1974	2.200	0.000	0.000	32.541	0.000	0.000	34.741
1975	0.000	0.000	0.000	32.399	0.000	0.000	32.399
1976	0.000	0.000	0.000	107.652	0.000	0.000	107.652

EXPORTS
WOOD PULP (2510)
ALASKA

VALUE--MILLION CURRENT DOLLARS

YEAR	INDIA	KOR REP	TAIWAN	JAPAN	OTH ASIA	OTHERS	TOTAL
1967	3.007	0.383	0.000	22.664	0.000	0.032	26.087
1968	3.911	0.250	0.313	19.140	0.277	0.000	23.891
1969	4.961	0.000	0.808	22.239	0.104	0.000	28.112
1970	4.903	0.554	1.476	22.860	0.567	1.560	31.939
1971	2.196	1.137	1.854	23.295	0.444	2.489	31.416
1972	0.936	1.154	2.258	26.553	1.843	2.402	35.147
1973	2.798	1.368	1.963	21.170	1.381	2.180	30.880
1974	0.000	2.803	3.969	50.855	2.784	4.879	65.290
1975	0.939	3.639	3.106	38.315	4.241	4.851	55.091
1976	1.558	9.471	4.484	42.883	3.336	5.186	66.919

QUANTITY--1000 STN

YEAR	INDIA	KOR REP	TAIWAN	JAPAN	OTH ASIA	OTHERS	TOTAL
1967	20.432	2.497	0.000	162.087	0.000	0.258	185.274
1968	24.865	1.859	2.124	136.972	2.829	0.000	169.650
1969	30.242	0.000	5.319	152.214	0.975	0.000	188.750
1970	29.153	4.222	9.123	140.082	4.332	9.817	196.729
1971	12.054	6.692	10.853	137.996	3.047	14.206	184.847
1972	5.196	6.520	13.149	153.958	10.793	13.358	202.973
1973	15.155	6.912	9.578	102.850	7.439	10.897	152.830
1974	0.000	8.319	12.026	147.392	7.982	12.626	188.343
1975	2.487	9.614	8.193	99.911	11.256	12.335	143.796
1976	4.680	25.051	12.652	127.588	9.883	17.644	197.497

IMPORTS
SOFTWOOD LUMBER (2432)
ALASKA

VALUE--MILLION CURRENT DOLLARS

YEAR	CANADA	OTH N AM	S AMERIC	JAPAN	OTH ASIA	OTHERS	TOTAL
1967	0.159	0.000	0.000	0.000	0.000	0.000	0.159
1968	0.218	0.000	0.000	0.000	0.000	0.000	0.218
1969	0.400	0.000	0.000	0.000	0.000	0.000	0.400
1970	0.360	0.000	0.000	0.005	0.000	0.000	0.365
1971	0.448	0.000	0.000	0.000	0.000	0.000	0.448
1972	0.361	0.000	0.000	0.000	0.000	0.000	0.361
1973	0.205	0.000	0.000	0.000	0.000	0.000	0.205
1974	0.321	0.000	0.000	0.000	0.000	0.000	0.321
1975	0.655	0.000	0.000	0.044	0.000	0.000	0.699
1976	0.837	0.000	0.000	0.000	0.000	0.000	0.837

QUANTITY--MILLION BF

YEAR	CANADA	OTH N AM	S AMERIC	JAPAN	OTH ASIA	OTHERS	TOTAL
1967	1.929	0.000	0.000	0.000	0.000	0.000	1.929
1968	2.166	0.000	0.000	0.000	0.000	0.000	2.166
1969	3.003	0.000	0.000	0.003	0.000	0.000	3.006
1970	3.609	0.000	0.000	0.028	0.000	0.000	3.637
1971	3.833	0.000	0.000	0.000	0.000	0.000	3.833
1972	3.015	0.000	0.000	0.000	0.000	0.000	3.015
1973	1.207	0.000	0.000	0.000	0.000	0.000	1.207
1974	2.060	0.000	0.000	0.000	0.000	0.000	2.060
1975	4.683	0.000	0.000	0.054	0.000	0.000	4.737
1976	4.552	0.000	0.000	0.000	0.000	0.000	4.552

IMPORTS
PULPWOOD CHIPS (631.8320)
ALASKA

VALUE--MILLION CURRENT DOLLARS

YEAR	CANADA	OTH N AM	S AMERIC	EUROPE	ASIA	OTHERS	TOTAL
1967	0.138	0.000	0.000	0.000	0.000	0.000	0.138
1968	0.067	0.000	0.000	0.000	0.000	0.000	0.067
1969	0.000	0.000	0.000	0.000	0.000	0.000	0.000
1970	0.283	0.000	0.000	0.000	0.000	0.000	0.283
1971	0.071	0.000	0.000	0.000	0.000	0.000	0.071
1972	0.000	0.000	0.000	0.000	0.000	0.000	0.000
1973	0.000	0.000	0.000	0.000	0.000	0.000	0.000
1974	0.000	0.000	0.000	0.000	0.000	0.000	0.000
1975	0.000	0.000	0.000	0.000	0.000	0.000	0.000
1976	0.000	0.000	0.000	0.000	0.000	0.000	0.000

QUANTITY--1000 STN

YEAR	CANADA	OTH N AM	S AMERIC	EUROPE	ASIA	OTHERS	TOTAL
1967	18.400	0.000	0.000	0.000	0.000	0.000	18.400
1968	8.200	0.000	0.000	0.000	0.000	0.000	8.200
1969	0.000	0.000	0.000	0.000	0.000	0.000	0.000
1970	35.532	0.000	0.000	0.000	0.000	0.000	35.532
1971	3.053	0.000	0.000	0.000	0.000	0.000	3.053
1972	0.000	0.000	0.000	0.000	0.000	0.000	0.000
1973	0.000	0.000	0.000	0.000	0.000	0.000	0.000
1974	0.000	0.000	0.000	0.000	0.000	0.000	0.000
1975	0.000	0.000	0.000	0.000	0.000	0.000	0.000
1976	0.000	0.000	0.000	0.000	0.000	0.000	0.000

IMPORTS
NEWSPRINT (6411)
ALASKA

VALUE--MILLION CURRENT DOLLARS

YEAR	CANADA	S AMERIC	EUROPE	ASIA	OCEANIA	OTHERS	TOTAL
1967	0.006	0.000	0.000	0.000	0.000	0.000	0.006
1968	0.088	0.000	0.000	0.000	0.000	0.000	0.088
1969	0.313	0.000	0.000	0.000	0.000	0.000	0.313
1970	0.517	0.000	0.000	0.000	0.000	0.000	0.517
1971	0.614	0.000	0.000	0.000	0.000	0.000	0.614
1972	0.741	0.000	0.000	0.000	0.000	0.000	0.741
1973	0.683	0.000	0.000	0.000	0.000	0.000	0.683
1974	1.053	0.000	0.000	0.000	0.000	0.000	1.053
1975	1.312	0.000	0.000	0.000	0.000	0.000	1.312
1976	0.928	0.000	0.000	0.000	0.000	0.000	0.928

QUANTITY--1000 STN

YEAR	CANADA	S AMERIC	EUROPE	ASIA	OCEANIA	OTHERS	TOTAL
1967	0.050	0.000	0.000	0.000	0.000	0.000	0.050
1968	0.810	0.000	0.000	0.000	0.000	0.000	0.810
1969	2.739	0.000	0.000	0.000	0.000	0.000	2.739
1970	4.567	0.000	0.000	0.000	0.000	0.000	4.567
1971	5.350	0.000	0.000	0.000	0.000	0.000	5.350
1972	4.822	0.000	0.000	0.000	0.000	0.000	4.822
1973	4.144	0.000	0.000	0.000	0.000	0.000	4.144
1974	5.302	0.000	0.000	0.000	0.000	0.000	5.302
1975	5.536	0.000	0.000	0.000	0.000	0.000	5.536
1976	3.527	0.000	0.000	0.000	0.000	0.000	3.527

Appendix M

HAWAII AND PUERTO RICO

Exports

Regional Overview

The insular regions of Hawaii and Puerto Rico were insignificant forest product exporters during 1967-76. In no year did their combined exports constitute even 1 percent (by value) of U.S. forest product exports. The 1976 exported amounted to $6 million.

Throughout the decade, paper and paperboard was the most important export, accounting for an average of 82 percent of the annual value of the regions' combined forest product exports. Puerto Rico, in 1976, supplied 99 percent of the $3.4 million worth of paper and paperboard shipped by the two regions. Nearly all of the Puerto Rican paper and paperboard was shipped to Caribbean and South American countries. In 1976 Venezuela was the leading importer, accounting for 57 percent of the value of Puerto Rico's exports.

Softwood lumber and plywood were the regions' other important exports during 1967-76. The two regions' combined exports of softwood lumber in 1976 were 613,000 board-feet (worth $407,000). Jamaica imported 48 percent of the value of the lumber exports in that year.

Plywood exports from the islands totaled 2.3 million square feet (valued at $570,000). Caribbean countries, especially in the Leeward and Windward Islands and the Netherlands Antilles, were again the largest importers.

Hawaii began exporting pulpwood chips to Japan in 1975, and by 1976 the $1.3 million worth of chip exports represented 22 percent of the two region's combined forest product exports.

Imports

Regional Overview

In an average year during 1967-76, the combined imports of Hawaii and Puerto Rico represented about 1 percent (by value) of all U.S. forest product imports. In 1976 the regions' imports were valued at $43.7 million, which was only 90 percent of the 1967 nominal import value.

Newsprint, softwood lumber, and plywood were the regions' most important imports throughout the period. Those three commodities represented, on average, 89 percent of the value of the regions' total forest product imports.

Newsprint imports into the two regions totaled $19 million in 1976. About 56 percent was imported by Puerto Rico and 44 percent by Hawaii. Nearly all (97 percent by value) of the regions' imports originated in Canada, but there was some newsprint shipped from Finland to Puerto Rico. The nominal value of newsprint imports decreased from 1967 to 1976 by 166 percent, or an average of 11 percent per year.

Softwood lumber represented 27 percent of the value of the 1976 forest product imports of Hawaii and Puerto Rico combined. The 100.7 million board-feet imported in that year were valued at $11.9 million. Of that amount, more than 99 percent was shipped to Puerto Rico. Canada was the most important exporter of lumber to Puerto Rico, annually providing an average of 63 percent of the value of the island country's softwood lumber imports. Honduras and Brazil supplied most of the remainder of Puerto Rico's lumber imports. Over the 1967-76 period, Puerto Rico's imports of softwood lumber declined by 13 percent in volume, and increased in nominal value by only 4 percent per year.

Puerto Rico's plywood imports in 1976 amounted to 56.7 million square feet (worth $8.6 million), while Hawaii's imports were about 5 percent of that--2.4 million square feet (valued at $400,000). In that year, Brazil supplied 29 percent and Taiwan 35 percent of the value of Puerto Rico's plywood imports. While Hawaii's imports of plywood declined during the study period, Puerto Rico's trade more than doubled in both volume and value. The annual import growth rates were 10 percent in volume and 15 percent in nominal value.

EXPORTS
HARDWOOD LOGS (2423)
HAWAII

VALUE--MILLION CURRENT DOLLARS

YEAR	ITALY	REP KOR	JAPAN	OTH ASIA	TR PAC I	OTHERS	TOTAL
1967	0.000	0.000	0.000	0.000	0.000	0.000	0.000
1968	0.000	0.000	0.000	0.000	0.000	0.000	0.000
1969	0.000	0.000	0.000	0.000	0.000	0.000	0.000
1970	0.000	0.000	0.000	0.000	0.000	0.000	0.000
1971	0.000	0.000	0.000	0.000	0.000	0.000	0.000
1972	0.001	0.000	0.011	0.000	0.000	0.000	0.011
1973	0.000	0.000	0.071	0.000	0.000	0.000	0.071
1974	0.000	0.002	0.150	0.000	0.000	0.000	0.152
1975	0.000	0.005	0.036	0.000	0.000	0.000	0.040
1976	0.000	0.000	0.102	0.000	0.004	0.000	0.106

QUANTITY--MILLION BF

YEAR	ITALY	REP KOR	JAPAN	OTH ASIA	TR PAC I	OTHERS	TOTAL
1967	0.000	0.000	0.000	0.000	0.000	0.000	0.000
1968	0.000	0.000	0.000	0.000	0.000	0.000	0.000
1969	0.000	0.000	0.000	0.000	0.000	0.000	0.000
1970	0.000	0.000	0.000	0.000	0.000	0.000	0.000
1971	0.000	0.000	0.000	0.000	0.000	0.000	0.000
1972	0.003	0.000	0.022	0.000	0.000	0.000	0.025
1973	0.000	0.000	0.081	0.000	0.000	0.000	0.081
1974	0.000	0.026	0.142	0.000	0.000	0.000	0.168
1975	0.000	0.009	0.025	0.000	0.000	0.000	0.034
1976	0.000	0.000	0.060	0.000	0.003	0.000	0.062

EXPORTS
PULPWOOD CHIPS (631.8320)
HAWAII

VALUE--MILLION CURRENT DOLLARS

YEAR	N AMERIC	S AMERIC	JAPAN	OTH ASIA	OCEANIA	OTHERS	TOTAL
1967	0.000	0.000	0.000	0.000	0.000	0.000	0.000
1968	0.000	0.000	0.000	0.000	0.000	0.000	0.000
1969	0.000	0.000	0.000	0.000	0.000	0.000	0.000
1970	0.000	0.000	0.000	0.000	0.000	0.000	0.000
1971	0.000	0.000	0.000	0.000	0.000	0.000	0.000
1972	0.000	0.000	0.000	0.000	0.000	0.000	0.000
1973	0.000	0.000	0.000	0.000	0.000	0.000	0.000
1974	0.000	0.000	0.000	0.000	0.000	0.000	0.000
1975	0.000	0.000	0.577	0.000	0.000	0.000	0.577
1976	0.000	0.000	1.330	0.000	0.000	0.000	1.330

QUANTITY--1000 STN

YEAR	N AMERIC	S AMERIC	JAPAN	OTH ASIA	OCEANIA	OTHERS	TOTAL
1967	0.000	0.000	0.000	0.000	0.000	0.000	0.000
1968	0.000	0.000	0.000	0.000	0.000	0.000	0.000
1969	0.000	0.000	0.000	0.000	0.000	0.000	0.000
1970	0.000	0.000	0.000	0.000	0.000	0.000	0.000
1971	0.000	0.000	0.000	0.000	0.000	0.000	0.000
1972	0.000	0.000	0.000	0.000	0.000	0.000	0.000
1973	0.000	0.000	0.000	0.000	0.000	0.000	0.000
1974	0.000	0.000	0.000	0.000	0.000	0.000	0.000
1975	0.000	0.000	9.339	0.000	0.000	0.000	9.339
1976	0.000	0.000	33.488	0.000	0.000	0.000	33.488

EXPORTS
SOFTWOOD LUMBER (2432)
PUERTO RICO

VALUE--MILLION CURRENT DOLLARS

YEAR	JAMAICA	DOM REP	LW WW IS	N ANTILL	FR W IND	OTHERS	TOTAL
1967	0.000	0.000	0.060	0.012	0.005	0.003	0.080
1968	0.000	0.002	0.061	0.011	0.004	0.000	0.078
1969	0.000	0.049	0.067	0.005	0.011	0.000	0.132
1970	0.032	0.055	0.054	0.003	0.007	0.000	0.152
1971	0.141	0.023	0.021	0.034	0.005	0.006	0.230
1972	0.262	0.029	0.019	0.017	0.023	0.012	0.361
1973	0.492	0.087	0.028	0.049	0.006	0.023	0.685
1974	0.345	0.210	0.076	0.106	0.050	0.095	0.882
1975	0.410	0.162	0.100	0.004	0.007	0.035	0.717
1976	0.195	0.013	0.054	0.016	0.004	0.061	0.343

QUANTITY--MILLION BF

YEAR	JAMAICA	DOM REP	LW WW IS	N ANTILL	FR W IND	OTHERS	TOTAL
1967	0.000	0.000	0.445	0.056	0.035	0.021	0.557
1968	0.000	0.013	0.333	0.057	0.008	0.000	0.410
1969	0.000	0.096	0.323	0.026	0.062	0.000	0.506
1970	0.112	0.172	0.330	0.028	0.011	0.000	0.653
1971	0.244	0.119	0.115	0.129	0.009	0.043	0.658
1972	1.162	0.090	0.086	0.083	0.109	0.069	1.599
1973	2.444	0.212	0.110	0.241	0.023	0.130	3.159
1974	0.717	0.671	0.242	0.253	0.122	0.246	2.251
1975	0.278	0.221	0.179	0.016	0.015	0.032	0.740
1976	0.132	0.015	0.127	0.039	0.003	0.083	0.398

EXPORTS
HARDWOOD LUMBER (2433)
PUERTO RICO

VALUE--MILLION CURRENT DOLLARS

YEAR	JAMAICA	DOM REP	LW WW IS	OTH C AM	EUROPE	OTHERS	TOTAL
1967	0.000	0.009	0.000	0.000	0.000	0.000	0.010
1968	0.000	0.000	0.001	0.000	0.000	0.000	0.001
1969	0.000	0.017	0.003	0.000	0.000	0.000	0.020
1970	0.001	0.000	0.023	0.000	0.002	0.000	0.026
1971	0.000	0.000	0.003	0.001	0.000	0.000	0.004
1972	0.000	0.029	0.009	0.005	0.000	0.000	0.042
1973	0.051	0.028	0.001	0.044	0.000	0.000	0.123
1974	0.066	0.003	0.023	0.050	0.046	0.000	0.186
1975	0.165	0.000	0.037	0.011	0.000	0.000	0.214
1976	0.012	0.004	0.012	0.030	0.000	0.000	0.057

QUANTITY--MILLION BF

YEAR	JAMAICA	DOM REP	LW WW IS	OTH C AM	EUROPE	OTHERS	TOTAL
1967	0.000	0.034	0.001	0.001	0.000	0.000	0.037
1968	0.000	0.000	0.002	0.000	0.000	0.000	0.002
1969	0.000	0.027	0.009	0.000	0.000	0.000	0.036
1970	0.002	0.000	0.073	0.000	0.011	0.000	0.086
1971	0.000	0.000	0.010	0.005	0.000	0.000	0.015
1972	0.000	0.094	0.028	0.013	0.000	0.000	0.135
1973	0.186	0.122	0.002	0.104	0.000	0.000	0.414
1974	0.077	0.007	0.057	0.110	0.115	0.000	0.367
1975	0.235	0.000	0.076	0.025	0.000	0.000	0.336
1976	0.021	0.030	0.010	0.125	0.000	0.000	0.186

EXPORTS
PLYWOOD (6312)
PUERTO RICO

VALUE--MILLION CURRENT DOLLARS

YEAR	HAITI	DOM REP	LW WW IS	N ANTILL	OTH C AM	OTHERS	TOTAL
1967	0.000	0.043	0.047	0.012	0.004	0.000	0.106
1968	0.000	0.014	0.049	0.013	0.001	0.000	0.077
1969	0.000	0.019	0.073	0.003	0.001	0.000	0.097
1970	0.003	0.009	0.067	0.013	0.003	0.000	0.094
1971	0.005	0.032	0.028	0.012	0.008	0.000	0.084
1972	0.009	0.047	0.033	0.059	0.030	0.000	0.179
1973	0.118	0.049	0.086	0.053	0.024	0.000	0.330
1974	0.097	0.061	0.124	0.089	0.028	0.000	0.399
1975	0.038	0.013	0.174	0.032	0.053	0.000	0.309
1976	0.000	0.019	0.169	0.154	0.104	0.053	0.499

QUANTITY--MIL SQ FT

YEAR	HAITI	DOM REP	LW WW IS	N ANTILL	OTH C AM	OTHERS	TOTAL
1967	0.000	0.127	0.376	0.057	0.003	0.000	0.564
1968	0.000	0.061	0.232	0.054	0.009	0.000	0.357
1969	0.000	0.080	0.351	0.011	0.003	0.000	0.444
1970	0.018	0.037	0.414	0.056	0.007	0.000	0.532
1971	0.022	0.159	0.131	0.038	0.018	0.000	0.369
1972	0.039	0.155	0.112	0.211	0.113	0.000	0.630
1973	0.418	0.189	0.341	0.185	0.058	0.000	1.191
1974	0.282	0.147	0.486	0.343	0.070	0.000	1.328
1975	0.061	0.042	0.527	0.123	0.187	0.000	0.940
1976	0.000	0.072	0.435	0.484	0.231	0.027	1.249

EXPORTS
WOOD PULP (2510)
PUERTO RICO

VALUE--MILLION CURRENT DOLLARS

YEAR	JAMAICA	HAITI	DOM REP	TRINIDAD	N ANTILL	OTHERS	TOTAL
1967	0.000	0.000	0.000	0.000	0.000	0.000	0.000
1968	0.000	0.000	0.002	0.000	0.000	0.000	0.002
1969	0.000	0.000	0.003	0.000	0.000	0.000	0.003
1970	0.000	0.000	0.006	0.000	0.000	0.025	0.031
1971	0.000	0.007	0.005	0.000	0.000	0.000	0.011
1972	0.011	0.000	0.006	0.000	0.000	0.000	0.017
1973	0.060	0.002	0.016	0.050	0.000	0.000	0.127
1974	0.154	0.000	0.034	0.076	0.000	0.000	0.264
1975	0.063	0.000	0.009	0.056	0.000	0.000	0.127
1976	0.012	0.000	0.000	0.000	0.005	0.000	0.016

QUANTITY--1000 STN

YEAR	JAMAICA	HAITI	DOM REP	TRINIDAD	N ANTILL	OTHERS	TOTAL
1967	0.000	0.000	0.000	0.000	0.000	0.000	0.000
1968	0.000	0.000	0.016	0.000	0.000	0.000	0.016
1969	0.000	0.000	0.111	0.000	0.000	0.000	0.111
1970	0.000	0.000	0.027	0.000	0.000	0.243	0.270
1971	0.000	0.051	0.027	0.000	0.000	0.000	0.078
1972	0.087	0.000	0.027	0.000	0.000	0.000	0.114
1973	0.316	0.029	0.091	0.188	0.000	0.000	0.623
1974	1.359	0.000	0.080	0.198	0.000	0.000	1.637
1975	0.101	0.000	0.021	0.085	0.000	0.000	0.208
1976	0.025	0.000	0.000	0.000	0.011	0.000	0.036

EXPORTS
PAPER & PAPERBOARD (6410)
PUERTO RICO

VALUE--MILLION CURRENT DOLLARS

YEAR	PANAMA	JAMAICA	DOM REP	TRINIDAD	FR W IND	OTHERS	TOTAL
1967	0.909	0.250	0.309	0.000	0.739	0.713	2.920
1968	1.163	0.341	0.551	0.047	0.896	1.005	4.002
1969	0.835	0.382	0.402	0.000	0.731	1.811	4.160
1970	0.365	1.204	0.629	0.211	0.451	0.872	3.730
1971	0.136	1.150	0.575	0.623	0.547	0.972	4.003
1972	0.021	1.098	0.544	0.543	0.628	0.462	3.295
1973	0.155	2.103	1.595	0.734	0.072	0.709	5.368
1974	0.625	2.302	3.402	2.091	0.998	2.508	11.926
1975	0.000	1.598	0.640	0.876	0.101	0.689	3.905
1976	0.000	0.231	0.448	0.252	0.002	2.387	3.320

QUANTITY--1000 STN

YEAR	PANAMA	JAMAICA	DOM REP	TRINIDAD	FR W IND	OTHERS	TOTAL
1967	8.805	2.296	2.647	0.000	6.078	6.812	26.639
1968	10.281	4.438	4.218	0.456	7.290	9.588	36.270
1969	7.332	3.944	3.504	0.000	6.009	19.987	40.777
1970	2.536	10.788	5.364	2.064	4.191	8.544	33.487
1971	1.214	10.249	4.198	5.223	4.864	9.680	35.428
1972	0.225	8.921	2.894	4.340	5.358	4.992	26.731
1973	0.577	10.137	5.618	4.023	0.519	3.377	24.252
1974	1.418	6.015	7.881	7.009	4.784	11.678	38.785
1975	0.000	5.157	1.034	2.832	0.503	2.004	11.531
1976	0.000	0.945	0.692	0.723	0.001	23.386	25.746

IMPORTS
SOFTWOOD LUMBER (2432)
HAWAII

VALUE--MILLION CURRENT DOLLARS

YEAR	CANADA	PHIL REP	JAPAN	AUSTRALA	N ZEALND	OTHERS	TOTAL
1967	0.128	0.000	0.000	0.000	0.000	0.002	0.129
1968	0.137	0.003	0.000	0.000	0.008	0.001	0.150
1969	0.510	0.000	0.000	0.000	0.034	0.000	0.544
1970	0.461	0.000	0.000	0.000	0.028	0.000	0.490
1971	3.131	0.000	0.000	0.000	0.087	0.000	3.218
1972	4.139	0.000	0.090	0.000	0.221	0.000	4.450
1973	6.218	0.000	0.000	0.006	0.223	0.000	6.448
1974	1.966	0.000	0.000	0.000	0.110	0.000	2.076
1975	0.509	0.005	0.000	0.000	0.106	0.000	0.621
1976	0.000	0.000	0.000	0.000	0.021	0.000	0.021

QUANTITY--MILLION BF

YEAR	CANADA	PHIL REP	JAPAN	AUSTRALA	N ZEALND	OTHERS	TOTAL
1967	1.265	0.000	0.000	0.000	0.000	0.003	1.268
1968	1.051	0.020	0.000	0.000	0.069	0.002	1.142
1969	3.858	0.000	0.000	0.000	0.236	0.000	4.094
1970	4.327	0.000	0.000	0.000	0.269	0.000	4.596
1971	25.797	0.000	0.000	0.000	0.770	0.000	26.567
1972	30.011	0.000	0.744	0.000	1.621	0.000	32.376
1973	35.537	0.000	0.000	0.032	1.900	0.000	37.469
1974	10.849	0.000	0.000	0.000	0.685	0.000	11.534
1975	3.096	0.026	0.000	0.000	0.517	0.000	3.639
1976	0.000	0.000	0.000	0.000	0.140	0.000	0.140

IMPORTS
HARDWOOD LUMBER (2433)
HAWAII

VALUE--MILLION CURRENT DOLLARS

YEAR	THAILAND	PHIL REP	JAPAN	OTH ASIA	OCEANIA	OTHERS	TOTAL
1967	0.009	0.147	0.126	0.019	0.021	0.000	0.323
1968	0.036	0.155	0.270	0.020	0.012	0.000	0.493
1969	0.046	0.081	0.212	0.020	0.010	0.000	0.368
1970	0.014	0.083	0.135	0.041	0.014	0.004	0.290
1971	0.002	0.092	0.068	0.034	0.008	0.000	0.203
1972	0.005	0.085	0.000	0.028	0.007	0.004	0.129
1973	0.013	0.217	0.000	0.007	0.020	0.010	0.267
1974	0.000	0.145	0.000	0.000	0.000	0.000	0.145
1975	0.006	0.104	0.000	0.007	0.003	0.000	0.120
1976	0.010	0.102	0.000	0.003	0.001	0.000	0.116

QUANTITY--MILLION BF

YEAR	THAILAND	PHIL REP	JAPAN	OTH ASIA	OCEANIA	OTHERS	TOTAL
1967	0.011	0.673	0.503	0.178	0.065	0.000	1.429
1968	0.074	0.705	0.936	0.153	0.056	0.000	1.925
1969	0.115	0.396	0.651	0.171	0.051	0.000	1.384
1970	0.039	0.252	0.435	0.246	0.110	0.011	1.093
1971	0.012	0.393	0.191	0.156	0.034	0.000	0.785
1972	0.023	0.342	0.000	0.112	0.023	0.019	0.520
1973	0.044	0.639	0.000	0.035	0.048	0.020	0.787
1974	0.000	0.314	0.000	0.002	0.000	0.000	0.316
1975	0.004	0.253	0.000	0.015	0.002	0.000	0.275
1976	0.015	0.209	0.000	0.013	0.001	0.000	0.238

IMPORTS
PLYWOOD (6312)
HAWAII

VALUE--MILLION CURRENT DOLLARS

YEAR	FINLAND	PHIL REP	TAIWAN	JAPAN	OTH ASIA	OTHERS	TOTAL
1967	0.091	0.093	0.122	0.080	0.000	0.000	0.386
1968	0.044	0.105	0.140	0.082	0.022	0.000	0.393
1969	0.551	0.212	0.295	0.118	0.000	0.012	1.188
1970	0.127	0.231	0.076	0.125	0.008	0.003	0.570
1971	0.067	0.238	0.100	0.149	0.003	0.000	0.558
1972	0.058	0.215	0.119	0.174	0.006	0.000	0.573
1973	0.126	0.505	0.206	0.174	0.003	0.011	1.025
1974	0.000	0.262	0.426	0.149	0.003	0.001	0.841
1975	0.000	0.138	0.180	0.040	0.000	0.002	0.360
1976	0.000	0.207	0.181	0.012	0.001	0.000	0.400

QUANTITY--MIL SQ FT

YEAR	FINLAND	PHIL REP	TAIWAN	JAPAN	OTH ASIA	OTHERS	TOTAL
1967	0.660	0.958	1.151	0.499	0.000	0.000	3.268
1968	0.242	0.432	0.586	0.250	0.092	0.000	1.602
1969	3.313	1.254	2.232	0.379	0.000	0.059	7.237
1970	0.573	1.989	0.581	0.487	0.091	0.010	3.731
1971	0.385	2.522	0.820	0.528	0.029	0.000	4.282
1972	0.243	1.777	0.796	0.652	0.015	0.000	3.483
1973	0.373	3.561	0.925	0.350	0.005	0.034	5.249
1974	0.000	1.117	1.793	0.247	0.025	0.002	3.184
1975	0.000	0.753	0.821	0.092	0.000	0.004	1.670
1976	0.000	0.899	0.681	0.047	0.003	0.000	1.630

IMPORTS
NEWSPRINT (6411)
HAWAII

VALUE--MILLION CURRENT DOLLARS

YEAR	CANADA	SWEDEN	FINLAND	OTH EURO	ASIA	OTHERS	TOTAL
1967	3.151	0.000	0.010	0.000	0.000	0.000	3.161
1968	3.279	0.002	0.029	0.000	0.000	0.000	3.310
1969	3.861	0.000	0.070	0.000	0.000	0.000	3.931
1970	4.407	0.000	0.223	0.000	0.000	0.000	4.630
1971	5.158	0.000	0.000	0.000	0.000	0.000	5.158
1972	4.949	0.000	0.000	0.000	0.000	0.000	4.949
1973	5.835	0.000	0.000	0.000	0.000	0.000	5.835
1974	4.534	0.000	0.000	0.000	0.000	0.000	4.534
1975	1.636	0.000	0.000	0.000	0.000	0.000	1.636
1976	8.403	0.000	0.000	0.000	0.000	0.000	8.403

QUANTITY--1000 STN

YEAR	CANADA	SWEDEN	FINLAND	OTH EURO	ASIA	OTHERS	TOTAL
1967	27.436	0.000	0.099	0.000	0.000	0.000	27.535
1968	27.997	0.022	0.269	0.000	0.000	0.000	28.288
1969	32.020	0.000	0.649	0.000	0.000	0.000	32.669
1970	35.405	0.000	2.018	0.000	0.000	0.000	37.423
1971	40.574	0.000	0.000	0.000	0.000	0.000	40.574
1972	36.139	0.000	0.000	0.000	0.000	0.000	36.139
1973	39.819	0.000	0.000	0.000	0.000	0.000	39.819
1974	24.524	0.000	0.000	0.000	0.000	0.000	24.524
1975	6.550	0.000	0.000	0.000	0.000	0.000	6.550
1976	31.428	0.000	0.000	0.000	0.000	0.000	31.428

IMPORTS
PAPER & PAPERBOARD (6410)
HAWAII

VALUE--MILLION CURRENT DOLLARS

YEAR	CANADA	EUROPE	JAPAN	OTH ASIA	OCEANIA	OTHERS	TOTAL
1967	0.026	0.007	0.005	0.003	0.000	0.000	0.041
1968	0.008	0.014	0.011	0.005	0.001	0.000	0.040
1969	0.004	0.008	0.021	0.032	0.019	0.000	0.084
1970	0.030	0.005	0.023	0.020	0.030	0.000	0.108
1971	0.060	0.001	0.027	0.001	0.036	0.000	0.125
1972	0.047	0.000	0.019	0.009	0.002	0.000	0.076
1973	0.026	0.000	0.021	0.000	0.018	0.000	0.065
1974	0.000	0.000	0.032	0.013	0.000	0.000	0.046
1975	0.000	0.000	0.044	0.006	0.000	0.000	0.050
1976	0.000	0.000	0.008	0.008	0.005	0.000	0.022

QUANTITY--1000 STN

YEAR	CANADA	EUROPE	JAPAN	OTH ASIA	OCEANIA	OTHERS	TOTAL
1967	0.120	0.039	0.014	0.001	0.000	0.000	0.174
1968	0.211	0.073	0.007	0.002	0.002	0.000	0.294
1969	0.015	0.040	0.013	0.012	0.134	0.000	0.213
1970	0.122	0.025	0.046	0.007	0.199	0.000	0.400
1971	0.331	0.004	0.020	0.001	0.215	0.000	0.571
1972	0.284	0.000	0.014	0.003	0.003	0.000	0.304
1973	0.175	0.000	0.017	0.000	0.071	0.000	0.264
1974	0.000	0.000	0.018	0.002	0.000	0.000	0.020
1975	0.000	0.000	0.012	0.001	0.000	0.000	0.012
1976	0.000	0.000	0.004	0.005	0.009	0.000	0.018

IMPORTS
SOFTWOOD LUMBER (2432)
PUERTO RICO

VALUE--MILLION CURRENT DOLLARS

YEAR	CANADA	HONDURAS	OTH N AM	BRAZIL	OTH S AM	OTHERS	TOTAL
1967	4.946	1.859	0.071	1.331	0.000	0.000	8.207
1968	6.408	1.803	0.037	1.164	0.318	0.000	9.730
1969	8.242	2.373	0.093	2.110	0.614	0.000	13.432
1970	6.988	2.163	0.232	1.569	0.218	0.000	11.170
1971	8.854	2.414	0.683	1.296	0.220	0.051	13.518
1972	11.364	4.369	0.899	1.622	0.233	0.008	18.494
1973	13.702	4.011	1.806	1.260	0.171	0.008	20.958
1974	11.277	2.778	2.118	0.431	0.582	0.000	17.186
1975	7.120	3.840	0.860	0.524	0.007	0.016	12.366
1976	7.415	2.854	1.230	0.354	0.031	0.000	11.884

QUANTITY--MILLION BF

YEAR	CANADA	HONDURAS	OTH N AM	BRAZIL	OTH S AM	OTHERS	TOTAL
1967	75.766	24.540	0.633	14.401	0.000	0.000	115.340
1968	84.959	21.568	0.301	13.254	4.003	0.000	124.085
1969	86.150	26.668	1.055	16.320	6.121	0.000	136.314
1970	87.516	22.635	2.277	10.546	1.669	0.000	124.644
1971	109.882	25.911	7.463	8.331	1.928	0.784	154.299
1972	119.091	40.720	9.415	10.330	1.816	0.020	181.393
1973	92.961	27.591	13.437	6.024	0.820	0.047	140.880
1974	81.758	34.373	11.508	1.327	1.878	0.000	130.844
1975	63.899	31.089	6.824	1.713	0.031	0.155	103.711
1976	62.875	27.636	8.694	1.166	0.168	0.000	100.538

IMPORTS
HARDWOOD LUMBER (2433)
PUERTO RICO

VALUE--MILLION CURRENT DOLLARS

YEAR	HONDURAS	NICARAGA	OTH N AM	BRAZIL	OTH S AM	OTHERS	TOTAL
1967	0.342	0.451	0.264	0.240	0.067	0.108	1.471
1968	0.459	0.211	0.377	0.281	0.129	0.019	1.476
1969	0.291	0.240	0.309	0.339	0.215	0.012	1.405
1970	0.228	0.185	0.103	0.956	0.239	0.004	1.714
1971	0.087	0.150	0.168	1.134	0.124	0.007	1.669
1972	0.126	0.009	0.167	1.184	0.139	0.000	1.624
1973	0.097	0.000	0.289	0.717	0.415	0.000	1.518
1974	0.204	0.107	0.202	1.339	0.604	0.047	2.502
1975	0.417	0.066	0.123	0.894	0.171	0.000	1.672
1976	0.000	0.015	0.029	0.775	0.034	0.000	0.853

QUANTITY--MILLION BF

YEAR	HONDURAS	NICARAGA	OTH N AM	BRAZIL	OTH S AM	OTHERS	TOTAL
1967	1.172	1.720	1.430	1.569	0.257	0.496	6.645
1968	1.808	0.944	1.595	1.458	0.737	0.065	6.608
1969	1.029	0.936	1.711	1.940	1.226	0.057	6.900
1970	0.956	0.485	0.554	4.079	1.367	0.009	7.451
1971	0.366	0.695	0.680	5.627	0.716	0.016	8.100
1972	0.551	0.024	0.662	4.667	0.764	0.000	6.668
1973	0.299	0.000	0.957	2.366	1.610	0.000	5.233
1974	0.541	0.241	0.577	3.295	2.633	0.006	7.294
1975	0.931	0.183	0.396	2.385	0.466	0.000	4.362
1976	0.000	0.039	0.117	2.386	0.102	0.000	2.645

IMPORTS
PLYWOOD (6312)
PUERTO RICO

VALUE--MILLION CURRENT DOLLARS

YEAR	BRAZIL	SPAIN	TAIWAN	JAPAN	OTH ASIA	OTHERS	TOTAL
1967	0.192	0.428	0.412	0.283	0.173	0.938	2.426
1968	0.467	0.636	0.546	0.365	0.075	0.883	2.972
1969	0.504	0.611	0.772	0.438	0.023	1.078	3.426
1970	0.592	0.787	0.524	0.514	0.090	0.864	3.371
1971	1.061	0.869	1.093	0.686	0.114	0.771	4.594
1972	1.260	0.752	1.632	0.721	0.355	1.565	6.285
1973	2.094	0.357	1.644	0.691	0.320	1.178	6.283
1974	2.985	0.688	2.585	0.091	0.265	0.845	7.459
1975	1.018	0.100	1.466	0.012	0.511	1.427	4.534
1976	2.495	0.054	3.025	0.000	1.210	1.816	8.601

QUANTITY--MIL SQ FT

YEAR	BRAZIL	SPAIN	TAIWAN	JAPAN	OTH ASIA	OTHERS	TOTAL
1967	1.500	2.587	3.778	1.388	1.291	5.513	16.057
1968	2.569	3.437	4.036	1.301	0.417	4.307	16.068
1969	3.340	3.397	6.033	2.146	0.127	6.184	21.227
1970	4.185	3.892	4.337	2.210	0.433	4.761	19.819
1971	7.988	4.255	8.728	2.619	0.670	3.925	28.185
1972	8.551	3.515	13.519	2.630	2.128	9.120	39.463
1973	9.106	1.447	10.091	1.406	1.531	4.615	28.196
1974	10.553	1.739	12.338	0.197	1.246	2.059	28.133
1975	4.168	0.228	8.455	0.021	2.455	7.222	22.549
1976	9.049	0.162	15.244	0.000	5.886	7.460	37.802

IMPORTS
RECONSTITUTED WOOD (6314)
PUERTO RICO

VALUE--MILLION CURRENT DOLLARS

YEAR	N AMERIC	SURINAM	OTH S AM	PORTUGAL	OTH EURO	OTHERS	TOTAL
1967	0.000	0.088	0.000	0.004	0.006	0.000	0.098
1968	0.000	0.100	0.000	0.000	0.000	0.002	0.102
1969	0.003	0.263	0.002	0.016	0.001	0.000	0.285
1970	0.009	0.173	0.000	0.041	0.005	0.001	0.230
1971	0.052	0.155	0.008	0.000	0.006	0.000	0.221
1972	0.000	0.101	0.000	0.000	0.001	0.000	0.103
1973	0.008	0.253	0.000	0.000	0.022	0.000	0.284
1974	0.000	0.194	0.006	0.000	0.039	0.000	0.238
1975	0.000	0.000	0.000	0.001	0.000	0.000	0.001
1976	0.000	0.000	0.000	0.009	0.000	0.000	0.009

QUANTITY--MIL SQ FT

YEAR	N AMERIC	SURINAM	OTH S AM	PORTUGAL	OTH EURO	OTHERS	TOTAL
1967	0.000	0.895	0.000	0.030	0.063	0.002	0.990
1968	0.000	0.968	0.000	0.004	0.000	0.009	0.981
1969	0.032	2.153	0.012	0.148	0.007	0.000	2.352
1970	0.080	1.531	0.000	0.364	0.040	0.011	2.026
1971	0.491	1.057	0.089	0.000	0.058	0.000	1.695
1972	0.000	0.776	0.000	0.000	0.012	0.000	0.788
1973	0.071	1.553	0.000	0.000	0.197	0.000	1.822
1974	0.000	0.943	0.030	0.000	0.329	0.000	1.302
1975	0.000	0.000	0.000	0.005	0.000	0.000	0.005
1976	0.000	0.000	0.000	0.014	0.000	0.000	0.014

IMPORTS
WOOD PULP (2510)
PUERTO RICO

VALUE--MILLION CURRENT DOLLARS

YEAR	CANADA	S AMERIC	EUROPE	ASIA	AFRICA	OTHERS	TOTAL
1967	0.327	0.000	0.000	0.000	0.000	0.000	0.327
1968	0.290	0.000	0.000	0.000	0.000	0.000	0.290
1969	0.426	0.000	0.000	0.000	0.000	0.000	0.426
1970	0.345	0.000	0.000	0.000	0.000	0.000	0.345
1971	0.240	0.000	0.000	0.000	0.000	0.000	0.240
1972	0.525	0.000	0.000	0.000	0.000	0.000	0.525
1973	0.159	0.000	0.000	0.000	0.000	0.000	0.159
1974	0.000	0.000	0.000	0.000	0.000	0.000	0.000
1975	0.000	0.000	0.000	0.000	0.000	0.000	0.000
1976	0.000	0.000	0.000	0.000	0.000	0.000	0.000

QUANTITY--1000 STN

YEAR	CANADA	S AMERIC	EUROPE	ASIA	AFRICA	OTHERS	TOTAL
1967	2.893	0.000	0.000	0.000	0.000	0.000	2.893
1968	2.561	0.000	0.000	0.000	0.000	0.000	2.561
1969	3.805	0.000	0.000	0.000	0.000	0.000	3.805
1970	2.533	0.000	0.000	0.000	0.000	0.000	2.533
1971	1.565	0.000	0.000	0.000	0.000	0.000	1.565
1972	3.501	0.000	0.000	0.000	0.000	0.000	3.501
1973	1.244	0.000	0.000	0.000	0.000	0.000	1.244
1974	0.000	0.000	0.000	0.000	0.000	0.000	0.000
1975	0.000	0.000	0.000	0.000	0.000	0.000	0.000
1976	0.000	0.000	0.000	0.000	0.000	0.000	0.000

IMPORTS
NEWSPRINT (6411)
PUERTO RICO

VALUE--MILLION CURRENT DOLLARS

YEAR	CANADA	S AMERIC	FINLAND	OTH EURO	AFRICA	OTHERS	TOTAL
1967	2.723	0.000	0.028	0.000	0.000	0.000	2.750
1968	2.400	0.000	0.035	0.000	0.000	0.000	2.435
1969	3.218	0.000	0.030	0.000	0.000	0.000	3.248
1970	4.260	0.000	0.025	0.000	0.000	0.000	4.285
1971	5.221	0.000	0.014	0.000	0.000	0.000	5.235
1972	4.471	0.000	0.032	0.000	0.000	0.000	4.503
1973	4.451	0.000	0.000	0.000	0.000	0.000	4.451
1974	7.555	0.000	0.000	0.000	0.000	0.000	7.555
1975	9.053	0.000	0.000	0.000	0.000	0.000	9.053
1976	10.040	0.000	0.526	0.000	0.000	0.000	10.566

QUANTITY--1000 STN

YEAR	CANADA	S AMERIC	FINLAND	OTH EURO	AFRICA	OTHERS	TOTAL
1967	22.550	0.000	0.252	0.000	0.000	0.000	22.802
1968	18.743	0.000	0.318	0.000	0.000	0.000	19.061
1969	24.390	0.000	0.305	0.000	0.000	0.000	24.695
1970	32.023	0.000	0.248	0.000	0.000	0.000	32.271
1971	38.028	0.000	0.111	0.000	0.000	0.000	38.139
1972	32.283	0.000	0.269	0.000	0.000	0.000	32.552
1973	48.518	0.000	0.000	0.000	0.000	0.000	48.518
1974	35.603	0.000	0.000	0.000	0.000	0.000	35.603
1975	35.173	0.000	0.000	0.000	0.000	0.000	35.173
1976	32.129	0.000	1.978	0.000	0.000	0.000	34.107

IMPORTS
BUILDING BOARD (6416)
PUERTO RICO

VALUE--MILLION CURRENT DOLLARS

YEAR	BRAZIL	OTH S AM	SCANDINA	OTH EURO	ASIA	OTHERS	TOTAL
1967	0.003	0.026	0.118	0.028	0.000	0.006	0.179
1968	0.000	0.084	0.135	0.012	0.000	0.009	0.240
1969	0.007	0.093	0.095	0.010	0.000	0.012	0.217
1970	0.062	0.103	0.071	0.005	0.000	0.014	0.254
1971	0.104	0.081	0.034	0.002	0.000	0.017	0.238
1972	0.060	0.134	0.037	0.230	0.004	0.037	0.501
1973	0.067	0.116	0.019	0.091	0.000	0.047	0.339
1974	0.070	0.032	0.011	0.019	0.133	0.012	0.278
1975	0.174	0.187	0.004	0.000	0.043	0.014	0.421
1976	0.027	0.103	0.040	0.001	0.035	0.012	0.219

QUANTITY--1000 STN

YEAR	BRAZIL	OTH S AM	SCANDINA	OTH EURO	ASIA	OTHERS	TOTAL
1967	0.021	0.490	2.038	0.599	0.000	0.049	3.198
1968	0.000	1.532	2.337	0.251	0.000	0.088	4.208
1969	0.056	1.766	1.485	0.174	0.000	0.074	3.555
1970	0.922	1.984	0.960	0.055	0.000	0.074	3.995
1971	1.578	1.232	0.400	0.035	0.000	0.083	3.328
1972	0.923	2.248	0.382	3.416	0.062	0.498	7.530
1973	0.611	1.552	0.166	1.161	0.000	0.479	3.969
1974	0.467	0.269	0.046	0.118	0.958	0.050	1.908
1975	1.428	1.867	0.008	0.000	0.251	0.038	3.593
1976	0.167	0.935	0.265	0.001	0.149	0.077	1.594

IMPORTS
PAPER & PAPERBOARD (6410)
PUERTO RICO

VALUE--MILLION CURRENT DOLLARS

YEAR	CANADA	VENEZUEL	SCANDINA	OTH EURO	ASIA	OTHERS	TOTAL
1967	0.042	0.000	0.399	0.023	0.002	0.002	0.469
1968	0.049	0.000	0.493	0.035	0.002	0.001	0.580
1969	0.056	0.000	0.380	0.037	0.000	0.001	0.473
1970	0.061	0.000	0.331	0.043	0.000	0.000	0.435
1971	0.232	0.000	0.242	0.108	0.005	0.010	0.596
1972	0.345	0.000	0.242	0.672	0.043	0.326	1.627
1973	1.370	0.000	0.156	1.200	0.000	0.141	2.866
1974	0.267	0.000	0.336	2.465	0.000	0.007	3.074
1975	0.019	0.773	0.267	3.764	0.000	0.026	4.848
1976	0.018	0.964	0.140	1.364	0.000	0.123	2.608

QUANTITY--1000 STN

YEAR	CANADA	VENEZUEL	SCANDINA	OTH EURO	ASIA	OTHERS	TOTAL
1967	0.372	0.000	2.085	0.032	0.006	0.000	2.496
1968	0.266	0.000	2.583	0.061	0.005	0.000	2.916
1969	0.467	0.000	1.837	0.083	0.000	0.000	2.386
1970	0.253	0.000	1.309	0.086	0.000	0.001	1.649
1971	1.129	0.000	0.892	0.344	0.012	0.039	2.416
1972	3.321	0.000	0.891	2.454	0.020	4.040	10.725
1973	11.406	0.000	0.424	3.739	0.000	1.323	16.891
1974	1.557	0.000	0.490	4.856	0.000	0.006	6.909
1975	0.005	1.890	0.275	5.787	0.000	0.027	7.984
1976	0.067	2.102	0.182	2.103	0.000	0.490	4.944

www.ingramcontent.com/pod-product-compliance
Lightning Source LLC
Chambersburg PA
CBHW060417220326
41598CB00021BA/2208

* 9 7 8 1 1 3 8 9 5 4 2 9 8 *